J. Kacgud
'91

AF324119

Selected Titles in This Series

Volume

9 **Shing-Tung Yau, Editor**
Mirror Symmetry I
1998

8 **Jürgen Jost, Wilfrid Kendall, Umberto Mosco,**
and Karl-Theodor Sturm
New Directions in Dirichlet Forms
1998

7 **D. A. Buell and J. T. Teitelbaum, Editors**
Computational Perspectives on Number Theory
1998

6 **Harold Levine**
Partial Differential Equations
1997

5 **Qi-keng Lu, Stephen S.-T. Yau, and Anatol**
Singularities and Complex Geometry
1997

4 **Vyjayanthi Chari and Ivan B. Penkov, Ed**
Modular Interfaces: Modular Lie Algebras, Qua
1997

3 **Xia-Xi Ding and Tai-Ping Liu, Editors**
Nonlinear Evolutionary Partial Differential Eq
1997

2.2 **William H. Kazez, Editor**
Geometric Topology
1997

2.1 **William H. Kazez, Editor**
Geometric Topology
1997

1 **B. Greene and S.-T. Yau, Editors**
Mirror Symmetry II
1997

Mirror Symmetry I

J. Kacyot

AMS/IP
Studies in Advanced Mathematics

Volume 9

Mirror Symmetry I

Shing-Tung Yau, Editor

American Mathematical Society   ·   International Press

Shing-Tung Yau, Managing Editor

1991 *Mathematics Subject Classification.* Primary 14–xx; Secondary 32–xx, 81–xx.

Library of Congress Cataloging-in-Publication Data

Mirror symmetry I / Shing-Tung Yau, editor.
  p. cm. — (AMS/IP studies in advanced mathematics, ISSN 1089-3288 ; v. 9)
  Includes bibliographical references.
  ISBN 0-8218-0665-3 (alk. paper)
  1. Mirror symmetry. 2. Manifolds (Mathematics) I. Yau. Shing-Tung, 1949- II. Series.
QC174.17.S9M562 1998   98-35520
516.3′62–dc21   CIP

Copying and reprinting. Material in this book may be reproduced by any means for educational and scientific purposes without fee or permission with the exception of reproduction by services that collect fees for delivery of documents and provided that the customary acknowledgment of the source is given. This consent does not extend to other kinds of copying for general distribution, for advertising or promotional purposes, or for resale. Requests for permission for commercial use of material should be addressed to the Assistant to the Publisher, American Mathematical Society, P. O. Box 6248, Providence, Rhode Island 02940-6248. Requests can also be made by e-mail to `reprint-permission@ams.org`.

 Excluded from these provisions is material in articles for which the author holds copyright. In such cases, requests for permission to use or reprint should be addressed directly to the author(s). (Copyright ownership is indicated in the notice in the lower right-hand corner of the first page of each article.)

© 1998 by the American Mathematical Society and International Press. All rights reserved.
The American Mathematical Society and International Press retain all rights
except those granted to the United States Government.
Printed in the United States of America.

♾ The paper used in this book is acid-free and falls within the guidelines
established to ensure permanence and durability.
Visit the AMS home page at URL: `http://www.ams.org/`
Visit the International Press home page at URL: `http://www.intlpress.com/`
10 9 8 7 6 5 4 3 2 1   03 02 01 00 99 98

Contents

Preface ix

Introduction xi

An introduction to mirror manifolds
B. R. Greene and M. R. Plesser 1

A pair of Calabi-Yau manifolds as an exactly soluble superconformal theory
Philip Candelas, Xenia C. de la Ossa, Paul S. Green,
and Linda Parkes 31

Topological mirrors and quantum rings
Cumrun Vafa 97

Mirror manifolds and topological field theory
Edward Witten 121

Rational curves and classification of algebraic varieties
Yujiro Kawamata 161

Rational curves on Calabi-Yau threefolds
Sheldon Katz 169

Picard-Fuchs equations and mirror maps for hypersurfaces
David R. Morrison 185

Kähler classes on Calabi-Yau threefolds—An informal survey
P. M. H. Wilson 201

Automorphic functions and special Kähler geometry
S. Ferrara, C. Kounnas, D. Lüst, and F. Zwirner 215

Picard-Fuchs equations and flat holomorphic connections from $N = 2$
supergravity
S. Ferrara and J. Louis 237

A new geometry from superstring theory
P. S. Aspinwall and C. A. Lütken 253

The geometry of Calabi-Yau orbifolds
Shi-Shyr Roan 279

Properties of superstring vacua from (topological) Landau-Ginzburg models
Amit Giveon and Dirk-Jan Smit 287

An $SL(2,\mathbb{C})$ action on certain Jacobian rings and the mirror map
TRISTAN HÜBSCH AND SHING-TUNG YAU
311

A generalized construction of mirror manifolds
PER BERGLUND AND TRISTAN HÜBSCH
327

New constructions of mirror manifolds: Probing moduli space far from Fermat points
B. R. GREENE, M. R. PLESSER, AND S.-S. ROAN
347

Deformations of Calabi-Yau Kleinfolds
ZIV RAN
391

Smoothing 3-folds with trivial canonical bundle and ordinary double points
GANG TIAN
399

Modified Calabi-Yau manifolds with torsion
MARTIN ROČEK
421

Calabi-Yau threefolds and complex multiplication
CIPRIAN BORCEA
431

Preface to the New Edition

This volume represents a new edition of the papers which were originally published in *Essays on Mirror Manifolds*. That volume has continued to be quite popular, and this new edition is supplemented by an additional volume, Mirror Symmetry 2, edited by B. Greene and myself.

In this volume, we would like to address a number of errors which were found in the original: for example, the workshop on Mirror Manifolds at the Mathematical Sciences Research Institute took place in 1991, not 1911. The paper by Geir Ellingsrud and Stein A. Stromme has been withdrawn at the authors' request. Meanwhile, P.M.H. Wilson's paper on "Kahler Classes on Calabi-Yau Threefolds– an Informal Survey"; Sheldon Katz's paper on "Rational Curves on Calabi-Yau Threefolds"; Ziv Ran's paper on "Deformations of Calabi-Yau Kleinfolds"; and David R. Morrison's paper on "Picard-Fuchs Equations and Mirror Maps for Hypersurfaces" have all been corrected per new drafts from the authors.

We are quite pleased with the continuing interest in each of the papers in this volume and look forward to the continuing development of this field in the new century.

S.-T. Yau
Cambridge, MA

Introduction

Irrespective of the ultimate verdict regarding the physical relevance of string thoery, there is no doubt about its mathematical richness. Bodies of knowledge involving aspects of algebraic geometry, Riemann surfaces, modular forms, infinite dimensional Lie algebras and numerous others — once exclusively the domain of pure mathematics — have come to be commonplace tools of physicists working in string theory and conformal field theory. Mathematics has concurrently benefited from this cross fertilization as well. One striking example of this is the central focus of essays collected in this volume.

In passing from its initial formulation in ten dimensional spacetime to a physically realistic description in four dimension, the consistency of string theory requires the 'additional' six dimensional space to be a compact Kähler manifold with vanishing first Chern class — a so called 'Calabi–Yau' (CY) manifold. More precisely, these 'additional' degrees of freedom must constitute a $N = 2$ super–conformal field theory — one way of achieving this is by making use of CY manifolds. Two CY manifolds are said to be a mirror pair if their respective Hodge numbers satisfy $h^{p,q} = \tilde{h}^{3-p,q}$ and they yield the same conformal field theory. In fact, one needs not restrict attention to the string motivated case of complex dimension three and apply this definition (after replacing 3 by D) to CY manifolds of arbitrary complex dimension D. By using the fact that they corrspond to the same conformal field theory, highly nontrivial identities can be established which relate topological and geometrical properties of the mirror pair.

The existence of such mirror manifolds was boldly conjectured by a number of physicists including Dixon and Lerche, Vafa and Warner based on physical intuition. Namely, there is a physically unnatural ambiguity in associating CY manifolds to conformal field theories which is elegantly resolved via the hypothesized mirror pair. Although a compelling possibility of existence of mirror, it seems unlikely from the mathematics vantage point because, for example, it was not clear how to construct CY manifolds whose Hodge number $h^{1,1}$ would be as large as previously known values of $h^{2,1}$. The belief and interest in mirror manifolds dramatically changed with the appearance of two papers. The first, by Greene and Plesser, established the existence of mirror manifolds for a particular subclass of CY manifolds by direct construction and deduced from these some compelling mathematical identities. The second, by Candelas, de La Ossa, Greene and Parkes, made use of one such construction to argue convincingly that they could determine the number of rational curves of arbitrary degree on the quintic threefold, thus establishing the computational power of mirror manifolds. Additionally, the computer study of CY hypersurfaces in weighted projective four space of Candelas, Lynker and Schimmrigk gave a numerical hint of the existence of mirror manifolds.

As with any important new development which straddles traditional academic disciplines, two pervasive obstacles facing prospective adherents are the differences in language and assumed knowledges between the respective fields of inquiry. The rapidly growing number of both mathematicians and physicists seeking to understand the basics and most recent results in mirror manifolds thus made manifest the need for the present volume. Both the purpose and compilation of these essays were further facilitated by a workshop on mirror manifolds held at the Mathematical Sciences Research Institute in Berkeley from May 6–8, 1911. Numerous lectures each day were followed by evening question sessions allowing detailed discussions of various points covered during the day. There was an unusually productive flow of ideas between mathematicians and physicists as each attempted to grasp the vantage point and conceptual framework of the other.

This volume contains both introductory discussions on the theory of mirror manifolds as well as papers describing more recent results. Aspects of some of these papers were also presented at the MSRI workshop. The papers of Greene, Plesser and Candelas, et. al. provide an introduction to mirror manifolds for mathematicians, review the known constructions and describe with explicit examples the uses of mirror manifolds. The latter paper also describes the mirror symmetry calculation of rational curves. The papers of Vafa and of Witten provide discussions of what may be the most natural and productive physical framework for studying mirror manifolds: topological field theory. We are familiar with the mathematical power of topological field theory and hence this approach is likely to be fundamental to mathematical understanding of mirror manifolds.

The papers of Roan and of Morrison are concerned with placing the analysis of Greene, Plesser and Candelas et. al, respectively, into purely mathematical language. Roan finds a natural description of certain aspects of mirror manifolds using toric geometry while Morrison systematizes and extends the result of Candelas et. al. calculating the number of rational curves on an algebraic variety and the contribution of Ellingsrud and Stromme presents a mathematical calculation of the number of degree three curves on the quintic hypersurface. An initial discrepancy of this work with that of the mirror symmetry calculation of Candelas et. al. (since resolved) led to some of the most detailed and productive discussions of the workshop. Wilson's paper focuses on the nature of the Kähler cone of CY manifolds, while Kawamata's contribution discusses the birational geometry of algebraic varieties. Aspinwall and Lutken review their calculation with Ross which explicitly verifies some predictions arising from mirror manifolds and also call attention to as yet ill understood subtleties surrounding resolution of singularity. Ferrara–Louis and Ferrara, Kounnas, Lust and Zwirner discuss the geometry of the moduli space which is important in understanding the structure of the mirror map. Hübsch and myself made an attempt to understand the mirror of the Kähler classes.

Tian and Ran's paper addresses the smoothing of certain kinds of singularities on CY manifolds. The papers of Greene, Plesser and Roan as well as that by Berglund and Hubsch address extending the domain of CY manifolds for which a mirror can be constructed.

Rocek discusses the inclusion of CY manifolds with nonzero torsion while Giveon and Smit pursue aspects of topological Landau–Ginsberg models. Borcea gave more examples of CY manifolds.

There are numerous important unsolved problems surrounding this new structure of mirror symmetry in algebraic geometry. This volume should serve as an introduction to the subject as well as a compendium of the most recent advances for both mathematicians and physicists.

Gratitude is due to all of the contributors to this volume for their prompt submission of a wide range of papers dealing with aspects of mirror manifolds. It is also a sincere pleasure to thank I. Singer for suggesting that an MSRI workshop on mirror manifolds. The conference was organized by Candelas and myself. I would like to thank him for important contributions which are essential for the success of the conference.

The workshop facilitated the productive flow of ideas between mathematicians and physicists and helped initiate some of the work presented in this volume. We thank the MSRI and the Paul and Gabriella foundation for making this opportunity possible. Finally, thanks are due to E. Archer for her tireless assistance in preparing this volume for publication. I should add that I am pleased to participate in all these activities on mirror manifolds and proud to say that both Greene and Hübsch were supported by my research grant DOE DE–FG02–88ER–25065 as postdoctoral fellows during their research on the subject.

AMS/IP Studies in Advanced Mathematics
Volume **9**, 1998

An Introduction to Mirror Manifolds

B.R. Greene

F.R. Newman Laboratory of Nuclear Studies

Cornell University

Ithaca, NY 14853

M.R. Plesser

Department of Physics

Yale University

New Haven, CT 06511

Abstract

Mirror symmetry establishes an unexpected and powerful link between *a priori* unrelated topologically distinct Calabi–Yau manifolds. We give an introduction to the subject of mirror manifolds intended for a mathematical audience. We discuss the explicit construction of mirror pairs which established their existence and also summarize their basic mathematical properties. Some important applications and verifications of mirror symmetry are also discussed.

© 1998 American Mathematical Society
and International Press

1. Introduction

Mirror symmetry is a property of Calabi–Yau spaces which relates topologically distinct manifolds in a surprising and powerful way [1] [2] [3] [4] [5] [6]. At present, mirror symmetry finds its most potent expression in the language of string theory in which the symmetry ensures that propagation of strings on the two manifolds comprising a mirror pair leads to identical physical theories.[1] From a purely mathematical point of view, mirror symmetry leads to equalities between geometrical quantities on the two members of a mirror pair. As will be clear, these equalities are remarkably powerful. For example, in some instances they provide a dictionary for translating difficult questions of interest on one manifold into far simpler questions on the other. These results may be stated without reference to the underlying physical theory although at present there is no means of establishing their validity without recourse to physical methods. That is, although recent work [8] has taken some interesting steps towards a mathematical understanding of mirror symmetry, the full justification of the only known constructions of mirror manifolds relies on arguments from conformal field theory [2]. A program of certain importance is to find the full mathematical framework which embodies mirror symmetry and hence ascertain the complete set of mathematical implications.

This paper is designed as an introduction to mirror symmetry for an audience whose interest in the subject originates in its mathematical aspects. We will thus avoid using detailed concepts from conformal field theory whenever possible, and emphasize those implications of the result for which the geometric interpretation is most manifest. However, we also provide a sketch of the arguments used to prove the mirror conjecture for a particular class of Calabi–Yau spaces, and in presenting these we will be forced to appeal to conformal field theory methods. These methods are standard and fully understood from the physics viewpoint but may be unfamiliar to some readers.[2] Our intent is not to provide a detailed derivation, but rather to convey those elements of the construction which we believe may be relevant to a reformulation of the result

---

[1] In fact, from the point of view of string theory, the existence of mirror manifolds is quite natural, and had been conjectured by several groups [7] [1].

[2] In this context, we point out that over the last few years a substantial amount of progress has been made in putting conformal field theory on solid mathematical footing.

in purely mathematical terms. Details can be found in the listed references.

Before entering in upon the mathematical details of our discussion, we want to briefly describe mirror symmetry in heuristic physical terms. This informal description will facilitate our later discussions of the arguments establishing explicit examples of mirror symmetry, as well as introduce the physical intuition which leads to the mirror symmetry conjecture; it is not required for an understanding of the more technical description which follows.

Studies in string theory over the last few years have established a correspondence between certain manifolds (and, in fact, more general spaces, as we shall discuss) and objects known as conformal field theories.[3] Roughly, the correspondence involves considering the manifold as the spacetime in which strings propagate. Many of the basic mathematical properties of a given one of these manifolds (such as its cohomology ring) find direct expression in the associated conformal field theory.[4] It turns out, however, that this association of manifolds and conformal field theories is not one-to-one: two and possibly more distinct manifolds may, in fact, correspond to the same conformal theory. We are thus led to partition the space of all manifolds which have conformal field theoretic realizations into classes of those which have the *same* realization. Such distinct manifolds which correspond to the same conformal theory shall be called *string equivalent manifolds*. A pair of Calabi–Yau spaces M and $\widetilde{M}$ are said to form a *mirror pair* if they are string equivalent and have Hodge diamonds satisfying the relation $h^{p,q}(M) = h^{d-p,q}(\widetilde{M})$ where $h^{p,q}$ is the dimension of the (p,q) Dolbeault cohomology group of M, and d is the complex dimension of the manifold.[5] Because the manifolds are string equivalent, the conformal theory associated to a mirror pair provides a previously unsuspected link between these topologically distinct manifolds. By passing through this link, intrinsic properties of one manifold of the pair are reflected in the other. Of much importance is the fact that some such intrinsic properties are easier to understand and calculate when rephrased on

---

[3] In the appendix we briefly outline this Calabi–Yau/conformal field theory correspondence.

[4] In principle, *all* properties of the conformal field theory are expressible as geometric quantities. This translation has been studied in detail only for the three-point functions we describe below, so the geometric interpretation of our results will be most manifest in terms of these.

[5] The two Hodge diamonds are thus related by a reflection about a diagonal axis – hence the name *mirror manifolds* [2].

the mirror. This is the origin of the recent success of [5] in calculating the number of rational curves of arbitrary degree on the quintic threefold – a seemingly impossible calculation becomes tractable when rephrased on the mirror manifold. We will discuss the essential ideas underlying these calculations and illustrate them with explicit examples of mirror manifolds.

In section two we give a more formal discussion of the basic points covered in this introduction and state the mirror conjecture. In section three we prove a less ambitious version of this mirror conjecture and present explicit examples of mirror pairs. In section four we describe some implications, applications and verifications of mirror symmetry. Finally, in section five we offer our conclusions and mention a few important outstanding problems. In an appendix we briefly outline the correspondence between Calabi–Yau manifolds and conformal field theories which underlies the justification for most of the results discussed.

2. Mirror Symmetry Conjecture

The manifolds of interest to our discussion are of the so-called *Calabi–Yau* type. These were introduced in the context of string theory in [9] [10]. For our purposes we take the following definition:

Definition. *A Calabi–Yau manifold is a compact, complex Kähler manifold with vanishing first Chern class and zero first Betti number.*[6]

We note at the outset that it is insufficient for our purposes to limit attention to smooth Calabi-Yau manifolds. Rather, our discussion will also make use of *orbifolds* [10] of smooth Calabi–Yau manifolds by discrete group actions that generically have fixed points. We shall only need to consider, though, V-manifolds with at most abelian quotient singularities. It is known that these admit resolutions to smooth manifolds [11] and it is properties of the latter manifolds which will be of interest to us.

We also note that when we speak of a Calabi–Yau manifold, we generally assume that

---

[6]The requirement of vanishing first Betti number is a convenience for our discussion, but essentially all that we describe applies to tori as well.

a particular choice of polarization (i.e. Kähler structure) and complex structure has been made. Different choices of either the complex or the Kähler structure on the same underlying topological space are said to constitute different points in the moduli space of the Calabi–Yau space. Infinitesimal deformations of the complex structure of a Calabi–Yau manifold are parameterized by the cohomology group $H^{2,1}$ while $H^{1,1}$ parameterizes Kähler deformations. From the point of view of string theory, as explained in the appendix, the specification of a geometrical solution of the string consistency equations requires not only the topology of the Calabi–Yau manifold, but also a choice of Ricci-flat metric (the existence of which is guaranteed by Yau's theorem [12]). The space of all Ricci-flat metrics is parameterized by the possible choices of complex and Kähler structures, thus matching the mathematical data specified above. Furthermore, although the cohomology group $H^{1,1}(M)$ is naturally real, string theory leads one to *complexify* the Kähler cone of M. Concretely, what we mean by this is as follows. If the Kähler metric on M is denoted by $g_{i\bar{j}}$, then in local coordinates the Kähler form can be written $J = ig_{i\bar{j}}dX^i \wedge dX^{\bar{j}}$. In considering string propagation it is important to include an additional structure on the manifold, which leads us to consider $J = (ig_{i\bar{j}} + b_{i\bar{j}})dX^i \wedge dX^{\bar{j}}$ where $b_{i\bar{j}}$ is some chosen element of $H^{1,1}(M)$.[7]

Having given the definition of Calabi–Yau manifolds (and their orbifolds), we would like to give a precise definition of mirror manifolds. The most precise definition, which we give below, relies upon conformal field theory. To motivate the definition we first spend a moment on a descriptive précis of the Calabi–Yau/conformal field theory connection. Immediately after giving the precise definition of mirror manifolds we give two mathematical implications which, as we mention, can be taken as a definition of mirror manifolds for the purposes of this introduction to the subject.

As explained in the appendix, string theory leads us to associate with a chosen Calabi–Yau manifold an object known as a conformal field theory. Many of the geometrical properties of the manifold find a natural expression in the conformal field theory associated to it. In particular, let us specialize to the case of complex dimension three for the clarity of the discussion, though

---

[7]Note that the usual positivity conditions on the imaginary part of J are still expected to hold.

most of the results will generalize to other dimensions. The Hodge diamond for a Calabi–Yau threefold takes the following form:

$$
\begin{array}{ccccccc}
 & & & 1 & & & \\
 & & 0 & & 0 & & \\
 & 0 & & h^{1,1} & & 0 & \\
1 & & h^{1,2} & & h^{2,1} & & 1 \\
 & 0 & & h^{2,2} & & 0 & \\
 & & 0 & & 0 & & \\
 & & & 1 & & &
\end{array}
$$

We see that the only Hodge numbers which are not uniquely determined are $h^{1,1} = h^{2,2}$ and $h^{2,1} = h^{1,2}$. Recall that deformations of M preserving the condition of Ricci-flatness are parameterized by precisely these cohomology groups: $H^{1,1}$ parameterizes infinitesimal deformations of the Kähler class while $H^{2,1}$ parameterizes infinitesimal deformations of the complex structure. The fact that $h^{3,0} = 1$ reflects the property that the canonical line bundle on M (more precisely, canonical sheaf) is trivial and hence admits a global holomorphic section Ω. The three-form Ω allows us to establish a canonical isomorphism between $H^{2,1}(M)$ and $H^1(M,T)$, where T is the holomorphic tangent bundle. Concretely, we have $\omega : H^1(M,T) \to H^{2,1}(M)$ given in local coordinates by

$$
b^j_{\bar{i}} dX^{\bar{i}} \frac{\partial}{\partial X^j} \to \Omega_{ijl} b^j_{\bar{i}} dX^i dX^l dX^{\bar{i}} . \tag{2.2}
$$

As described in the appendix, these cohomology groups are related to a special class of fields in the conformal field theory, which generate deformations of the theory preserving its superconformal structure. These fields are labeled by their transformation properties under the symmetry algebra. Further, the difference between the two kinds of fields generating the two kinds of deformations lies in the *sign* of their charge (eigenvalue) under a $U(1)$ subalgebra, the choice of sign being a matter of convention. In other words, whereas elements of $H^{1,1}$ and $H^{2,1}$ are vastly different geometrical objects, their conformal field theory counterparts have no intrinsic differentiating characteristics. This led several physicists [7] [1] to conjecture the existence of a second Calabi–Yau manifold, which would correspond to the *same* conformal field theory, but such that the identifications of fields with geometrical objects would follow the opposite convention for the sign of the $U(1)$ eigenvalue.

We can now define what is meant by a mirror pair.

Definition. *Two Calabi–Yau manifolds M and $\widetilde{M}$ constitute a mirror pair if they correspond to the same conformal field theory, and the association of geometrical objects on the two manifolds to fields in the conformal field theory differs by a reversal of the charges under the left-moving $U(1)$ of all fields.*

Although this is technically the most precise definition, as well as the strongest, it may not be very illuminating in its present form. Hence, we now give two particular implications which follow from this definition. These embody the properties of mirror manifolds which have been used in mathematical applications to date.[8]

Implication: If M and $\widetilde{M}$ are a mirror pair then

1) $h^{p,q}(M) = h^{3-p,q}(\widetilde{M})$

2) $H^{p,q}(M) \cong H^{3-p,q}(\widetilde{M})$ under an isomorphism preserving the quantum triple products.

The first of these implies that the Hodge diamond of $\widetilde{M}$ is related to that of M by a reflection about a diagonal axis; this is the origin of the term "mirror manifolds" [2]. In particular, except in very special cases, M and $\widetilde{M}$ are topologically distinct. The quantum trilinear products are deformations of natural mathematical triple products based on the ordinary cup product of $H^*(M)$. Their motivation from the point of view of string theory is described in the appendix; here we give the explicit expressions for the nontrivial cases. To do so, let us first define the natural mathematical triple products referred to above. These are [13] [14] :
$I^{1,1} : H^{1,1} \times H^{1,1} \times H^{1,1} \to \mathbb{C}$ given by

$$I^{1,1}(B_{(i)}, B_{(j)}, B_{(k)}) = \int_M B_{(i)} \wedge B_{(j)} \wedge B_{(k)} \tag{2.3}$$

and $I^{2,1} : H^{2,1} \times H^{2,1} \times H^{2,1} \to \mathbb{C}$ given by

$$I^{2,1}(A_{(i)}, A_{(j)}, A_{(k)}) = \int_M \Omega \wedge A^l_{(i)} \wedge A^m_{(j)} \wedge A^n_{(k)} \Omega_{lmn} \tag{2.4}$$

---

[8] Thinking of these implications as constituting the definition of mirror manifolds will suffice for the present discussion. However, it is crucial to emphasize that in the language of physics, mirror manifolds, as stated, correspond to the same conformal field theory. Thus, not only are their respective three point functions equal (which is the content of implication 2), but their n-point functions are also equal, for arbitrary n. The implications we give emphasize the three point functions as these have the most direct geometrical interpretation; it would be of interest to study the geometrical content of the other n-point functions (both in three and also higher dimensions).

where Ω_{lmn} is he holomorphic three-form on M and $A_{(i)}$ are elements of $H^{2,1}$ (expressed as elements of $H^1(M,T)$ with their subscripts being tangent space indices). Notice that (2.3) is nothing but the cup product giving rise to the triple intersection matrix. Equation (2.4) is most easily described as the natural map from $H^1(T) \times H^1(T) \times H^1(T) \to \mathbb{C}$ arising from the triviality of the canonical bundle. As written, (2.4) is this map after the isomorphism (2.2) has been employed.

The quantum triple products are deformations of these natural mathematical constructs in a manner dictated by string theory as outlined in the appendix. In particular, these are $I_Q^{1,1} : H^{1,1} \times H^{1,1} \times H^{1,1} \to \mathbb{C}$ given by [15]

$$
\begin{aligned}
I_Q^{1,1}(B_{(i)}, B_{(j)}, B_{(k)}) &= I^{1,1}(B_{(i)}, B_{(j)}, B_{(k)}) \\
&+ \sum_{I_{n,m}} m^{-3} e^{(-\int_{I_{n,m}} \iota^*(J))} \int_{I_{n,m}} \iota^*(B_{(i)}) \int_{I_{n,m}} \iota^*(B_{(j)}) \int_{I_{n,m}} \iota^*(B_{(k)})
\end{aligned}
\tag{2.5}
$$

where $I_{n,m}$ is an m-fold cover of a rational curve on M of degree n, $\iota : I_{n,m} \hookrightarrow M$ the inclusion, and J is the Kähler form of M. The factor m^{-3} in (2.5) was discovered phenomenologically in [5] by an application of mirror symmetry (see section four) and has recently been put on a firm footing in [16]. The definition of the quantum deformation of (2.4) is much simpler : $I_Q^{2,1} : H^{2,1} \times H^{2,1} \times H^{2,1} \times \to \mathbb{C}$ is given by

$$
I_Q^{2,1}(A_{(i)}, A_{(j)}, A_{(k)}) = \int_M \Omega \wedge A_{(i)}^l \wedge A_{(j)}^m \wedge A_{(k)}^n \Omega_{lmn} .
\tag{2.6}
$$

In other words, the relevant triple product arising in string theory is precisely the natural mathematical construct in this case [17].

Having defined what we mean by the quantum triple products, let us briefly explain why the implications given above follow from the definition of mirror manifolds. To do so, let M and $\widetilde{M}$ be a mirror pair of Calabi–Yau manifolds which correspond to the conformal field theory K. This implies that every construct in K has *two geometrical interpretations* – one on M and one on $\widetilde{M}$. In particular, this is true of the correlation functions in K. As discussed in the appendix, there are two special classes of correlation functions in K. One such class has the geometrical interpretation on M as $I_Q^{1,1}$ while the other is given geometrically by $I_Q^{2,1}$. Now, the crucial point of mirror symmetry is that fields in K which correspond to $(1,1)$ forms $((2,1)$ forms) on M correspond to $(2,1)$ forms $((1,1)$ forms) on $\widetilde{M}$. Thus, correlation functions in K which are geometrically interpretable as $I_Q^{1,1}$ on M necessarily are geometrically interpretable as $I_Q^{2,1}$ on $\widetilde{M}$. Similarly, correlation functions in K which are geometrically interpretable as $I_Q^{2,1}$ on M are geometrically interpretable as $I_Q^{1,1}$ on $\widetilde{M}$. Hence, we arrive at the equality stated in implication 2. In summary then, the identities which arise from mirror symmetry have their origin in the fact that a given object in the conformal field theory has two vastly different geometrical interpretations on the manifolds of a mirror pair.

We now state the *mirror conjecture* and return shortly to a less robust version that can be elevated to the level of a theorem.

Conjecture: For every[9] Calabi–Yau manifold M there exists a *mirror* Calabi–Yau manifold $\widetilde{M}$. In particular, for $d = 3$

$$h_M^{p,q} = h_{\widetilde{M}}^{3-p,q} \tag{2.7a}$$

and

$$I_Q^{1,1}(M) = I_Q^{2,1}(\widetilde{M}) \tag{2.7b}$$

$$I_Q^{2,1}(M) = I_Q^{1,1}(\widetilde{M}) . \tag{2.7c}$$

We note that by the latter equalities of triple products we mean that there exist suitable bases of the relevant cohomology groups such that the equalities hold. In the next section we will show that for a certain subclass of Calabi-Yau spaces we can prove the mirror conjecture and explicitly construct the mirror manifold $\widetilde{M}$.

3. Mirror Manifolds

In this section we focus on a subclass of Calabi-Yau spaces for which we can prove the mirror conjecture described previously. The proof will involve an explicit construction of the

---

[9]The precise meaning of 'every' in this context is not completely clear at this time. For example, a natural question to ask is what happens in examples in which one member of a supposed pair is rigid, i.e. $h^{d-1,1} = 0$? Then the supposed mirror would appear to be non–Kähler. It may well be that a suitable generalization of Calabi-Yau manifolds (that is, more general string solutions – possibly with torsion [18] [2] [19]) can yield the mirror in such situations.

mirror manifold pair; we will see in section four that this allows the extraction of nontrivial information about both members of the pair. We again specialize the discussion to the case of complex dimension three for definiteness, although most of what we say is independent of this choice. We begin with the main result of [2]:

Theorem. *Let M be a Calabi–Yau hypersurface in weighted projective four space of Fermat type. Let G be the group of coordinate scaling symmetries which preserve the holomorphic three form on M. Then, the mirror $\widetilde{M}$ of M exists and is given by the quotient M/G.*

We note that M/G is a singular space since G generally acts with fixed points. However, due to [11] we know that M/G admits a desingularization to a smooth Calabi–Yau manifold; we in fact refer to the cohomology of the latter. (From a string theory point of view, this is simply the well known statement that the string is sensitive to the cohomology of resolved rather than singular orbifolds [10].)

A related point and one that must be emphasized is that although we perform the explicit construction of the mirror for a specific manifold representing a very special point in the parameter space (the Fermat point) the theorem in fact allows us to construct the mirror manifold in (at least) a neighborhood of this point. This is because mirror symmetry yields an isomorphism of the tangents to the paired moduli spaces, allowing us to relate deformations of one manifold away from the Fermat point to deformations of the mirror manifold from the Fermat mirror. The isomorphism relates deformations of the complex structure of M to deformations of the (complexified) Kähler structure of $\widetilde{M}$ and vice versa. Note that in general the deformed version of M and the correspondingly deformed version of $\widetilde{M}$ are not related by an orbifolding operation (as are M and $\widetilde{M}$) and are not of Fermat type – nonetheless they are a mirror pair. The result that these corresponding deformations preserve the mirror symmetry property is, at present, best understood from the conformal field theory point of view. It reflects the fact that we have *one* conformal theory with two geometric interpretations. Deforming the Kähler structure on $\widetilde{M}$ corresponds to a particular deformation of the conformal field theory (by turning on a vacuum expectation value for some truly marginal operator). This deformation of the conformal theory (by this particular marginal operator) is interpretable on M as a complex structure deformation (by the corresponding complex structure modulus). After deformation, then, we again have one conformal theory (a deformation of the original) with two geometrical interpretations (each a deformation of one of the original spaces).

We believe that there is no obstruction to extending this local deformation argument to the whole of moduli space thus yielding the result: *if a Calabi–Yau space lies in the same moduli space as a Fermat hypersurface then it has a mirror.* We do not have a rigorous proof of this assertion, but note that the applications of mirror symmetry discussed in the next section in fact rely on such an extension. Specifically, asymptotic properties at boundaries of the two moduli spaces which are infinitely far from the Fermat (resp. Fermat mirror) point are compared. The success of these calculations reinforces the intuition from conformal field theory that the identification is in fact global.

An immediate corollary of the theorem which underlies its mathematical applications thus far is:

Corollary.

$$I_Q^{1,1}(M) = I_Q^{2,1}(\widetilde{M}) \tag{3.1a}$$

$$I_Q^{2,1}(M) = I_Q^{1,1}(\widetilde{M}) \ . \tag{3.1b}$$

As noted in [2], there is an interesting limit of this corollary – namely, if the 'radius' of M, say, should tend to infinity, then $I_Q^{1,1}(M)$ tends to the intersection matrix on M, i.e. $I^{1,1}(M)$. Then, (3.1) relates this topological quantity on M to a quasi–topological (depends on the complex structure as well) quantity on the topologically distinct space $\widetilde{M}$. We will return to these ideas shortly. Before moving on to a description of the proof of the theorem, we illustrate the result with one example; we will return to this example in the final section, as the object of study in the first applications of mirror symmetry.

Consider the quintic Fermat hypersurface in $\mathbb{CP}^4$. The coordinate scaling symmetries constitute a $(\mathbb{Z}_5)^5$, generated by $Z_i \to \alpha Z_i$ with $\alpha^5 = 1$. One of these (the diagonal) is trivial because we work in projective space. Another does not preserve the holomorphic three form.

Thus, G is the group $(\mathbb{Z}_5)^3$. Hence,

$$Z_1^5 + Z_2^5 + Z_3^5 + Z_4^5 + Z_5^5 = 0 \tag{3.2a}$$

and

$$\frac{Z_1^5 + Z_2^5 + Z_3^5 + Z_4^5 + Z_5^5}{(\mathbb{Z}_5)^3} = 0 \tag{3.2b}$$

constitute a mirror pair. We note that the former has $h^{1,1} = 1$ and $h^{2,1} = 101$ while the latter has $h^{1,1} = 101$ and $h^{2,1} = 1$. In particular, $\chi(M) = -200$ and $\chi(\widetilde{M}) = 200$, where χ denotes the Euler characteristic. In table 1 we list some other explicit examples of the construction.

We now turn to a brief summary of how we establish the mirror symmetry theorem stated above. It is at this point in the discussion that conformal field theory ideas play a crucial role. At present, we are not aware of any means of establishing the results without recourse to these techniques. As alluded to earlier, it is of considerable interest to determine if more standard mathematical methods can be used instead of conformal field theory methods. Our goal here is to communicate the most skeletal description of our arguments, referring the interested reader to [2] for details. Hopefully, the following will indicate the basic phenomenon that possibly will have a mathematical analogue. We also suggest that the reader unfamiliar with these ideas consult the appendix for a brief description of the connection between Calabi–Yau manifolds and conformal field theories.

Our strategy for constructing mirror manifolds is as follows: We seek a nontrivial *automorphism* $\mathcal{O}$ of the conformal field theory K which realizes an operation on the corresponding geometrical construction which is *not* an automorphism of the Calabi–Yau hypersurface M. Rather, we seek an $\mathcal{O}$ such that $\mathcal{O} : M \to \widetilde{M}$ with $\widetilde{M}$ being Calabi–Yau and not isomorphic to M. In this way, M and $\widetilde{M}$ would be intimately linked by their common conformal field theory. In particular, as mentioned above, if $\mathcal{O}$ acts on the conformal field theory by reversing the sign of all charges under the left-moving $U(1)$ symmetry, we have the following situation : Both M and $\mathcal{O}(M)$ correspond to the same conformal field theory since K and $\mathcal{O}(K)$ are isomorphic. Furthermore, since the explicit isomorphism between K and $\mathcal{O}(K)$ is the reversal of the sign of the left moving $U(1)$ charges, the identification of conformal fields with differential forms is *reversed* on $\mathcal{O}(M)$ relative to M. Thus M and $\mathcal{O}(M)$ would constitute a mirror pair, if such an operation $\mathcal{O}$ could be found. In particular, we would have

$$h_M^{2,1} = h_{\mathcal{O}(M)}^{1,1} \tag{3.3a}$$

$$h_M^{1,1} = h_{\mathcal{O}(M)}^{2,1} \ . \tag{3.3b}$$

In [2] such an operation $\mathcal{O}$ was constructed. Its existence relies on the notion of orbifolding. If a manifold M is invariant under a group of symmetries G (in our case these will generally be holomorphic automorphisms) we can consider the quotient space $\widetilde{M} = M/G$. We restrict attention to groups G which ensure that $\widetilde{M}$ is also Calabi–Yau. Such G are called 'allowable'.[10]Similarly, if a conformal theory K respects a symmetry group G, we can consider the quotient theory K/G. The crucial property of these operations for our argument, is that the quotient conformal theory describes propagation on the quotient of the underlying Calabi–Yau manifold by the same allowable group actions G.[11] The correspondence between group actions in the two pictures is furnished by the identification of homogeneous coordinates on the manifold with primary fields in the conformal field theory given by the arguments of [20] [21].

The correspondence between conformal field theories and Calabi–Yau manifolds is best established for what are known as 'minimal model' conformal field theories and represent Calabi–Yau hypersurfaces in weighted projective space [22] [20] [21]. The proof of the theorem given in [2] relies on the algebraic construction of these 'minimal model' conformal field theories (derived from simple representations of the superconformal algebra). Using this formulation, one can show that the operation $\mathcal{O}$ defined by taking the quotient by the group of 'allowable' scaling symmetries G is an automorphism of the conformal field theory satisfying the conditions mentioned. Namely, the explicit isomorphism involves reversing the $U(1)$ eigenvalues of all fields. This statement and its proof comprise the essence of the mirror manifold construction. Let us

---

[10]We note that G is allowable if it preserves the holomorphic d-form present on the initial Calabi–Yau d-fold.

[11]To be more precise, the quotient will have singularities at the fixed points of the G action. The orbifold conformal field theory corresponds to a singular limit of an appropriate desingularization of the quotient. The condition on G described above assures us of the existence of a desingularization which is itself a Calabi–Yau manifold.

note here the points in the argument at which the conformal field theory enters in a crucial way. Initially, arguments from conformal field theory motivated us to construct the particular quotient described above; in a sense this is inessential, since in principle the construction itself can be carried out with no reference to the physical interpretation. One simply considers a given Fermat Calabi–Yau hypersurface together with the appropriate orbifolds thereof. One then finds two manifolds with Hodge diamonds related in the requisite way. The crucial contribution of the conformal field theory argument is in showing that the relation goes quite a bit deeper, giving the isomorphism between the relevant cohomology classes and the equality of the quantum triple products (and more generally quantum n-products) as well as the extension of these equalities through deformation arguments.

Before discussing the implications of mirror symmetry, we pause here to emphasize three important points. First, our construction works equally well on any orbifold of the theories under discussion. That is, the space of all orbifolds of a given theory can be partitioned into mirror pairs. We illustrate this with the quintic hypersurface in table 2. We note that the mirror of a theory M/H with $H \subset G$ is given by M/H^* where H^* is the complement of H in G (that is, the smallest group containing H and H^* is G). Second, our arguments are not specific to complex dimension three and immediately generalize to other dimensions. We illustrate this with a few examples in table three. Third, as discussed earlier, although our discussion to this point has been tied to very special points in the respective Calabi–Yau moduli spaces by deformation arguments we can immediately extend our results to more general points in moduli space.

We also note that even further generalizations can be made since for example, as shown in [2] and illustrated by the third line of table 1, the hypersurface constraint can be relaxed. As mentioned in the introduction, we have reason to believe that even then the class of models for which the above argument works is not the largest for which mirror manifolds exist. Concrete evidence for this can be found in [3]. Simultaneous with the above work on Fermat hypersurfaces and their deformations, Candelas, Lynker and Schimmrigk [3] completed a thorough computer search of Calabi–Yau manifolds in weighted projective four space (not restricted to Fermat type). They found that the set of all such constructions is almost invariant under the interchange of $h^{1,1}$ and $h^{2,1}$. While finding two manifolds differing by the interchange of these Hodge numbers does not assure that the manifolds are a mirror pair (to be a mirror pair they must correspond to the same conformal field theory) this data indicates that mirror duality probably extends beyond the particular class of models treated in [2]. We have subsequently shown [23] that at least some of the candidate mirror pairs found in [3] do in fact correspond to isomorphic conformal field theories, although there are unexpected subtleties in the argument.

Having outlined the arguments by which mirror symmetry was shown to exist, let us briefly turn to some consequences. In [2] we described in some detail a number of important implications of mirror symmetry. In the next section we review the arguments that lead to these claims, as well as some subsequent work by other groups which lead to a rather spectacular verification of their veracity and to some new applications.

4. Applications of Mirror Symmetry

In the previous section we have briefly summarized the arguments which establish the existence of a class of mirror manifolds. In this section we will review some applications of this construction. These will center on the consequences of (2.7b), which we reiterate here explicitly:

$$
\begin{aligned}
\int_{\widetilde{M}} \widetilde{\Omega} \wedge \widetilde{B}_{(i)}^l \wedge \widetilde{B}_{(j)}^m \wedge \widetilde{B}_{(k)}^n \widetilde{\Omega}_{lmn} &= \int_M B_{(i)} \wedge B_{(j)} \wedge B_{(k)} \\
&+ \sum_{I_{n,m}} m^{-3} e^{(-\int_{I_{n,m}} \iota^*(J))} \int_{I_{n,m}} \iota^*(B_{(i)}) \int_{I_{n,m}} \iota^*(B_{(j)}) \int_{I_{n,m}} \iota^*(B_{(k)})
\end{aligned}
\tag{4.1}
$$

where we remind the reader that the $\widetilde{B}_{(i)}$ are $(2,1)$ forms (expressed as elements of $H^1(\widetilde{M}, T)$ with their subscripts being tangent space indices), $\widetilde{\Omega}$ is the holomorphic three form, the $B_{(i)}$ are $(1,1)$ forms corresponding[12] to the $\widetilde{B}_{(i)}$, $I_{n,m}$ is an m-fold cover of a rational curve on M of degree n, $\iota : I_{n,m} \hookrightarrow M$ the inclusion, and J is the Kähler form of M. In the physics literature, the terms in the second line of (4.1) are called instanon corrections. We recall the crucial role

---

[12] A mathematical description of how one can explicitly determine this correspondence is discussed in [8].

played by the underlying conformal field theory in deriving this equation. If we simply had two manifolds whose Hodge numbers were interchanged we could not make any such statement. This result [2] is rather surprising and clearly very powerful. We have related expressions on *a priori* unrelated manifolds which probe rather intimately the structure of each. Furthermore, the left hand side of (4.1) is directly calculable while the right hand side requires, among other things, knowledge of the rational curves of every degree on the space.[13]

Since written, (4.1) has been verified in several illuminating examples [5] [4] [24]. We briefly discuss the results of these papers and some interesting implications. The authors of [4] sought to verify (4.1) in a particular example through the direct computation of both sides. For ease of computation the authors chose to work with the mirror pair having $\chi = \pm 40$ in table 2. They also chose to work at large radius on the mirror manifold (the $\chi = +40$ member of the pair) so as to suppress the instanton contributions in (4.1) and leave only the topological intersection number on the right hand side [2]. By direct computation, they were able to explicitly verify (4.1) in this limit. This work also uncovered an interesting and as yet incompletely understood subtlety: If a manifold has $h^{1,1} > 1$, one must be careful regarding precisely how the 'large radius' limit in this multidimensional moduli space is defined. It is possible that different limits will pick out different theories, possibly corresponding to the inequivalent ways in which orbifold singularities can be repaired. These ideas and further elaborations on them are discussed in [25], to which the reader is referred to for details.

The other work we would like to describe is [5]. In this paper, Candelas, de La Ossa, Green and Parkes analyze (4.1) for the case of the quintic–mirror-quintic pair (3.2).[14] In particular, they consider a one parameter family of mirror manifolds given by deforming along the single Kähler modulus (complex structure modulus) on the quintic (mirror quintic), as we discussed in the general context in section two. Through a careful analysis of the map between Kähler and complex structure deformations, the authors are able to extract information about the

---

[13]We recall a remark made earlier – (4.1) is but one of an infinite series of equalities implied by having a pair of mirror manifolds. As yet, the full geometrical content of any of the other equalities has not been studied in much detail.

[14]For a recent mathematical presentation of this work see [26].

right hand side of (4.1) from direct calculation of the left hand side. In practice, the integral on the left hand side is computed in terms of the periods of the holomorphic three form $\widetilde{\Omega}$. A comparison to the right hand side then yields a relation between the numbers of rational curves of various degrees on M and the coefficients in an asymptotic expansion of solutions to the Picard-Fuchs equations of $\widetilde{M}$. This enabled these authors to find the *number* of rational curves (holomorphic instantons) of any desired degree on the quintic hypersurface. This is to be compared with the fact that mathematicians have to this point only succeeded in computing these numbers up to degree three, and then only with great effort [27] [28] [29]. Using mirror symmetry in the guise of (4.1), the number of curves of any degree can be calculated. This work has been extended to a number of other examples in [24].

5. Conclusions

In this paper we have sought to introduce the basic ideas and results of recent work that has established the existence of mirror manifolds. In summary, we have established that Calabi-Yau hypersurfaces of Fermat type in weighted projective space (and deformations thereof) have associated mirror manifolds. The latter have Hodge diamonds which are 'mirror images' of the former: $H^{p,q}$ and $H^{d-p,q}$ are interchanged, where d is the complex dimension of the Calabi-Yau spaces. For odd values of d this implies, in particular, that the Euler characteristics of a mirror pair differ by a minus sign. The power of mirror manifolds becomes apparent when we study various triple products on the cohomology of the manifolds, making use of modifications to the classical geometrical structures dictated by string theory. We have explicitly illustrated the identities which follow in the best studied case of $d = 3$. The veracity and power of such identities has recently been exemplified by Candelas, et. al. [5] in which they have been used to determine the number of rational curves of arbitrary degree on the quintic threefold – a problem impervious to standard mathematical methods.

There are a number of open issues which deserve further study. Amongst these are:

1. Can one establish the mirror symmetry results described above using standard mathe-

matical methods? Some progress along this line has been taken by Roan, although his approach has not dealt with the trilinear forms.

2. Does mirror symmetry hold for 'all' Calabi–Yau spaces? As discussed, the naturality arguments based upon the physical interpretation suggest that this should hold. Of course, this will require some flexibility in the class of spaces we consider since there are rigid Calabi–Yau's ($h^{2,1} = 0$) whose mirrors would appear to not be Kähler ($h^{1,1} = 0$). It is possible that Calabi–Yau spaces with torsion [18] are the correct direction to study for such mirrors [2] [19], although no concrete results along this line have been achieved.

3. The construction of mirror manifolds involves building singular spaces from orbifolding smooth varieties. Often there are topologically distinct desingularizations (which nonetheless have the same Hodge numbers) of the latter orbifolds. These distinct desingularizations are all related by flopping [30]. An important question is to determine the conformal field theoretic interpretation of flopping [25] and hence clarify the physical interpretation of these different resolutions. For example, is one particular desingularization singled out or are they all on equal footing? If the latter is true (in the sense of conformal field theory) we might have many 'mirrors' for a given Calabi–Yau manifold.

4. Conformal field theories are relatively complicated constructs. On the other hand, most of our geometrical discussion focuses attention on a small subsector of the full conformal theory, namely those fields with direct geometrical interpretation in terms of cohomology elements. One might wonder whether the results of mirror symmetry might have a natural formulation in terms of a 'truncated version' of conformal field theory that similarly focuses on the geometrically interpretable fields. Topological field theories [31] are the natural candidates for such truncated conformal theories and hence it seems worthwhile to understand the formulation of mirror symmetry in this context. It may well be that topological field theory provides the link required for a fully mathematical formulation of the results we have discussed. We remark once more that such a formulation would, at least superficially, lead to less powerful results limited to the topological subsector of the theory.

These question are currently under active investigation and are sure to ultimately shed much light on the full mathematical and physical ramifications of mirror symmetry.

We thank P. Candelas, I.M. Singer, S.-T. Yau for organizing a productive and enjoyable workshop and G.G. Ross and C. Vafa for a number of important discussions. B.R.G. is supported by the National Science Foundation and M.R.P. is supported in part by DOE grant DE-AC-02-76ERO3075.

Appendix A. Conformal Field Theory and Algebraic Geometry

This appendix attempts to describe the correspondence between conformal field theories and Calabi–Yau manifolds suggested by string theory. The intent is more to convey a glossary of terms and the spirit of the construction than to present a full derivation of the results mentioned.

A1. From Manifolds to Conformal Field Theories

String theory describes the physics of one-dimensional extended objects. As they propagate, strings sweep out a two dimensional "worldsheet", and calculations in string theory lead us to consider a quantum field theory on this two-dimensional surface. A quantum field theory on a surface Σ is constructed from two essential ingredients: *fields* and an *action*. The former are sections of chosen bundles on Σ while the latter is a real valued functional of the sections. For the case at hand, the fields are simply coordinates parameterizing the embedding of Σ in the chosen Calabi–Yau manifold M, and the action is basically the area of the image of Σ in the induced metric (more precisely, this is true after solving the Euler–Lagrange equations for the metric on Σ).[15] Explicitly, the action S is defined by

$$S = \frac{1}{2\pi\alpha'} \int_\Sigma d^2\sigma \sqrt{|h|}(h^{\alpha\beta}G_{ij}(X)\partial_\alpha X^i\partial_\beta X^j + \epsilon^{\alpha\beta}B_{ij}(X)\partial_\alpha X^i\partial_\beta X^j) \, . \qquad (A.1)$$

In (A.1) the 'fields' X^i comprise the map $X : \Sigma \to M$ embedding the worldsheet in a (spacetime) manifold M, $h_{\alpha\beta}$ is a metric on Σ, $d^2\sigma\sqrt{|h|}$ an invariant integration measure on Σ, and G_{ij} a metric on M; the antisymmetric tensor B_{ij} is an additional term, of crucial importance in

---

[15] For all of our discussion, the reader can safely take Σ to have the topology of a sphere.

string theory but absent in discussions of particle propagation; α' is a dimensionful parameter related to the 'tension' of the string. The fundamental objects in a quantum field theory are the *correlation functions* of field ϕ_j (constructed for example from X and its derivatives) inserted at points z_j on Σ which are formally calculated using the action S as

$$\langle \phi_1(z_1) \cdots \phi_N(z_N) \rangle = \int DX e^{-S(X)} \phi_1(z_1) \cdots \phi_N(z_N) \, . \tag{A.2}$$

A rigorous definition of the integral (A.2) over maps X is a very difficult and in general unsolved problem, though in the situations we discuss it is unusually well-behaved. For our purposes, to motivate the construction presented later, it will suffice to consider it merely as a formal device. For example, for small α' we can evaluate the integral as an asymptotic series by a saddle-point calculation.[16] In the theories of interest to us, there are additional (fermionic) degrees of freedom, and the action (A.1) contains other terms, but for our heuristic purposes the explicit form of these will not be necessary.

The consistency conditions of string theory lead us to study a specific class of these "sigma models", namely those which lead to a field theory with $N = 2$ superconformal symmetry. We note that superconformal symmetry embodies conformal symmetry (invariance under conformal transformations on Σ) and supersymmetry. The latter symmetry is not evident in our action since the fermionic degrees of freedom crucial to this symmetry have been suppressed for clarity of exposition. Applying the requirements that (A.1) give rise to a theory respecting superconformal invariance we find a number of conditions.[17] To lowest order in α' these are as follows. The manifold M is required to be a three dimensional complex Kähler manifold, and the metric G to have vanishing Ricci tensor. This is the origin of the interest in Calabi–Yau manifolds on the part of string theorists. In other words, the quantum field theory action (A.1) is conformally invariant if the manifold M is chosen to be Calabi–Yau.[18] The antisymmetric

---

[16] Note that α' can be absorbed into G by a rescaling so the true (dimensionless) expansion parameter is $\frac{\alpha'}{R^2}$. This limit thus corresponds to large (nearly flat) manifolds.

[17] We are also imposing an additional requirement that the central charge of the representation have a prescribed value – this is required for a consistent string theory.

[18] We note that maintaining conformal invariance in the presence of quantum corrections forces adjustments to be made to the Ricci-flat metric on M but they do not change the topological property of M having vanishing first Chern class.

tensor B_{ij} is required to be such that $B_{ij}dX^i dX^j$ is a harmonic $(1,1)$ form on M. The space of solutions to these consistency equations is the space of consistent string theories – loosely referred to in this context as the "moduli space" of the physical problem. Locally, then, this space is parameterized by the inequivalent Ricci-flat metrics on M and by an element of $H^{1,1}$ representing B. The former space is well known to decompose locally as a product of deformations of the complex structure of M parameterized by harmonic $(2,1)$ forms and deformations of the Kähler class of G parameterized by harmonic $(1,1)$ forms. The way the two terms appear in the action leads us to consider the variation of B_{ij} as *complexifying* the Kähler cone.

From this discussion, we see that starting from a Calabi–Yau manifold with the additional structure of the harmonic $(1,1)$ form B, we can build a unique conformal field theory. The point of our discussion in this paper is that going from a conformal field theory to a corresponding Calabi–Yau manifold is *not* unique. Rather, more than one Calabi–Yau manifold can yield the same conformal field theory.

The fields in our theory can be described by their transformation properties under the superconformal symmetry. In the models of interest, they transform as irreducible highest weight representations of this algebra. The detailed structure of the algebra will not be needed here but we note that the highest weight vectors are labeled by their conformal weight h and their charge q under a $U(1)$ symmetry. In fact, in a conformal field theory we find two copies of this algebra which act on holomorphic (resp. antiholomorphic) degrees of freedom, in terms of complex coordinates on Σ. There is a sector of the theory which is of special interest in string theory. It is composed of those highest weight fields with conformal dimension and $U(1)$ charge $(h,q) = (1,\pm 1)$. These "exactly marginal" fields parameterize deformations of the theory which preserve the consistency conditions mentioned above, thus representing tangents to the space of consistent theories. These deformations may be related to deformations of the action, and thus to the geometrically constructed moduli space discussed above. In particular, the space of fields with $q = 1$ under both the holomorphic and the antiholomorphic algebras is related to $H^{2,1}$, and the space of fields with $(q = -1)$ $q = 1$ under the (anti-)holomorphic algebras to $H^{1,1}$. This identification can be made quite explicit using the construction of the fields in terms of the coordinates X^i (the form structure is encoded in the fermionic degrees of freedom we have omitted here), and one may thus translate the evaluation of correlation functions of these fields to geometrical calculations on M. This is not a trivial calculation in general but for a certain set of correlators, the results are especially simple. These are the socalled "three-point functions" involving three operators of either type.[19]Because of an exact $SL(2, \mathbb{C})$ symmetry of the theory, these are in fact independent of the points z_i on Σ at which the fields are inserted (see (A.2)) and thus lead to a triple product structure on the relevant cohomology groups which appears quite natural: We find $I^{2,1} : H^{2,1} \times H^{2,1} \times H^{2,1} \to \mathbb{C}$ given by [13]

$$I^{2,1}(A_{(i)}, A_{(j)}, A_{(k)}) = \int_M \Omega \wedge A_{(i)}^l \wedge A_{(j)}^m \wedge A_{(k)}^n \Omega_{lmn} \tag{A.3}$$

where Ω_{lmn} is the holomorphic three-form on M and $A_{(i)}$ are elements of $H^{2,1}$ (expressed as elements of $H^1(M, T)$ with their subscripts being tangent space indices), and $I^{1,1} : H^{1,1} \times H^{1,1} \times H^{1,1} \to \mathbb{C}$ given by [14]

$$I^{1,1}(B_{(i)}, B_{(j)}, B_{(k)}) = \int_M B_{(i)} \wedge B_{(j)} \wedge B_{(k)} \ . \tag{A.4}$$

We thus see that both the topology of M and its complex structure are very naturally embodied in properties of the conformal field theory it defines.

The considerations above were limited to the first nontrivial order in an expansion in small α'. In string theory, the correlation functions (A.2) are understood to be well-defined as functions of α', and the first term in an asymptotic series contains information only about some limit of these functions. However, in the case we are discussing, the existence of the superconformal symmetry places severe restrictions on the dependence of correlation functions on α'. These are embodied in "nonrenormalization theorems", which guarantee that the statements made above at lowest order in α' are in fact true to *all* orders in the expansion [32].[20]Since the

---

[19]We are not being quite precise here; the correlation functions of the fields we discussed in fact vanish (for example by conservation of the $U(1)$ charge). What we are really referring to are the correlation functions of other fields in the theory which are canonically related to the marginal fields by the action of elements of the superconformal algebra.

[20]An exception are the conditions on G. These receive corrections at every order; the convergence of this series has not been proved.

series is asymptotic, "nonperturbative" contributions to the exact answers can still appear. In fact, they do appear, and one may obtain an explicit form for them [15] [17]

$$I^{1,1} \to I_Q^{1,1} = I^{1,1} + \sum_{I_{n,m}} m^{-3} e^{\left(\frac{-1}{\alpha'} \int_\Sigma X^*(J)\right)}$$

$$\int_\Sigma X^*(B^{(i)}) \int_\Sigma X^*(B^{(j)}) \int_\Sigma X^*(B^{(k)}) \tag{A.5a}$$

$$I^{2,1} \to I_Q^{2,1} = I^{2,1} \tag{A.5b}$$

where $I_{n,m}$ is an m-fold cover of a rational curve on M of degree n, $X : \Sigma \to I_{n,m}$, and J is the Kähler form of M. The factor m^{-3} in (A.5a) was discovered phenomenologically in [5] by an application of mirror symmetry (see section four) and has recently been put on a firm footing in [16]. The fact that $I^{2,1}$ is not corrected was shown in [17]. Heuristically, the nonrenormalization theorems tell us that the saddle-point approximation to (A.2) is exact. The equations (A.5) express the contribution (or its absence) from nontrivial extrema.

A2. Conformal Field Theories to Manifolds[21]

We have described how string theory associates a conformal field theory to a geometrical situation. The construction, however, is quite complicated and does not seem like a very promising approach to studying the geometric situation itself. Additionally, various questions of convergence and rigour are left unsatisfactorily addressed. For a particular class of Calabi–Yau spaces, this situation was dramatically improved in [22]. In this work, Gepner constructed superconformal field theories which are exactly solvable in the sense that the correlation functions (A.2) can be explicitly evaluated. This is achieved by constructing a twisted product of simple, well-studied representations of the $N = 2$ superconformal algebra. What is remarkable is that this purely algebraic construction leads to exactly the same field theory as the geometrical construction described above. More precisely, one can obtain in this way the conformal field theory corresponding to any Calabi–Yau manifold given by the vanishing locus of a polynomial of Fermat type in a weighted projective space, with a prescribed Kähler structure corresponding to

---

[21]To avoid any confusion we emphasize that only a limited subclass of conformal theories have a geometrical interpretation.

a "radius" of order 1 (in units of $\sqrt{\alpha'}$). This correspondence was conjectured by Gepner on the basis of the representation content of the theories under large discrete symmetry groups, and further strengthened by the work of [17] in which the values of $I^{2,1}$ were shown to agree exactly between the two approaches (to within an undetermined normalization). A formal argument proving the equivalence based upon path integrals was later given in [20] [21] and a more rigorous derivation of some of the results using more traditional methods in [33]. For these specified points in certain moduli spaces we thus have exact formulas for the correlation functions of interest. As mentioned above, we also have an explicit identification of deformations of the conformal field theory with deformations of the geometrical structure. This allows us to extend the solution to a neighborhood of the exactly solved point.

Globally, the validity of the identification of stringy moduli space with the geometric construction is not guaranteed, and one may expect some additional identifications between inequivalent points in the geometrical parameter space when considered as string theories. For example, one readily sees that (A.2) is unchanged by adding to B_{ij} an element of $H^{1,1}(M, \mathbb{Z})$ so this variable is to be thought of as periodic, and there are other such identifications. Mirror symmetry may be considered a highly nontrivial generalization of these, relating two manifolds which are in fact topologically distinct.

A3. A simple Example

An example may serve to clarify some of the preceding discussion. The simplest example of a compact Ricci-flat manifold is a torus (this is not three-dimensional but we can imagine a tensor product of this with a "trivial" theory). Since M is in fact flat, the discussion simplifies considerably and the conformal field theory is exactly solvable throughout the moduli space. In the discussion above, there are no higher-order corrections to G in this case. The infinitesimal deformations of the conformal field theory are generated by two (complex) fields – one corresponding to a deformation of the complex structure and one to a Kähler deformation. Indeed, in this example the global moduli space is known: the flat metrics G are parameterized by the complex structure labelled by a point τ in the fundamental domain of $SL(2, \mathbb{Z})$, and the volume $|G|$. When we incorporate the B field we are led to the complex combination $\rho = B_{12} + i\sqrt{|G|}$

parameterizing the upper half plane. In fact, the conformal field theory is invariant under $B \to B + 1$ as mentioned above, but also under $\rho \to -1/\rho$ so that inequivalent conformal field theories on a torus are parameterized by two copies of the fundamental domain. The "modular group" [5] is thus $SL(2, \mathbb{Z}) \times SL(2, \mathbb{Z})$. Finally, there is an additional identification, representing mirror symmetry,which interchanges the roles of complex structure and Kähler structure deformations. In this example the symmetry is simply $\tau \leftrightarrow \rho$.

References

[1] W. Lerche, C. Vafa, and N.P. Warner, Nucl. Phys. **B324** (1989) 427.

[2] B.R. Greene and M.R. Plesser, Nucl. Phys. **B338** (1990) 15.

[3] P. Candelas, M. Lynker, and R. Schimmrigk, Nucl. Phys. **B341** (1990) 383.

[4] P.S. Aspinwall, C.A. Lütken, and G.G. Ross, Phys. Lett. **241B** (1990) 373.

[5] P. Candelas, X.C. de la Ossa, P.S. Green, and L. Parkes, Nucl. Phys. **B359** (1991) 21; Phys. Lett. **258B** (1991) 118.

[6] S. Ferrara, M. Bodner, and A.C. Cadavid, Phys. Lett. **247B** (1990) 25.

[7] L.J. Dixon, in *Superstrings, Unified Theories, and Cosmology 1987*, (G. Furlan et. al., eds.), World Scientific, 1988, p.67.

[8] S.-S. Roan, Int. Jour. of Mathematics 2 (1991) 439.

[9] P. Candelas, G. Horowitz, A. Strominger and E. Witten, Nucl. Phys. **B258** (1985) 46.

[10] L. Dixon, J. Harvey, C. Vafa and E. Witten, Nucl. Phys. **B261** (1985) 620; **B274** (1986) 285.

[11] S.-S. Roan and S.-T. Yau, Acta Math. Sinica, New Series, 3 (1989) 256.

[12] S.-T Yau, Proc. Nat. Acad. Sci USA **74** (1977) 1798.

[13] A. Strominger, Phys. Rev. Lett. **55** (1985) 2547.

[14] A. Strominger and E. Witten, Comm. Math. Phys. **101** (1985) 341.

[15] M. Dine, N. Seiberg, X.G. Wen, and E. Witten, Nucl. Phys. **B278** (1987) 769; **B289** (1987) 319.

[16] P.S. Aspinwall and D.R. Morrison, *Topological field theory and rational curves*, Oxford and Duke Preprint OUTP-91-32P DUK-M-91-12.

[17] J. Distler and B.R. Greene, Nucl. Phys. **B309** (1988) 295.

[18] A. Strominger, Nucl. Phys. **B274** (1986) 253.

[19] M. Roček, *Modified Calabi–Yau Manifolds with Torsion*, in this volume.

[20] B.R. Greene, C. Vafa and N.P. Warner, Nucl. Phys. **B324** (1989) 371.

[21] E. Martinec, in *V.G. Knizhnik Memorial Volume* (L. Brink et. al., eds.); Phys. Lett **B217** 431.

[22] E. Gepner, Phys. Lett. **199B** (1987) 380; Nucl. Phys. **B296** (1987) 380.

[23] M.R. Plesser, talk presented at this workshop; B.R. Greene and M.R. Plesser, in preparation.

[24] D.R. Morrison, *Picard-Fuchs equations and mirror maps for hypersurfaces*, Duke Preprint DUK-M-91-14.

[25] P.S. Aspinwall and C.A. Lütken, Nucl. Phys. **B355** (1991) 482.

[26] D. Morrison, *Mirror Symmetry and Rational Curves on Quintic Threefolds: A Guide for Mathematicians*, Duke Preprint DUK-M-91-01.

[27] J. Harris, Duke Math. Jour. **46** (1979) 685.

[28] S. Katz, Compositio Math. **60** (1986) 151.

[29] G. Ellingsrude and S.A. Strømme, *The Number of Twisted Cubic Curves on the General Quintic Threefold*, Preprint, 1991.

[30] S.-S. Roan, Int. Jour. Math. **1** (1990) 211.

[31] See the contributions of C. Vafa and E. Witten to this volume, and references therein.

[32] E. Witten, Nucl. Phys. **B268** (1986) 79.

[33] S. Cecotti, Nucl. Phys. **B355** (1991) 755.

M	$h^{2,1}$	$h^{1,1}$	S	$\widetilde{M}$
$z_1^5 + z_2^5 + z_3^5 + z_4^5 + z_5^5 = 0$	101	1	$\mathbb{Z}_5^5$	$M/(\mathbb{Z}_5^3)$
$z_1^5 + z_2^{10} + z_3^{10} + z_4^{10} + z_5^2 = 0$	145	1	$\mathbb{Z}_5 \times \mathbb{Z}_{10}^3 \times \mathbb{Z}_2$	$M/(\mathbb{Z}_{10}^2)$
$z_1^3 + z_2^3 + z_3^3 + z_4^3 = 0$ $z_1 x_1^3 + z_2 x_2^3 + z_3 x_3^3 = 0$	35	8	$\mathbb{Z}_3 \times \mathbb{Z}_9^3$	$M/(\mathbb{Z}_3 \times \mathbb{Z}_9)$
$z_1^3 + z_2^4 + z_3^5 + z_4^6 + z_5^{20} = 0$	47	23	$\mathbb{Z}_3 \times \mathbb{Z}_4 \times \mathbb{Z}_6 \times \mathbb{Z}_{20}$	$M/(\mathbb{Z}_2)$
$z_1^3 + z_2^4 + z_3^4 + z_4^{12} + z_5^{12} = 0$	89	5	$\mathbb{Z}_3 \times \mathbb{Z}_4^2 \times \mathbb{Z}_{12}^2$	$M/(\mathbb{Z}_3 \times \mathbb{Z}_4^2)$
$z_1^4 + z_2^6 + z_3^{21} + z_4^{28} + z_5^2 = 0$	52	28	$\mathbb{Z}_4 \times \mathbb{Z}_6 \times \mathbb{Z}_{21} \times \mathbb{Z}_{28}$	$M/(\mathbb{Z}_2)$
$z_1^5 + z_2^6 + z_3^{10} + z_4^{30} + z_5^2 = 0$	91	7	$\mathbb{Z}_5 \times \mathbb{Z}_6 \times \mathbb{Z}_{10} \times \mathbb{Z}_{30}$	$M/(\mathbb{Z}_2 \times \mathbb{Z}_{10})$

Table 1

Mirrors $\widetilde{M}$ of Some Minimal Model Compactifications M

The equations listed define M as a hypersurface in an appropriate weighted projective 4-space (such that the equation is homogeneous). The last column gives the mirror manifold $\widetilde{M}$ as a quotient of M. The third entry defines a submanifold of $\mathbb{CP}^3 \times \mathbb{CP}^2$, demonstrating that the restriction to minimal models sometimes extends beyond hypersurfaces.

Symmetries	$h^{2,1}$	$h^{1,1}$	χ
	101	1	-200
$[0,0,0,1,4]$	49	5	-88
$[0,1,2,3,4]$	21	1	-40
$[0,1,1,4,4]$ $[0,1,2,3,4]$	21	17	-8
$[0,1,1,4,4]$	17	21	8
$[0,1,3,1,0]$ $[0,1,1,0,3]$	1	21	40
$[0,1,4,0,0]$ $[0,3,0,1,1]$	5	49	88
$[0,1,0,0,4]$ $[0,0,1,0,4]$ $[0,0,0,1,4]$	1	101	200

Table 2

Orbifolds of $z_1^5 + z_2^5 + z_3^5 + z_4^5 + z_5^5 = 0$ in $\mathbf{CP}^4$

The generators of each symmetry group are represented by the powers of a fundamental fifth root of unity by which they multiply the homogeneous coordinates of $\mathbf{CP}^4$. Thus $[0,0,0,1,4]$, for example, represents

$$(z_1, z_2, z_3, z_4, z_5) \to (z_1, z_2, z_3, \alpha z_4, \alpha^4 z_5)$$

with $\alpha^5 = 1$.

d	M	$h^{d-p,p}(M) = h^{p,p}(\widetilde{M})$	G
2	$Y_{3;4}$	1 19 1	$\mathbb{Z}_4^2$
3	$Y_{4;5}$	1 101 101 1	$\mathbb{Z}_5^3$
4	$Y_{5;6}$	1 426 1752 426 1	$\mathbb{Z}_6^4$
5	$Y_{6;7}$	1 1667 18327 18327 1667 1	$\mathbb{Z}_7^5$

Table 3

Calabi–Yau Hypersurfaces of Fermat Type in various dimensions and their Mirrors

The indicated quotient of the manifold M (of complex dimension d) is its mirror $M/G \sim \widetilde{M}$. $Y_{x;y}$ denotes the vanishing locus of a Fermat polynomial of degree y in $\mathbb{CP}^x$. The middle column lists the Hodge numbers $h^{d-p,p}(M)$ (all others are fixed by the Lefschetz theorem to be $h^{p,q} = \delta_{p,q}$) which are identical to the Hodge numbers $h^{p,p}$ of $\widetilde{M}$. The first line represents the K3 manifold, which is its own mirror.

AMS/IP Studies in Advanced Mathematics
Volume **9**, 1998

A Pair of Calabi–Yau Manifolds As an Exactly Soluble Superconformal Theory[*]

Philip Candelas[1], Xenia C. de la Ossa[1†], Paul S. Green[2] and Linda Parkes[1]

1) Theory Group, Department of Physics, The University of Texas, Austin, Texas 78746

2) Department of Mathematics, University of Maryland, College Park, Maryland 20742

Abstract

We compute the prepotentials and the geometry of the moduli spaces for a Calabi–Yau manifold and its mirror. In this way we obtain all the sigma model corrections to the Yukawa couplings and moduli space metric for the original manifold. The moduli space is found to be subject to the action of a modular group which, among other operations, exchanges large and small values of radius though the action on the radius is not as simple as $R \to 1/R$. It is shown also that the quantum corrections to the coupling decompose into a sum over instanton contributions and moreover that this sum converges. In particular there are no 'sub-instanton' corrections. This sum over instantons points to a deep connection between the modular group and the rational curves of the Calabi–Yau manifold. The burden of the present work is that a mirror pair of Calabi–Yau manifolds is an exactly soluble superconformal theory, at least as

[*]Supported in part by the Robert A. Welch Foundation and N.S.F. Grants PHY-880637 and PHY-8605978.

© 1998 American Mathematical Society
and International Press

far as the massless sector is concerned. Mirror pairs are also more general than exactly soluble models that have hitherto been discussed since we here solve the theory for all points of the moduli space.

Contents

1. Introduction

2. The Mirror of $\mathbb{P}_4(5)$

    2.1   Rudiments of the Homology

    2.2   The Modular Group

3. The Periods

4. The Prepotential, Metric and Yukawa Coupling

    4.1   The Yukawa Coupling

5. $\mathbb{P}_4(5)$, The Mirror Map and Quantum Corrections

    5.1   The Loop Term

    5.2   The Mirror Map

    5.3   The Sum Over Instantons

    5.4   The Number of Rational Curves of Large Degree

    5.5   Some Further Remarks on the Modular Group

6. A Mechanism for Supersymmetry Breaking

A. A Second Look at the Homology of $\mathcal{W}$

    A.1   Cycles Corresponding to the Periods ϖ_j

B. Further Properties of the Periods

1. Introduction.

This discovery of mirror symmetry [1-3] among pairs of Calabi–Yau manifolds goes a long way towards resolving a long standing puzzle. A Calabi–Yau manifold $\mathcal{M}$ possesses a certain number of parameters. These are parameters associated with the structure of $\mathcal{M}$ as a complex manifold and parameters corresponding to the deformations of the Kähler metric of $\mathcal{M}$. These parameters, which are related to the cohomology of $\mathcal{M}$, give rise to families and antifamilies of particles in the effective low energy theory that results from compactification of the string. The parameters corresponding to deformations of the complex structure are related to the cohomology group H^{21} of (2,1)–forms while the parameters corresponding to deformations of the Kähler form correspond to the group H^{11} of (1,1)–forms. The Yukawa couplings of the low energy theory correspond to certain cubic forms on the cohomology ring [4]. There are no couplings between the two different sorts of parameters, so the Yukawa couplings come in two types. The puzzle has been that the two types of couplings are very different both at a mathematical level and with regard to renormalization. The couplings corresponding to the complex structure parameters vary with the parameters and are not renormalized either in loops or by instantons. By contrast the couplings corresponding to the Kähler class are topological numbers that are integers in an appropriate basis and which *are* renormalized by instantons [5-7]. In this sense one might describe analysis based on a Calabi–Yau manifold as being 'half exact'. However, since both $\mathcal{M}$ and its mirror $\mathcal{W}$, for which the roles of the two types of parameters are exchanged, correspond to the same superconformal theory, one can combine the calculations and obtain exact results. One can compute both types of Yukawa couplings by calculating the couplings for the complex structure parameters of $\mathcal{M}$ and then computing the remaining couplings, complete with their sigma model corrections, by computing the couplings corresponding to the complex structure parameters of $\mathcal{W}$. The suggestion that Calabi–Yau manifolds should arise in mirror pairs was made by Dixon and Gepner [8], and by Lerche, Vafa and Warner [9] the latter paper also uncovered the chiral ring structure of the superconformal theories. It is the identification of the chiral ring of the superconformal theory with the cohomology ring of (2,1)–forms [10] on the Calabi–Yau manifold that enables us to perform exact calculations by means of geometrical methods.

Of course if the Yukawa couplings were all we could compute then the results would not be very significant since we need also to be able to compute the metric on the parameter space, which appears in the kinetic terms of the sigma model, in order to correctly normalize the fields. Fortunately an extension of the nonrenormalization theorem that ensures that the superpotential does not receive sigma model corrections enables us to compute also the kinetic terms. The observation that the existence of mirror manifolds permits the calculation of both types of couplings has been made independently by Greene and Plesser [2], and also in the interesting article by Aspinwall, Lütken and Ross [3] who consider a mirror pair of manifolds with $\chi = \pm 40$ that has several parameters and who show that, in an appropriately defined large complex structure limit, the Yukawa couplings for the complex structure parameters of the mirror manifold coincide with the topological couplings of the original manifold. One of the new elements of the present work is that we are able to solve, in the context of a particular example, for the Yukawa couplings and the metric on the parameter space for all values of the complex structure.

This article is devoted to a discussion of these issues in the context of the solution of the conformal field theory of a particular example of a mirror pair of Calabi–Yau manifolds. We take for the manifold $\mathcal{M}$ the quintic threefold $\mathbb{P}_4(5)$, which has $b_{11} = 1$, $b_{21} = 101$ and Euler number -200. The mirror $\mathcal{W}$ of this manifold is known in virtue of a construction due to Greene and Plesser [2]. The mirror $\mathcal{W}$ has $b_{11} = 101$, $b_{21} = 1$ and Euler number $+200$. What we do here is calculate the prepotential for the complex structure parameter of $\mathcal{W}$. By mirror symmetry this yields the fully corrected prepotential for the original manifold $\mathcal{M}$. Although we concentrate on a specific and simple case we believe that many features of our results are of general validity.

Since the present work is somewhat beset by detail we list here the salient results:

(i) The metric and the Yukawa couplings are computed complete with all sigma model corrections for all points in the parameter space. There is a particular value of the parameter for which the appropriate conformal field theory is the Gepner model 3^5 and for this value we find agreement with the known coupling for this model.

(ii) It is found that a modular group, Γ, acts on the parameter space. Among other operations Γ exchanges large and small values of the radius. This is of interest because it is relevant to the conjectured existence of a minimum fundamental length in string theory. The existence of a modular group has been noted previously for the case of orbifold [11] and some consequences of modular invariance and the possibility that Calabi–Yau manifolds would also be subject to a modular group has been examined in a number of papers [12]. The modular group Γ however is not the group $SL(2, \mathbb{Z})$, as had previously been anticipated in the literature, and the operation is not as simple as $R \to 1/R$.

(iii) The exact Yukawa coupling admits a decomposition into a sum over instantons. The sum converges to the exact value so there are no 'sub-instanton' contributions to the coupling or prepotential. Moreover the decomposition of the coupling into a sum over instanton contributions seems to indicate a deep connection between the automorphic functions of the modular group and the rational curves (instantons) of $\mathcal{M}$. As an illustration of this we seem to be able to read off from our results the numbers of rational curves of each degree. This is quite likely of mathematical interest.

(iv) We abstract from the particular pair of Calabi–Yau manifolds studied here an expression for the fully corrected Yukawa coupling, which we conjecture to be of general validity. The fundamental object from which the corrected coupling derives is not so much the 'bare' manifold but rather a 'quantum manifold' consisting of the bare manifold together with its rational curves. In physics–speak this is just the statement that the quantum manifold is the bare manifold together with all worldsheet instantons. However this statement is rendered more precise and seems to be in line with recent developments in mathematics [13].

The layout of this article is as follows: in §2 we review the construction of the mirror $\mathcal{W}$ and discuss the rudiments of the geometry of its space of complex structures. A feature that is important for the three dimensional case is the existence of finite points in the parameter space corresponding to singular Calabi–Yau manifolds. We describe here also the large complex structure limit of $\mathcal{M}$. The detailed structure of the metric and Yukawa couplings derive from a discussion of the homology of $\mathcal{W}$ and a computation of the holomorphic three-form from which the prepotential is constructed. This is done in §3 and the metric and couplings are derived from the prepotential in §4. We turn in §5 to a comparison of the metric and couplings with the 'bare' quantities appropriate to $\mathcal{M}$. For the couplings we identify the quantum corrections with the contributions of instantons. For the metric, in addition to the exponentially small terms, there is also a loop correction similar to the 'four loop term' found in another context by Grisaru, van de Ven and Zanon [14]. We also extend the standard nonrenormalization theroem to show that the four loop term is the only loop term to affect the prepotential. In §6 we present a speculative proposal for a mechanism to achieve a small breaking of supersymmetry at low energies. We have included this proposal here even though it is logically separate from the issue of mirror symmetry because the mechanism is based on a non-Kähler resolution of a conifold and the conifold that arises in the study of the mirror manifold of $\mathbb{P}_4(t)$ is of precisely the type to which such a non-Kähler resolution is appropriate. Finally two appendices deal with a more detailed description of the homology of $\mathcal{W}$ and with further properties of the periods of the holomorphic three-form.

2. The Mirror of $\mathbb{P}_4(5)$

Let $M = \mathbb{P}_4(5)$ be the family of manifolds that can be represented as quintic hypersurfaces in $\mathbb{P}_4$. There are 101 parameters associated with the complex structure of these manifolds, which in this case can all be thought of as the coefficients of the quintic polynomial (see for example [10], and references to the original literature cited therein). There is also one parameter associated with the choice of Kähler class which in this case can be thought of as the radius of the $\mathbb{P}_4$. To contruct the mirror manifold [2] we start with a one parameter subfamily M_1 of quintic hypersurfaces given by the polynomials

$$p = \sum_{k=1}^{5} x_k^5 - 5\psi \prod_{k=1}^{5} x_k \ . \tag{2.1}$$

These hypersurfaces are invariant under the symmetry group[1] generated by

$$g_0 = (1,0,0,0,4) \, ,$$
$$g_1 = (0,1,0,0,4) \, ,$$
$$g_2 = (0,0,1,0,4) \, ,$$
$$g_3 = (0,0,0,1,4) \, ,$$

$$(2.2)$$

where the i-th entry is the power of the fifth root of unity multiplying the i-th coordinate. For example, g_1 represents the $\mathbb{Z}_5$ action

$$(x_1, x_2, x_3, x_4, x_5) \to (x_1, \alpha x_2, x_3, x_4, \alpha^4 x_5)$$

where $\alpha = e^{2\pi i/5}$. In virtue of the fact that the product of all the g's multiplies the homogeneous coordinates by a common phase only three of these are independent so we can choose to work with g_1, g_2 and g_3, say. Note also that the quintic polynomial (2.1) is in fact the most general quintic invariant under these identifications. A family W of mirror manifolds is then obtained by taking the quotient of each Calabi–Yau manifold $\mathcal{M}_\psi$ in M_1 by the group $\mathbb{Z}_5^3$ generated by g_1, g_2, and g_3.

The procedure for dividing out by the $\mathbb{Z}_5$'s involves cutting out the curves and points of the manifold which are left invariant by the symmetries. After making the $\mathbb{Z}_5^3$ identifications, the curves and points are replaced by their smooth equivalents. The action of the $\mathbb{Z}_5^3$ has the 10 fixed curves

$$C_{ijk} : x_i^5 + x_j^5 + x_k^5 = 0 \, , \quad i,j,k \text{ distinct} \, .$$

Each of these curves is a $\mathbb{P}_2(5)$ and is invariant under a $\mathbb{Z}_5$ subgroup. These fixed curves meet in the 10 fixed points

$$p_{ij} : x_i^5 + x_j^5 = 0 \, , \quad i,j \text{ distinct}$$

(there are in fact only ten of these points owing to the identifications (2.2)) each being left invariant by a $\mathbb{Z}_5 \times \mathbb{Z}_5$ subgroup. Each fixed curve contains 3 fixed points and 3 fixed curves

---

[1] We owe this choice of generators to B.R. Greene.

meet in each of the fixed points. We will take care of the curves and the points separately, so we need to know the Euler number of the curves less the points. This is simply

$$\chi(\mathbb{P}_2(5)) - 15 = -25 \ .$$

Also, we need to know how the $\mathbb{Z}_5$'s act on the curves. One leaves the curve invariant and the other two identify it with itself. Thus we calculate the Euler number of the mirror manifold to be

$$\chi = \frac{-200 - 10 \times 5 - 10 \times (-25)}{5 \times 5 \times 5} + 10 \times \frac{(-25) \times 5}{25} + 10 \times \frac{5 \times 25}{5} = +200 \ .$$

A curious fact is that the Euler number of $\mathbb{P}_4(5)$ minus the fifty points and ten curves is zero[2]

One naturally wonders if there are other ways of presenting this manifold. There are in fact two ways of constructing the manifold as a hypersurface in weighted $\mathbb{P}_4$'s. If we set

$$(x_1, x_2, x_3, x_4, x_5) = (y_1 y_3^{1/5}, y_2^{4/5} y_5^{1/5}, y_3^{4/5} y_4^{1/5}, y_4^{4/5} y_2^{1/5}, y_5^{4/5}) \ ,$$

the transformation being well defined in virtue of the identifications (2.2), indeed the transformation was chosen so as to incorporate the identifications in a natural way, then the defining equation becomes

$$p = y_1^5 y_3 + y_2^4 y_5 + y_3^4 y_4 + y_4^4 y_2 + y_5^4 - 5\psi \prod_{k=1}^{5} y_k \ ,$$

on reflection we see that the manifold is $\mathbb{P}_4^{(41,48,51,52,64)}(256)$. Similarly, one can also show[3] that the mirror manifold is also $\mathbb{P}_4^{51,60,64,65,80)}(320)$ (thereby showing that the two weighted hypersurfaces are biholomorphic). In this case the coordinate transformation takes the form

$$(x_1, x_2, x_3, x_4, x_5) = (w_1 w_4^{1/5}, w_2^{4/5} w_5^{1/5}, w_3, w_4^{4/5} w_2^{1/5}, w_5^{4/5})$$

and the polynomial is

$$p = w_1^5 w_4 + w_2^4 w_5 + w_3^5 + w_4^4 w_2 + w_5^4 - 5\psi \prod_{k=1}^{5} w_k \ .$$

---

[2] P.S. Aspinwall informs us that this is a general feature of constructing a mirror manifold by starting with a manifold $\mathcal{M}$ and dividing by a symmetry group. The noncompact manifold that remains after removing the fixed points and fixed curves has Euler number zero.

[3] For a more general dicussion of these techniques see [15].

These weighted projective spaces appear in the tables of [16] and have Euler number $+200$ and $b_{11} = 101$. We find it most convenient to work with the original form (2.1) of the polynomial p though we have to bear in mind that this form refers to a covering space of $\mathcal{W}$. To get to $\mathcal{W}$ itself we must make the identifications (2.2).

One of our aims is to describe the space of ψ's, that is the space W of complex structures of $\mathcal{W}_\psi$. The first thing to note is that ψ and $\alpha\psi$ correspond to the same complex structure since the replacement $\psi \to \alpha\psi$ is equivalent to the coordinate transformation

$$(x_1, x_2, x_3, x_4, x_5) \to (\alpha^{-1}x_1, x_2, x_3, x_4, x_5) ,\tag{2.3}$$

in other words, we learn that the true coordinate is ψ^5.

To describe the geometry of W, it is important to note that there are special values of ψ for which $\mathcal{W}_\psi$ is singular. This occurs when the quintic (2.1) fails to be transverse. That is when the five equations

$$\frac{\partial p}{\partial x_k} = 0 , \quad k = 1, \ldots, 5 \tag{2.4}$$

are simultaneously satisfied. These equations imply that

$$x_1^5 = x_2^5 = \cdots = x_5^5 = \psi \prod_{k=1}^{5} x_k , \tag{2.5}$$

whence

$$\prod_{k=1}^{5} x_k^5 = \psi^5 \prod_{k=1}^{5} x_k^5 .$$

If ψ is finite then none of the x_i's may be zero for if one were zero then by (2.5) all would be zero, which is not allowed. It follows that (2.4) can only be satisfied if $\psi^5 = 1$. If we take $\psi = 1$ say, then returning to (2.5), we see that

$$x_k = \alpha^{n_k} , \quad \sum n_k = 0 .$$

These points are all identified under the identifications (2.2) so that $\mathcal{W}$ at $\psi = 1$ has only one singular point which we may as well take to be the point $(1, 1, 1, 1, 1)$. The singularity is a node that is a point for which p and dp both vanish but the matrix of second derivatives is nonsingular. For these values of ψ, the corresponding $\mathcal{W}$ is a conifold [17]. This type of singular Calabi–Yau manifold has been described in some detail in [17] and [18]. For our present purpose it suffices to recall that a neighborhood of the node is locally a cone with base $S^2 \times S^3$. For values of ψ near $\psi = 1$, say, the situation is as in Figure 2.1. There is an S^3 which shrinks to zero as $\psi \to 1$.

The value $\psi = \infty$ corresponds to the singular quintic

$$\mathcal{W}_\infty : x_1 x_2 x_3 x_4 x_5 = 0$$

which, before identification under the $\mathbb{Z}_5^3$, consists of 5 $\mathbb{P}_3$'s meeting in 10 $\mathbb{P}_2$'s meeting in 10 $\mathbb{P}_1$'s meeting in 5 points. A lower dimensional hewzistic sketch is given in Figure 2.2. We shall see later that $\mathcal{W}_\infty$ is the large complex structure limit of $\mathcal{W}_\psi$ and is the mirror of the large radius limit of $\mathcal{M}$.

2.1. Rudiments of the Homology

The structure of the moduli space of a Calabi–Yau manifold reflects the homology of the manifold. Recall that the complex structure of the manifold can be described by giving the periods of the holomorphic three-form over a canonical homology basis [19] [17] [20]. More precisely we can proceed, for the case under consideration for which $b_{21}(\mathcal{W}) = 1$ and $b_3(\mathcal{W}) = 4$, as follows. We choose a symplectic basis (A^1, A^2, B_1, B_2) for $H_3(\mathcal{W}, \mathbb{Z})$ such that

$$A^a \cap B_b = \delta^a{}_b \,, \quad A^a \cap A^b = 0 \,, \quad B_a \cap B_b = 0 \,. \tag{2.6}$$

Let (α_a, β^b) be the cohomology basis dual to that above so that

$$\int_{A^a} \alpha_b = \delta^a{}_b \,, \quad \int_{B_a} \beta^b = \delta^b_a \,,$$

with other integrals vanishing. Then it follows that

$$\int_{\mathcal{W}} \alpha_a \wedge \beta^b = \delta^b_a \,, \quad \int_{\mathcal{W}} \alpha_a \wedge \alpha_b = 0 \,, \quad \int_{\mathcal{W}} \beta^a \wedge \beta^b = 0 \,.$$

Important in the following is the holomorphic three-form Ω. Being a three-form Ω may be expanded in terms of the basis

$$\Omega = z^a \alpha_a - \mathcal{G}_a \beta^a \ .$$

The coefficients $(z^a, \mathcal{G}_b)$ are the periods of Ω, so called because they are given by the integrals of Ω over the homology basis

$$z^a = \int_{A^a} \Omega \ , \quad \mathcal{G}_b = \int_{B_b} \Omega \ . \tag{2.7}$$

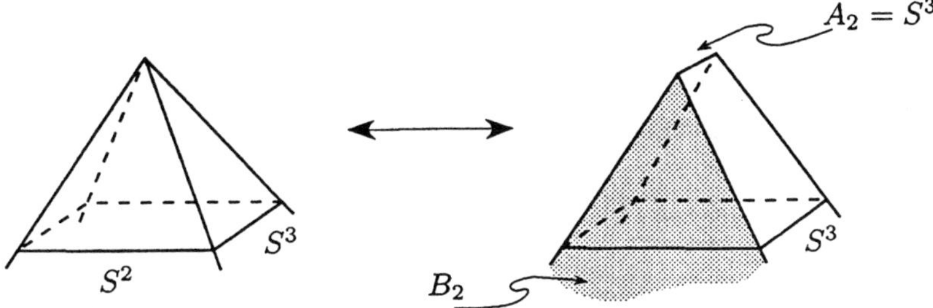

Figure 2.1: The singular point of the conifold has a neighborhood that is a cone with base $S^2 \times S^3$. For $\psi - 1$ small but nonzero the node is replaced by a sphere of radius $O((\psi - 1)^{\frac{1}{2}})$.

We have more periods than parameters. For the present case we have $b_3 = 4$ periods but we know there is only $1(= b_{21})$ parameter for the complex structure. We therefore choose to regard the $\mathcal{G}_b$ as being functions of the z^a, $\mathcal{G}_b = \mathcal{G}_b(z^a)$. This leaves us with the z^a as independent periods. We still have one too many parameters but this turns out to fit in quite well. The scale of Ω is not defined. In reality Ω is a section of a line bundle over the moduli space [21]. There is a gauge invariance associated with Ω due to the fact that Ω is undefined up to multiplication by an arbitrary holomorphic function of the parameter

$$\Omega(\psi) \to f(\psi)\Omega(\psi) \ , \tag{2.8}$$

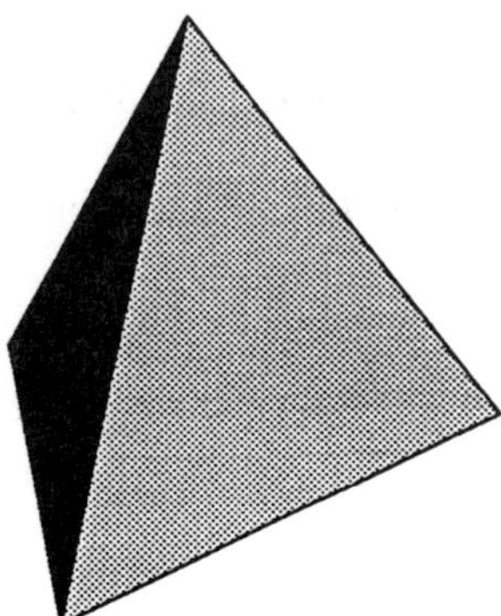

Figure 2.2: A heuristic sketch, drawn as a tetrahedron, of the large complex structure limit of $\mathcal{W}$. Before identification under the $\mathbb{Z}_5^3$ the space consists of 5 $\mathbb{P}_3$'s meeting in 10 $\mathbb{P}_2$'s meeting in 10 $\mathbb{P}_1$'s meeting in 5 points.

and we can regard the two z^a's as projective coordinates for Ω. The $\mathcal{G}_a$ and Ω are then homogeneous of degree one as functions of the z^a

$$\Omega(\lambda z) = \lambda \Omega(z) , \quad \mathcal{G}_b(\lambda z) = \lambda \mathcal{G}_b(z) ,$$

and it can be shown [17] [20] that the $\mathcal{G}_b$ are the gradients of a prepotential $\mathcal{G}(z)$ that is homogeneous of degree two

$$\mathcal{G}_a = \frac{\partial \mathcal{G}}{\partial z^a} , \quad \mathcal{G}(\lambda z) = \lambda^2 \mathcal{G}(z) .$$

For the situation to hand for which the manifold can be represented by a single polynomial there is a well known representation for the holomorphic 3-form [4] [10]. We make the specific gauge choice

$$\Omega = 5\psi \frac{x_5 dx_1 \wedge dx_2 \wedge dx_3}{\dfrac{\partial p}{\partial x_4}} . \tag{2.9}$$

We shall have more to say about the gauge invariance in the following. For the present however (2.9) is a convenient choice because the replacement $\psi \to \alpha\psi$ can be undone by the coordinate transformation (2.3). Thus

$$\Omega(\alpha\psi) = \Omega(\psi) \ .$$

As a choice of basis we take A^2 to be the S^3 that shrinks to zero as $\psi \to 1$. The cylce B_2 is then a three-cycle that intersects this S^3 in a point. The 'tip' of B_2, the rest of which lies outside the neighborhood, is indicated by the shading in Figure 2.1. The other basis cycles A^1 and B_1 do not intersect the S^3 and can be taken to lie outside the neighborhood. We are concerned about the monodromy of the basis about $\psi = 1$, say. Under transport about $\psi = 1$ the S^3, being unambiguously defined for each value of ψ, will return to the same cycle. The cycle B_2 on the other hand is really only defined as being the dual of the A^2 so nothing prevents B_2 from aquiring a multiple of A^2. Thus

$$A^2 \to A^2$$

$$B_2 \to B_2 + nA^2$$

for some integer n. The remaining cycles A^1 and B_1 are remote from the node and so are unaffected by this process. The monodromy of the basis induces a monodromy for the periods (2.7)

$$\begin{pmatrix} \mathcal{G}_1 \\ \mathcal{G}_2 \\ z^1 \\ z^2 \end{pmatrix} \to \begin{pmatrix} 1 & 0 & 0 & 0 \\ 0 & 1 & 0 & n \\ 0 & 0 & 1 & 0 \\ 0 & 0 & 0 & 1 \end{pmatrix} \begin{pmatrix} \mathcal{G}_1 \\ \mathcal{G}_2 \\ z^1 \\ z^2 \end{pmatrix} \ . \tag{2.10}$$

We shall see later that the periods, as functions of ψ, satisfy a linear differential equation which has a singularity at $\psi = 1$. The relation (2.10) can be understood to describe the monodromy of the solutions of the differential equation about the singular point. Such monodromy is familiar, expressing the fact that some of the solutions of a differential equation in the neighborhood of a singularity contain logarithms and hence are multivalued.

2.2. The Modular Group

It is convenient to adopt a matrix notation and to define a period vector

$$\Pi = \begin{pmatrix} \mathcal{G}_1 \\ \mathcal{G}_2 \\ z^1 \\ z^2 \end{pmatrix} \ . \tag{2.11}$$

We have just argued that under transport around $\psi = 1$ the period vector undergoes a transformation $\Pi \to T\Pi$. The matrix T is symplectic, since the transformation must preserve the intersection numbers (2.6), and it is also integral, since the homology basis is integral. We have observed also that ψ and $\alpha\psi$ correspond to the same manifold. It follows that

$$\Pi(\alpha\psi) = A\Pi(\psi) \; ,$$

with A an integral symplectic matrix such that $A^5 = 1$. We will obtain the precise form of A in the following section but for the present it suffices to remark that it is not the identity, neither is T, since we will also find that the integer in (2.10) has the value unity. The two matrices A and T generate a modular group Γ that acts on the period vectors, and also on the universal covering space of W, which may be identified with the upper half plane in such a way that Γ acts by hyperbolic isometries. We shall denote the operations of replacing ψ by $\alpha\psi$ and of analytic continuation about $\psi = 1$ by $\mathcal{A}$ and $\mathcal{T}$. These operations are represented by the matrices A and T. However care is required when composing the operations since the matrices compose 'backwards'. For example

$$\mathcal{T}(\mathcal{A}\Pi) = \mathcal{T}(A\Pi) = A(\mathcal{T}\Pi) = AT\Pi \; .$$

Transport about $\psi = \alpha^k$ corresponds to the operation $\mathcal{A}^k \mathcal{T} \mathcal{A}^{-k}$ and hence to the matrix

$$T_k = A^{-k} T A^k \; , \tag{2.12}$$

and the matrix T_∞ corresponding to transport about $\psi = \infty$ follows from the aboservation that a sum of loops around the fifth roots of unity and around infinity is contractible. Hence

$$T_\infty^{-1} = T_4 T_3 T_2 T_1 T = (AT)^5 \; . \tag{2.13}$$

Thus there are no new generators corresponding to these operations. ψ^5 is a modular invariant of Γ but we can pass to a modular parameter $\gamma = \gamma(\psi)$ such that the action of Γ on γ is represented by 2×2 matrices acting on the upper half γ plane. Since $\mathcal{A}^5 = 1$ we represent it as

$$A = - \begin{pmatrix} \cos\frac{2\pi j}{5} & \sin\frac{2\pi j}{5} \\ -\sin\frac{2\pi j}{5} & \cos\frac{2\pi j}{5} \end{pmatrix} \; ,$$

where the minus sign is of no consequence in virtue of the projectivity of the representation, and j is an integer, $1 \leq j \leq 4$. Since $\mathcal{T}^k$ is never the identity for any $k \neq 0$, $\mathcal{T}$ acts on the upper half plane as a 'deck transformation' of infinite order. Some composition of $\mathcal{T}$ with powers of $\mathcal{A}$ may therefore be represented by a matrix of the form

$$\begin{pmatrix} 1 & \delta \\ 0 & 1 \end{pmatrix} .$$

The magnitude of the translation is fixed by the fact that a fundamental region for the normal subgroup of Γ generated by $\mathcal{T}$ is a ten sided polygon whose sides are permuted by $\mathcal{A}$. This amounts to considering the images under $\mathcal{A}$, $\mathcal{A}^2$, $\mathcal{A}^3$ and $\mathcal{A}^4$ of the imaginary axis and then demanding compatibility with $\mathcal{T}$. Fundamental domains of Γ are illustrated in Figure 2.3.

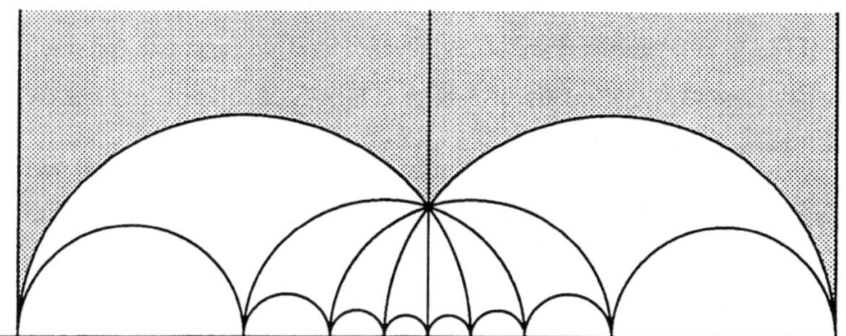

Figure 2.3: The fundamental regions for the group generated by $\mathcal{A}$ and $\mathcal{T}$ consist of a pair of triangles. The circles intersect the real axis where $\gamma = \tan\dfrac{r\pi}{10}$, $r = -4,\ldots,4$. If we take the shaded region to be the fundamental region the points $\psi = (0,1,\infty,\alpha)$ map to $\gamma = (i, \tan\dfrac{2\pi}{5}, \infty, -\tan\frac{2\pi}{5})$ respectively.

The explicit form of $\gamma(\psi)$ may be found by seeking a map that maps the entire ψ^5 plane into a pair of triangles. This is a standard procedure in the theory of automorphic functions involving triangle functions [22]. If the fundamental region for γ is taken to be the shaded region in Figure 2.3 then the relation is

$$\gamma = i\left(\frac{Z_1 - \alpha^2 Z_2}{Z_1 + \alpha^2 Z_2}\right) \tag{2.14}$$

$$= \frac{i}{\pi}\tan\frac{2\pi}{5}\left\{\log(\psi^5) - i\pi + \frac{\sum_{n=0}^{\infty}\dfrac{\Gamma(n+\frac{2}{5})\Gamma(n+\frac{3}{5})}{(n!)^2\,\psi^{5n}}[2\Psi(n+1) - \Psi(n+\frac{2}{5}) - \Psi(n+\frac{3}{5})]}{\sum_{n=0}^{\infty}\dfrac{\Gamma(n+\frac{2}{5})\Gamma(n+\frac{3}{5})}{(n!)^2\,\psi^{5n}}}\right\} ,$$

where the second equality is valid for $|\psi| > 1$, Ψ denotes the digamma function, and Z_1 and Z_2 are defined in terms of Gauss' series

$$Z_1 = \frac{\Gamma^2(\frac{2}{5})}{\Gamma(\frac{4}{5})}{}_2F_1\left(\frac{2}{5},\frac{2}{5},\frac{4}{5};\psi^5\right) , \qquad Z_2 = \psi\frac{\Gamma^2(\frac{3}{5})}{\Gamma(\frac{6}{5})}{}_2F_1\left(\frac{3}{5},\frac{3}{5},\frac{6}{5};\psi^5\right) .$$

We may now complete the specification of the matrices A and T. Recall that $\mathcal{A}$ maps ψ to $\alpha\psi$ and $\mathcal{T}$ transports ψ about $\psi = 1$. It is straightforward to compute the effect of these operations on γ in virtue of (2.14) and the standard analytic continuation formulae for the hypergeometric function.[4]. If we require that the action of A is to rotate fundamental regions by $\dfrac{2\pi}{5}$ about the fixed point $\gamma = i$ then it turns out that $j = 3$ and we find

$$A = \begin{pmatrix} \cos\frac{\pi}{5} & \sin\frac{\pi}{5} \\ -\sin\frac{\pi}{5} & \cos\frac{\pi}{5} \end{pmatrix} , \qquad AT = \begin{pmatrix} 1 & -2\tan\frac{2\pi}{5} \\ 0 & 1 \end{pmatrix} .$$

These relations define the representation. It is easy to check also that

$$A^3 T A^{-3} = \begin{pmatrix} 1 & 0 \\ -2\tan\frac{2\pi}{5} & 1 \end{pmatrix}$$

from which we see that the various monodromy matrices are conjugates of the matrix on the right.

---

[4]If we set

$$f(a;\zeta) = \frac{\Gamma^2(a)}{\Gamma(2a)}{}_2F_1(a,a;2a;\zeta)$$

then the essential relation is

$$\mathcal{T}f(a;\zeta) = f(a;\zeta) - i\tan\pi a\{f(a;\zeta) - \zeta^{1-2a}f(1-a;\zeta)\} .$$

The upper half plane may be mapped to the interior of the unit circle by the transformation

$$\xi = \alpha^3 \left(\frac{1 + i\gamma}{1 - i\gamma} \right) = \frac{Z_2}{Z_1} \, ,$$

the fundamental regions of Γ are then as illustrated in Figure 2.4. In this representation the action of $\mathcal{A}$ is simply that of multiplication by α.

We defer further discussion of Γ to §5 save to observe the Γ which might be termed the quantum modular group, is not $SL(2, \mathbb{Z})$ which had previously been suggested as the modular group based on its action on the 'bare' moduli space of $\mathbb{P}_4(5)$.

In order to find the explicit form of the metric we need to evaluate explicitly the periods of Ω. It is to this that we now turn.

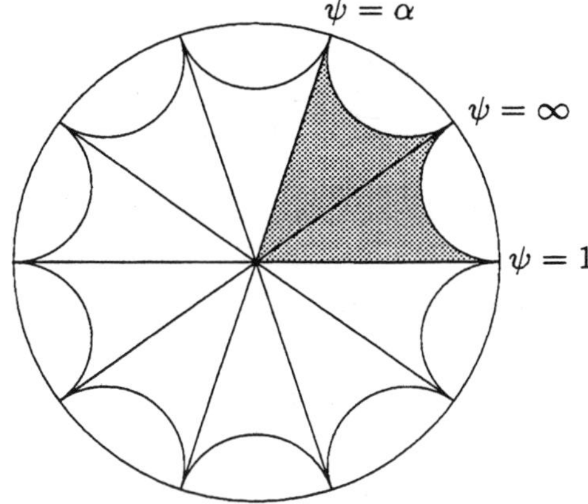

Figure 2.4:   Fundamental regions of Γ in the unit disk.

3. The Periods

We begin by specifying more carefully the A^2 and B_2 of our symplectic basis. A^2 is the S^3 that degenerates to zero at the conifold point $\psi = 1$ and B_2 is a certain torus associated with the degeneration of the manifold as $\psi \to \infty$.

$$A^2 = \{x_k | x_5 = 1 \ , \ x_i \text{ real} \ , \ i = 1, 2, 3,$$
$$x_4 \text{ given by the branch of } p(x) = 0 \text{ that is an } S^3 \text{ as } \psi \to 1\} \ . \tag{3.1}$$

Consider first the case that $\psi - 1$ is real, positive and small. Then by writing

$$\begin{aligned}
x_1 &= 1 + \frac{y_1}{\sqrt{10}} + \frac{y_2}{5} + \frac{y_4}{\sqrt{50}} \\
x_2 &= 1 + \frac{y_1}{\sqrt{10}} - \frac{y_2}{5} + \frac{y_4}{\sqrt{50}} \\
x_3 &= 1 + \frac{y_1}{\sqrt{10}} + \frac{y_3}{5} + \frac{y_4}{\sqrt{50}} \\
x_4 &= 1 + \frac{y_1}{\sqrt{10}} - \frac{y_3}{5} - \frac{y_4}{\sqrt{50}}
\end{aligned}$$

and keeping lowest order terms we see that

$$\sum_{k=1}^{4} y_k^2 = 5(\psi - 1) \ , \tag{3.2}$$

which is manifestly an S^3. The qualification in the definition of A^2 concerning the branch is necessary owing to the fact that the surface $p(x) = 0$ for x_k real has disconnected components as is easily seen by observing that the line

$$(x_1, -x_1, 0, -1, 1)$$

is remote from the point $(1, 1, 1, 1, 1)$ and yet identically satisfies $p(x) = 0$.

On integrating the three form Ω over A^2 we find

$$\begin{aligned}
z^2(\psi) &= \int_{A^2} \Omega \\
&= \frac{4\pi^2}{5^{3/2}}(\psi - 1) + \cdots \ .
\end{aligned} \tag{3.3}$$

The leading term may be obtained by integrating of the zero'th order locus (3.2) and the higher terms may be calculated systematically by means of an iteration scheme. Shortly we shall give an exact expression for $z^2(\psi)$ in terms of hypergeometric functions.

For B_2 we take the cycle

$$B_2 = \{x_k | x_5 = 1 \ , \ |x_1| = |x_2| = |x_3| = \delta \ , \tag{3.4}$$

x_4 given by the solution to $p(x) = 0$ that tends to zero as $\psi \to \infty\}$.

To understand the condition on the branch of the solution, set $x_4 = (\psi x_1 x_2 x_3)^{1/4}\xi$ and write the equation $p = 0$ in the form

$$\xi = \frac{(1 + x_1^5 + x_2^5 + x_3^5)}{5(\psi x_1 x_2 x_3)^{5/4}} + \frac{\xi^5}{5} \ , \tag{3.5}$$

from which it is clear that to leading order as $\psi \to \infty$, there is a solution for ξ given by the first term on the right hand side of (3.5). Thus there is a branch for x_4 with $x_4 = O(\psi^{-1})$ as $\psi \to \infty$ for fixed x_1, x_2, x_3. On the other hand, by rearranging the equation $p = 0$ into the form

$$\xi^4 = 5 - \frac{(1 + x_1^5 + x_2^5 + x_3^5)}{(\psi x_1 x_1 x_3)^{5/4}\xi} \ ,$$

we see that there are four branches for x_4 that are $O(\psi^{1/4})$ as $\psi \to \infty$ for fixed (x_1, x_2, x_3). At this stage it is far from clear that, as defined, B_2 meets A^2 in a single point. Showing that it does requires a more detailed discussion of the homology than we wish to give here so we defer the demonstration of this fact to Appendix A. We turn instead to a computation of the periods.

We have

$$\begin{aligned}
\mathcal{G}_2 &\overset{\text{def}}{=} \int_{B_2} \Omega \\
&= \int_{B_2} \frac{dx_1\, dx_2\, dx_3}{x_1 x_2 x_3 - \psi^{-1} x_4^4} \ .
\end{aligned} \tag{3.6}$$

A comment regarding the orientations of the cycles A^2 and B_2 is necessary here. The cycles do not possess an intrinsically defined orientation so a choice must be made. We have already implicitly fixed the orientation of A^2 by Eq. (3.3). This in turn fixes the orientation of B_2 since we require that $A^2 \cap B_2 = +1$. The choice made in the second of equations (3.6) is consistent with this. As $\psi \to \infty$ we see that the term involving ψ can be neglected and hence

$$\mathcal{G}_2 \to \left(\frac{2\pi i}{5}\right)^3 \ ,$$

the factor of 5^{-3} arising from the $\mathbf{Z}_5^3$ identifications. For ψ large the integrand can be expanded in powers of (ξ^4/ψ). ξ can itself be computed as a power series in ψ^{-1} in virtue of an iteration based on (3.5). The result is of the form

$$\mathcal{G}_2 = \sum_{n=0}^{\infty} a_n \int \frac{dx_1 dx_2 dx_3}{x_1 x_2 x_3} \frac{(1 + x_1^5 + x_2^5 + x_3^5)^{4n}}{(x_1^5 x_2^5 x_3^5 \psi^5)^n} \ .$$

We evaluate the integrals by residues. The only term in the quantity $(1 + x_1^5 + x_2^5 + x_3^5)^{4n}$ that contributes is the term $x_1^{5n} x_2^{5n} x_3^{5n}$ which appears with coefficient $(4n)!/(n!)^4$. Thus

$$\mathcal{G}_2 = \left(\frac{2\pi i}{5}\right)^3 \sum_{n=0}^{\infty} a_n \frac{(4n)!}{(n!)^4 \psi^{5n}} \ .$$

The surprise is that when the a_n's are calculated and substituted into this expression we find

$$\mathcal{G}_2 = \left(\frac{2\pi i}{5}\right)^3 \sum_{n=0}^{\infty} \frac{(5n)!}{(n!)^5 (5\psi)^{5n}} \ , \tag{3.7}$$

and the series on the right converges for $|\psi| \geq 1$.

The function that appears on the right hand side of Eq. (3.7)

$$\varpi_0(\psi) \stackrel{\text{def}}{=} \sum_{n=0}^{\infty} \frac{(5n)!}{(n!)^5 (5\psi)^{5n}} \ , \quad |\psi| > 1 \ , \quad 0 \leq \arg \psi < \frac{2\pi}{5} \ , \tag{3.8}$$

satisfies a linear differential equation of generalized hypergeometric type and it is useful to take advantage of this fact in order to write down a complete set of periods. The restriction on arg ψ anticipates that anlaytic continuation of $\varpi_0(\psi)$ will lead to branch cuts. The fact that periods of the type we are discussing satisfy linear differential equations with regular singular points is a very general property [23]. For the case at hand it is straightforward to check that

$$\left\{ \frac{d^4}{dz^4} - \frac{2(4z-3)}{z(1-z)} \frac{d^3}{dz^3} - \frac{(72z-35)}{5z^2(1-z)} \frac{d^2}{dz^2} - \frac{(24z-5)}{5z^3(1-z)} \frac{d}{dz} - \frac{24}{625z^3(1-z)} \right\} \varpi_0 = 0 \tag{3.9}$$

where $z = \psi^{-5}$. It is compelling to assume that all four periods satisfy this same equation, a fact that we shall assume for the present but verify shortly. Either directly from the equation or from the associated Riemann symbol

$$\mathcal{P} \left\{ \begin{array}{ccc} 0 & \infty & 1 \\ 0 & 1/5 & 0 \\ 0 & 2/5 & 1 \\ 0 & 3/5 & 2 \\ 0 & 4/5 & 1 \end{array} \ \psi^{-5} \right\} \ , \tag{3.10}$$

we observe that the equation is of generalized hypergeometric type. Recall that the generalized hypergeometric equation of fourth order is

$$\{\vartheta(\vartheta + c_1 - 1)(\vartheta + c_2 - 1)(\vartheta + c_3 - 1) - z(\vartheta + a_1)(\vartheta + a_2)(\vartheta + a_3)(\vartheta + a_4)\}w = 0 \ ,$$

where $\vartheta = z\dfrac{d}{dz}$ and the associated Riemann symbol is

$$\mathcal{P}\left\{\begin{matrix} 0 & \infty & 1 & \\ 0 & a_1 & 0 & \\ 1 - c_1 & a_2 & 1 & z \\ 1 - c_2 & a_3 & 2 & \\ 1 - c_3 & a_4 & \sum c_j - \sum a_k & \end{matrix}\right\} \ .$$

The solutions of the differential equation (3.9) can be written in terms of the generalized hypergeometric function [22] [24]

$$\begin{aligned} {}_4F_3(a_1, a_2, a_3, a_4; c_1, c_2, c_3; \zeta) &= \frac{\Gamma(c_1)\Gamma(c_2)\Gamma(c_3)}{\Gamma(a_1)\Gamma(a_2)\Gamma(a_3)\Gamma(a_4)} \times \\ &\times \sum_{n=0}^{\infty} \frac{\Gamma(a_1 + n)\Gamma(a_2 + n)\Gamma(a_3 + n)\Gamma(a_4 + n)}{\Gamma(c_1 + n)\Gamma(c_2 + n)\Gamma(c_3 + n)} \frac{\zeta^n}{n!} \ . \end{aligned}$$

In fact we have

$$\varpi_0(\psi) = {}_4F_3\left(\frac{1}{5}, \frac{2}{5}, \frac{3}{5}, \frac{4}{5}; 1, 1, 1; \frac{1}{\psi^5}\right) \ , \tag{3.11}$$

as is easily verified with the aid of the multiplication formula for the Γ-function in the form

$$\Gamma(z)\Gamma(z + 1/5)\Gamma(z + 2/5)\Gamma(z + 3/5)\Gamma(z + 4/5) = (2\pi)^2 5^{1/2 - 5z}\Gamma(5z) \ . \tag{3.12}$$

A standard maneuver for finding the other solutions to the differential equation in terms of the hypergeometric function amounts here to changing variables from $1/\psi^5$ to ψ^5 and to extracting a factor of ψ^k, with $k = 1, 2, 3,$ or 4, from the Riemann symbol. Thus a set of four linearly independent solutions are

$$\psi^k \mathcal{P}\left\{\begin{matrix} 0 & \infty & 1 & \\ 1/5 - k/5 & k/5 & 0 & \\ 2/5 - k/5 & k/5 & 1 & \psi^5 \\ 3/5 - k/5 & k/5 & 2 & \\ 4/5 - k/5 & k/5 & 1 & \end{matrix}\right\} = \psi^k {}_4F_3\left(\frac{k}{5}, \frac{k}{5}, \frac{k}{5}, \frac{k}{5}; \frac{\overbrace{k+1}}{5}, \frac{k+2}{5}, \frac{k+3}{5}, \frac{k+4}{5}; \psi^5\right) \ ,$$

$$\tag{3.13}$$

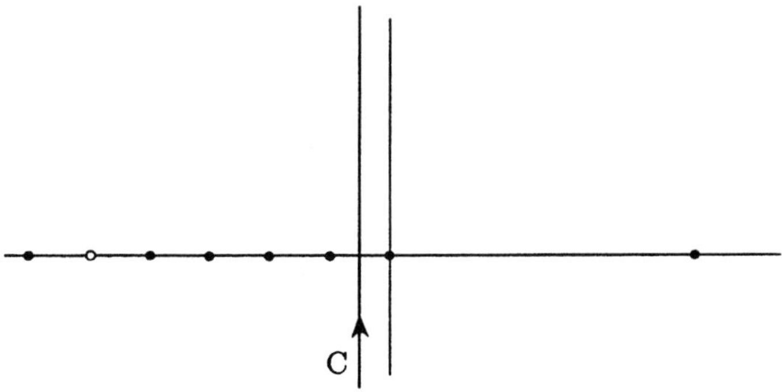

Figure 3.1:   The contour of integration for (3.14) showing the poles of $\Gamma(-s)$ and of $\Gamma(5s+1)$.

where the overbrace signifies that the parameter that is unity is to be omitted. This basis is useful for some purposes, however we find it more convenient to base our development on the functions $\varpi_0(\alpha^k\psi)$. At first sight one might be tempted to conclude from (3.11) that $\varpi_0(\psi)$ is a function of ψ^5 but this is not the case owing to the fact that there are branch cuts implicit in the definition of $\varpi_0(\psi)$. In order to clarify this and to find explicit expressions for $\varpi_0(\alpha^k\psi)$ we wish to analytically continue $\varpi_0(\psi)$ to a neighborhood of the origin. This is accomplished by introducing an integral representation of Barne's type. If $0 < \arg \psi < \dfrac{2\pi}{5}$ we have

$$\varpi_0(\psi) = \frac{1}{2\pi i} \int_C ds \frac{\Gamma(-s)\Gamma(5s+1)}{\Gamma^4(s+1)} e^{i\pi s}(5\psi)^{-5s} , \tag{3.14}$$

with the contour as in Figure 3.1. The integrand has poles on the positive real axis for $s = 0, 1, 2, \ldots$ due to the presence of the factor $\Gamma(-s)$. If in addition $|\psi| > 1$ the contour can be closed to the right and (3.11) is recovered as a sum over residues. If on the other hand $|\psi| < 1$ the contour can be closed to the left enclosing the poles of $\Gamma(5s+1)$. Thus we find

$$\varpi_0(\psi) = -\frac{1}{5} \sum_{m=1}^{\infty} \frac{\alpha^{2m}\Gamma(m/5)(5\psi)^m}{\Gamma(m)\Gamma^4(1-m/5)} , \quad |\psi| < 1 . \tag{3.15}$$

We shall show presently that $\varpi_0(\psi)$ contains logarithms when $\psi^5 = 1$ so we must specify cuts associated with these terms. It is convenient to take $\varpi_0(\psi)$ to be analytic in the neighborhood of the origin so we take the cuts to run radially outward from the points $\psi = \alpha^k$, $k = 0,\ldots,4$. It is clear from the form of the differential operator that the functions

$$\varpi_j(\psi) \stackrel{\text{def}}{=} \varpi_0(\alpha^j \psi) \, , \quad j = 0,\ldots,4 \tag{3.16}$$

all satisfy the differential equation. It is clear also from the series (3.15) that any four of these functions are linearly independent. The five $\varpi_j(\psi)$ being subject to the single relation

$$\sum_{j=0}^{4} \varpi_j(\psi) = 0 \, . \tag{3.17}$$

We wish next to examine further the monodromy of this basis about the point $\psi = 1$. We have argued that the basis $(\mathcal{G}_a, z^b)$ transforms according to the rule (2.10) under transport about the point $\psi = 1$. It follows that the ϖ_j have the transformation rule

$$\left(\frac{2\pi i}{5}\right)^3 \varpi_j(\psi) \longrightarrow \left(\frac{2\pi i}{5}\right)^3 \varpi_j(\psi) + c_j z^2(\psi) \, , \tag{3.18}$$

where the c_j are a set of numerical coefficients. This transformation rule is equivalent to the assertion that the ϖ_j have the structure

$$\left(\frac{2\pi i}{5}\right)^3 \varpi_j(\psi) = \frac{c_j}{2\pi i} z^2(\psi) \log(\psi - 1) + f_j(\psi) \, , \tag{3.19}$$

with $z^2(\psi)$ and $f_j(\psi)$ analytic for $|\psi - 1| < 1$. The period $z^2(\psi)$ corresponds to one of the indices which is unity in (3.10). The differential equation, as is easily verified, admits two solutions that are given by power series about $\psi = 1$. These are $z^2(\psi)$ and another series corresponding to the index 2. The other solutions contain logarithms owing to the fact that the index 1 is repeated. One might have anticipated a more complicated structure than (3.19) with the term multiplying the logarithm being a more general linear combination of the two solutions that have power series expansions. This however would not be consistent with (2.10). Checking that the transformation rule is indeed as in (3.19) amounts to checking the coefficient of $(\psi - 1)^2 \log(\psi - 1)$ in $\varpi_j(\psi)$. This will be subsumed in the following computation of the coefficient c_j.

Suppose $z^2(\psi)$ has the expansion

$$z^2(\psi) = \frac{4\pi^2}{5^{3/2}}\left\{(\psi - 1) + b(\psi - 1)^2 + \cdots\right\} \tag{3.20}$$

about $\psi = 1$. Then from (3.19) we have

$$\left(\frac{2\pi i}{5}\right)^3 \frac{d^2}{d\psi^2}\varpi_j(\psi) = -\frac{2\pi i}{5^{3/2}}c_j\left\{\frac{1}{\psi - 1} + 2b\log(\psi - 1) + \cdots\right\}, \tag{3.21}$$

where the terms indicated by the ellipsis have finite limits as $\psi \to 1$. Thus we can calculate c_j and b by differentiating the series (3.15) and computing its leading behaviour as $\psi \to 1$. It is convenient to set

$$m = 5N + k, \quad k = 0,\ldots,4, \quad N = 0,1,2,\ldots$$

for the variable of summation in (3.15). Then in virtue of Stirling's formula

$$\Gamma(z) = \sqrt{2\pi}z^{z-1/2}e^{-z}\left\{1 + \frac{1}{12z} + O(z^{-2})\right\},$$

we find that as $\psi \to 1$

$$\frac{d^2\varpi_j}{d\psi^2} \sim -\frac{5^{3/2}}{4\pi^2}\sum_{k=0}^{4}\alpha^{kj}(\alpha^k - 1)^4 \sum_{N}^{\infty}\left(\psi^{5N} + \frac{\psi^{5N}}{5N}\right).$$

By comparing this expression with (3.21) we find $b = 1/2$ and

$$c_j = (1, 1, -4, 6, -4), \quad \text{for } j = (0, 1, 2, 3, 4).$$

The next step is to express $z^2(\psi)$ in terms of the basis $\varpi_j(\psi)$. The result is

$$z^2(\psi) = -\left(\frac{2\pi i}{5}\right)^3\left(\varpi_1(\psi) - \varpi_0(\psi)\right). \tag{3.22}$$

To see this suppose x is real and $x > 1$. If ϵ is an infinitesimal, then from (3.19)

$$\left(\frac{2\pi i}{5}\right)^3\left(\varpi_1(x + i\epsilon) - \varpi_1(x - i\epsilon)\right) = -z^2(x).$$

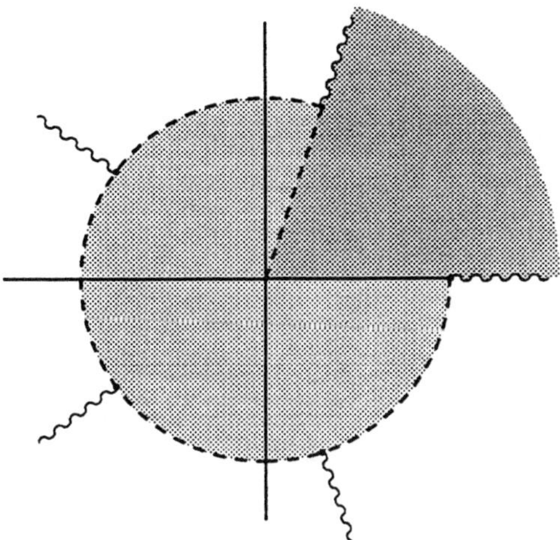

Figure 3.2: The cut ψ plane. The fundametal region is indicated by the heavier shading. The series for the $\varpi_j(\psi)$ given by (3.16) and (3.15) converge for $|\psi| < 1$.

However

$$\varpi_1(x - i\epsilon) = \varpi_0(\alpha(x - i\epsilon))$$

$$= \varpi_0(x + i\epsilon)$$

the last equality following from (3.8) which is valid for $0 \leq \arg \psi < \dfrac{2\pi}{5}$. Thus (3.22) holds for ψ's just above the cut that extends from 1 to ∞ and hence by analytic continuation for all ψ.

We now know z^2 and $\mathcal{G}_2$ in terms of the basis ϖ_j and we wish to find similar expressions for z^1 and $\mathcal{G}_1$. It turns out, rather surprisingly, that we do not have to explicitly describe A^1 and B_1 if we proceed somewhat indirectly. We choose a specific basis of linearly independent ϖ_j's and form a vector

$$\varpi \stackrel{\text{def}}{=} \left(\frac{2\pi i}{5} \right)^3 \begin{pmatrix} \varpi_2 \\ \varpi_1 \\ \varpi_0 \\ \varpi_4 \end{pmatrix} .$$

What we are seeking is a relation of the form

$$\Pi(\psi) = m\varpi(\psi) \, , \tag{3.23}$$

with Π as in (2.11) and with m a numerical matrix which in virtue of (3.22) and (3.7) is of the form

$$m = \begin{pmatrix} a & b & c & d \\ 0 & 0 & -1 & 0 \\ e & f & g & h \\ 0 & 1 & -1 & 0 \end{pmatrix} \, .$$

The periods z^1 and $\mathcal{G}_1$ are not as yet uniquely defined, since given any choice there is the freedom to make a second choice differing from the first by an $Sp(2; \mathbb{Z}) = SL(2; \mathbb{Z})$ transformation

$$\begin{pmatrix} \mathcal{G}_1 \\ z^1 \end{pmatrix} \longrightarrow \begin{pmatrix} p & q \\ r & 2 \end{pmatrix} \begin{pmatrix} \mathcal{G}_1 \\ z^1 \end{pmatrix} \, . \tag{3.24}$$

We can use this freedom to set $f = 0$ in m. We know also that the periods z^1 and $\mathcal{G}_1$ are free of logarithms at $\psi = 1$ which is the condition that the sums, weighted by the coefficients c_j, of the corresponding rows of m must vanish. Consider now how the basis ϖ changes under $\psi \to \alpha\psi$

$$\varpi(\alpha\psi) = a\varpi(\psi)$$

with

$$a = \begin{pmatrix} -1 & -1 & -1 & -1 \\ 1 & 0 & 0 & 0 \\ 0 & 1 & 0 & 0 \\ 0 & 0 & 1 & 0 \end{pmatrix} \, .$$

It is the simple form for a that motivated the choice of basis ϖ. In terms of the symplectic basis we have

$$\Pi(\alpha\psi) = A\Pi(\psi) \, , \quad A = mam^{-1} \, .$$

The matrix A must be integral and symplectic. This turns out to be a stringent condition with a solution for m that is unique up to the $Sp(2; \mathbb{Z})$ transformations (3.24)

$$m = \begin{pmatrix} -\frac{3}{5} & -\frac{1}{5} & \frac{21}{5} & \frac{8}{5} \\ 0 & 0 & -1 & 0 \\ -1 & 0 & 8 & 3 \\ 0 & 1 & -1 & 0 \end{pmatrix} \, , \tag{3.25}$$

and we find

$$A = \begin{pmatrix} -9 & -3 & 5 & 3 \\ 0 & 1 & 0 & -1 \\ -20 & -5 & 11 & 5 \\ -15 & 5 & 8 & -4 \end{pmatrix} .$$

Given m we can return now to the monodromy of the bases about $\psi = 1$.

$$\varpi \longrightarrow t\varpi , \quad \Pi \longrightarrow T\Pi .$$

Where the matrices t and T may be found from (3.18) and (3.25). Given the transformations T and A it is a simple matter to compute the monodromy of the bases about the other singularities in virtue of (2.12) and (2.13). We record these results in Table 3.1. Referring to the table we see that the integer n in our previous expression for the monodromy matrix (2.10) is in fact unity.

	ϖ	Π
$\psi \to \alpha\psi$	$a = \begin{pmatrix} -1 & -1 & -1 & -1 \\ 1 & 0 & 0 & 0 \\ 0 & 1 & 0 & 0 \\ 0 & 0 & 1 & 0 \end{pmatrix}$	$A = \begin{pmatrix} -9 & -3 & 5 & 3 \\ 0 & 1 & 0 & -1 \\ -20 & -5 & 11 & 5 \\ -15 & 5 & 8 & -4 \end{pmatrix}$
Monodromy about $\psi = 1$	$t = \begin{pmatrix} 1 & 4 & -4 & 0 \\ 0 & 0 & 1 & 0 \\ 0 & -1 & 2 & 0 \\ 0 & 4 & -4 & 1 \end{pmatrix}$	$T = \begin{pmatrix} 1 & 0 & 0 & 0 \\ 0 & 1 & 0 & 1 \\ 0 & 0 & 1 & 0 \\ 0 & 0 & 0 & 1 \end{pmatrix}$
Monodromy about $\psi = \infty$	$t_\infty = \begin{pmatrix} -34 & -55 & -310 & 50 \\ 10 & 16 & 90 & -15 \\ 0 & 0 & 1 & 0 \\ -15 & -25 & -155 & 21 \end{pmatrix}$	$T_\infty = \begin{pmatrix} 51 & 90 & -25 & 0 \\ 0 & 1 & 0 & 0 \\ 100 & 175 & -49 & 0 \\ -75 & -125 & 35 & 1 \end{pmatrix}$

Table 3.1: The matrices associated with the transformation $\psi \to \alpha\psi$ and monodromy about $\psi = 1$ and $\psi = \infty$. The matrices associated with monodromy about $\psi = \alpha^k$ are $\alpha^{-k}ta^k$ and $A^{-k}TA^k$.

4. The Prepotential, Metric and Yukawa Coupling

Given the periods $(\mathcal{G}_a, z^b)$ it is now straightforward to construct the prepotential and the metric. The prepotential can be defined by

$$\mathcal{G} = \frac{1}{2} z^a \mathcal{G}_a \ .$$

Recall that the holomorphic three-form Ω has the variational property

$$\int_W \Omega \wedge \frac{\partial \Omega}{\partial \psi} = 0$$

and that this property requires the $\mathcal{G}_a$ to be the derivatives of the prepotential

$$\mathcal{G}_a = \frac{\partial \mathcal{G}}{\partial z^a} \ , \quad a = 1, 2 \ .$$

As a consistency check we can verify that this relation is satisfied. To this end note that, in virtue of the fact that the prepotential is homogeneous of degree two as a function of the z^a,

$$\mathcal{G}(z^1, z^2) = (z^2)^2 \mathcal{G} \left(\frac{z^1}{z^2}, 1 \right) \ .$$

A short calculation now reveals that

$$\frac{\partial \mathcal{G}}{\partial z^a} = \mathcal{G}_a - \epsilon_{ab} z^b \frac{W[z^c, \mathcal{G}_c]}{W[z^1, z^2]}$$

where ϵ_{ab} is the permuation symbol and

$$W[u, v] = u \frac{dv}{d\psi} - v \frac{du}{d\psi} \ .$$

Thus for consistency the identity $W[z^c, \mathcal{G}_c] = 0$ must be satisfied. We have checked that it is indeed satisfied by the somewhtat brutish method of expanding the periods as power series in ψ by means of (3.15) and (3.25) and checking that $W[z^c, \mathcal{G}_c]$ vanishes order by order. Acutally, the vanishing of $W[z^c, \mathcal{G}_c]$ is a very natural condition as can be appreciated by noting that

$$\int_W \Omega(\psi) \wedge \Omega(\psi') = z^c(\psi)\mathcal{G}_c(\psi') - z^c(\psi')\mathcal{G}_c(\psi) \ . \tag{4.1}$$

By differentiating this expression with respect to ψ' and setting $\psi' = \psi$ we find that

$$W[z^c, \mathcal{G}_c] = \int_W \Omega \wedge \frac{\partial \Omega}{\partial \psi}$$

and hence must vanish.

The Kähler potential K is given by the relations

$$e^{-K} = i(\bar{z}^a \mathcal{G}_a - z^a \overline{\mathcal{G}}_a)$$

$$= -i\Pi^\dagger \Sigma \Pi \qquad (4.2)$$

$$= -i\varpi^\dagger \sigma \varpi$$

with

$$\Sigma = \begin{pmatrix} 0 & 0 & 1 & 0 \\ 0 & 0 & 0 & 1 \\ -1 & 0 & 0 & 0 \\ 0 & -1 & 0 & 0 \end{pmatrix} \quad \text{and} \quad \sigma = -\frac{1}{5}\begin{pmatrix} 0 & 1 & 3 & 1 \\ -1 & 0 & 3 & 3 \\ -3 & -3 & 0 & 1 \\ -1 & -3 & -1 & 0 \end{pmatrix} .$$

The metric on the moduli space follows from (4.2). A 3-D plot of $g_{\psi\bar{\psi}}$ against ψ is presented in Figure 4.1. The cusp at $\psi = 1$ (there is of course only one cusp owing to the identification $\psi \approx \alpha\psi$) corresponds to the value of ψ for which $\mathcal{W}$ is a conifold. The metric is mildly singular at the conifold. It can be shown that $g_{\psi\bar{\psi}}$ is asymptotically proportional to $\log|\psi - 1|$ so the conifold is at a finite distance from the smooth manifolds in agreement with general results [17]. In order to compare the metric $g_{\psi\bar{\psi}}$ with the corresponding metric for the moduli space of $\mathbb{P}_4(5)$ it is of interest to compute the asymptotic form of the metric as $\psi \to \infty$. In Appendix B series expansions, valid for $|\psi| > 1$, are derived for the ϖ_j

$$\varpi_j(\psi) = \sum_{r=0}^{3} \log^r(5\psi) \sum_{n=0}^{\infty} b_{jrn} \frac{(5n)!}{(n!)^5 (5\psi)^{5n}} , \quad |\psi| > 1 .$$

In order to compute the leading terms of the metric we retain only the terms with $n = 0$ and define vectors of coefficients similar to ϖ

$$b_r \stackrel{\text{def}}{=} -\left(\frac{2\pi i}{5}\right)^3 \begin{pmatrix} b_{2r0} \\ b_{1r0} \\ b_{0r0} \\ b_{4r0} \end{pmatrix}$$

so that

$$\varpi(\psi) \sim \sum_{r=0}^{3} b_r \log^r(5\psi) .$$

Most of the terms

$$\sigma_{rs} \stackrel{\text{def}}{=} -i b_r^\dagger \sigma b_s$$

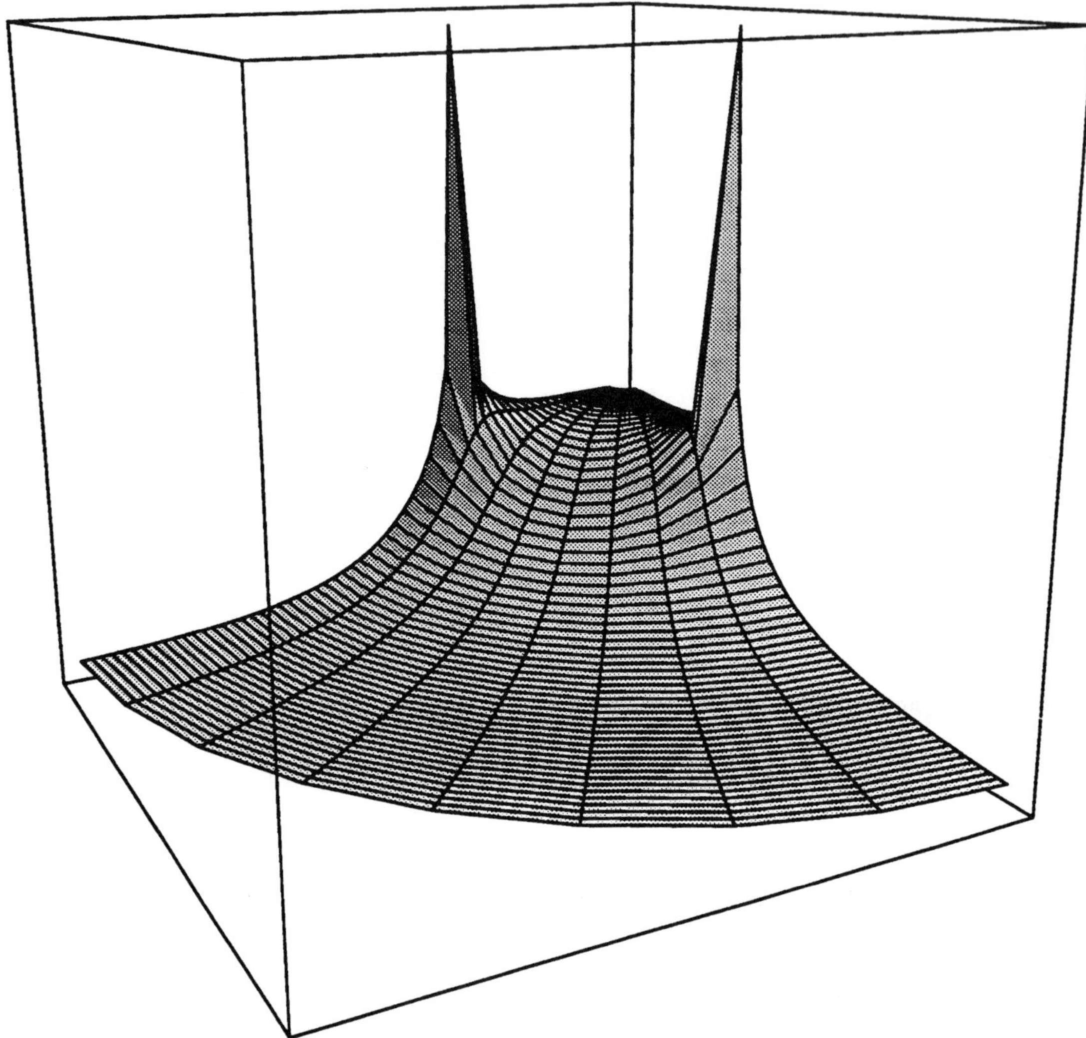

Figure 4.1: A plot of the metric $g_{\psi\bar{\psi}}$ against ψ for ψ in the fundamental region $0 \leq \arg \psi < \dfrac{2\pi}{5}$. The cusps correspond to the conifold at $\psi = 1$. As $\psi \to \infty$ the metric tends asymptotically to a metric of constant negative curvature.

vanish. The only nonzero products being

$$\sigma_{30} = \sigma_{03} = \frac{4\pi^3}{75} \;, \quad \sigma_{21} = \sigma_{12} = \frac{4\pi^3}{25}$$

$$\sigma_{00} = \frac{128}{625}\pi^3\zeta(3) \;,$$

where ζ is the Riemann ζ-function.

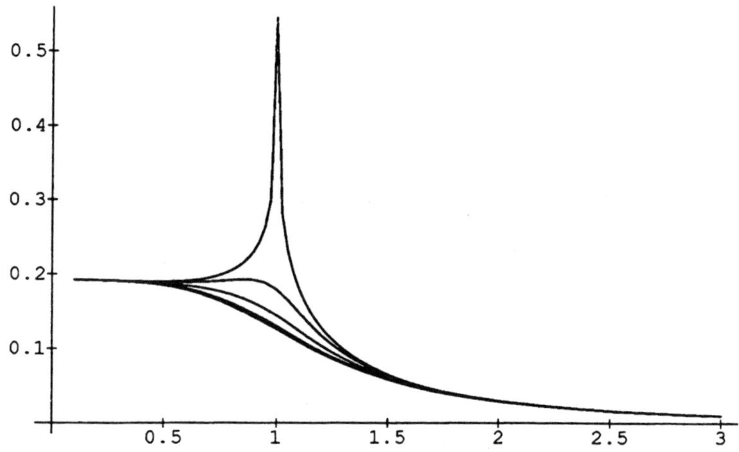

Figure 4.2: A plot of the metric, $g_{\psi\bar\psi}$ against $|\psi|$ for arg $\psi = \dfrac{k\pi}{20}$, $k = (0,1,2,3,4)$. Note the cusp at $\psi = 1$ where the metric has a logarithmic singularity.

Thus

$$e^{-K} \sim \left(\frac{2\pi}{5}\right)^3 \left\{ \frac{20}{3} \log^3 |5\psi| + \frac{16}{5}\zeta(3) \right\}$$

and so we find the leading terms of the metric to be

$$g_{\psi\bar\psi} = \frac{3}{4|\psi|^2 \log^2 |5\psi|} \left\{ 1 - \frac{48\zeta(3)}{25 \log^3 |5\psi|} + \cdots \right\} \;. \tag{4.3}$$

The leading term corresponds to a metric of uniform negative curvature. In fact on setting

$$t \sim -\frac{5}{2\pi i} \log(5\psi) \;,$$

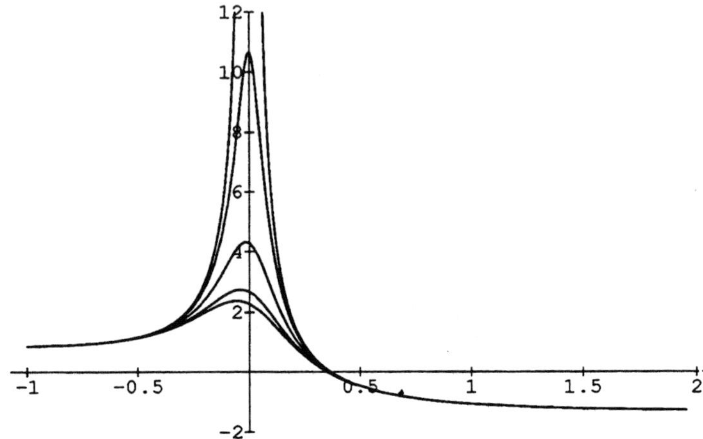

Figure 4.3: A plot of the Ricci–scalar against $\log_{10}|\psi|$ for $\arg\psi = \dfrac{k\pi}{20}$, $k = (0,1,2,3,4)$. The plot illustrates that the curvature scalar tends to the value $-\dfrac{4}{3}$ as $\psi \to \infty$ but does so logarithmically.

we find that

$$ds^2 \sim \frac{3}{2}\frac{|dt|^2}{(t_2)^2}\,,$$

where we have written t_2 for $\Im mt$, so the large complex structure limit of the geometry coincides, as we shall see in the following section, with the large radius limit of the moduli space for $\mathbf{P}_4(5)$. The subleading term in (4.3) corresponds to a loop correction to the metric which we shall discuss further in §5. Note finally that the singular manifold corresponding to $\psi = \infty$ is infinitely distant from the smooth manifolds.

4.1. The Yukawa Coupling

In order to make contact with the Yukawa coupling we introduce a set of 'Wronskians'

$$W_k \stackrel{\text{def}}{=} z^a \left(\frac{d}{d\psi}\right)^k \mathcal{G}_a - \mathcal{G}_a \left(\frac{d}{d\psi}\right)^k z_a\,.$$

Then, from (4.1), we find a relation between the Yukawa coupling and W_3

$$\kappa_{\psi\psi\psi} = \int \Omega \wedge \frac{\partial^3 \Omega}{\partial\psi^3}$$
$$= W_3 \ . \tag{4.4}$$

Now the form of W_3 follows straightforwardly from the differential equation (3.9) that governs the periods. Let us rewrite the equation in the form

$$\left\{ \left(\frac{d}{d\psi}\right)^4 + \sum_{k=0}^{3} C_k(\psi) \left(\frac{d}{d\psi}\right)^k \right\} \Pi = 0 \ .$$

From this relation it is immediate that

$$W_4 + \sum_{k=0}^{3} C_K W_k = 0 \ . \tag{4.5}$$

Note now that

$$W_0 = W_1 = W_2 = 0 \ .$$

The vanishing of W_0 is trivial. W_2 is the derivative of W_1 and we have verified that W_1 vanishes. Note also the elementary identity

$$W_4 - 2W_3' + W_2'' = 0 \ .$$

Putting these results together we find

$$W_3' + \frac{1}{2}C_3 W_3 = 0$$

and hence

$$\kappa_{\psi\psi\psi} = W_3 = \left(\frac{2\pi i}{5}\right)^3 \frac{5\psi^3}{1 - \psi^5} \tag{4.6}$$

the constant being determined by the computation of the specific form of W_3 as $\psi \to 0$. We could also have arrived at this same result by using the methods of [10].

Some comments are in order here concerning the gauge dependence of the Yukawa couplings (for a full account in the spirit of the present work see [21] and [20]). It was noted previously that the holomorphic three form is in reality undefined up to multiplication by a holomorphic function for ψ,

$$\Omega \to f\Omega \,,$$

and we refer to such a replacement as a gauge transformation. A gauge transformation induces a transformation of the period vector, $\Pi \to f\Pi$, and from (4.4) it follows that the coupling is gauge dependent

$$\kappa_{\psi\psi\psi} \to f^2 \kappa_{\psi\psi\psi} \,,$$

so is the Kähler potential which transforms according to the rule

$$e^{-K} \to |f|^2 e^{-K} \,.$$

The gauge choice is arbitrary and a physical quantity such as a decay rate cannot depend on the gauge. It must therefore be the case that the coupling enters into physical quantities only through the invariant combination

$$y_{\text{inv}} \stackrel{\text{def}}{=} g^{-\frac{3}{2}} e^K |\kappa| \,,$$

the factor $g^{-\frac{3}{2}}$ being included to remove the coordinate dependence of $|\kappa|$ (since the effect of a gauge transformation on the Kähler potential is $K \to K - \log f - \log \overline{f}$ the metric is gauge invariant).

Roughly speaking the mechanism that causes κ to be repalced by the invariant coupling is the following. The low energy Lagrangian contains terms of the form

$$g e^{-K} \overline{\chi} \, \slashed{D} \chi + \frac{1}{2} g |\partial\phi|^2 + (\kappa \chi\chi\phi + \text{conj.}) \,.$$

With a suitable choice of covariant derivative D, this is invariant under

$$\kappa \to f^2 \kappa \,, \quad e^{-K} \to |f|^2 e^{-K} \,,$$

if also

$$\chi \to f^{-1}\chi \,,$$

and ϕ is invariant. We can normalize the kinetic terms by defining

$$\tilde{\chi} = g^{\frac{1}{2}} e^{-\frac{K}{2}} \chi e^{i\frac{\theta}{2}} \quad \text{and} \quad \tilde{\phi} = g^{\frac{1}{2}} \phi \, ,$$

where $\theta = \arg \kappa$. Then the low energy terms become

$$\overline{\tilde{\chi}} \not{\!D} \tilde{\chi} + \frac{1}{2} |\partial \tilde{\phi}|^2 + g^{-\frac{3}{2}} e^{K} |\kappa| (\tilde{\chi}\tilde{\chi}\tilde{\phi} + \text{conj.}) \, .$$

We summarize the limiting forms of the metric, curvature and invariant Yukawa couplings with a Table. The reader will recognize the value obtained for the invariant coupling at $\psi = 0$ [7] as being the value corresponding to the Gepner model 3^5 [25]. This agreement is an inportant check on the mirror hypothesis. The $\psi = \infty$ row of the Table gives the 'bare' value of y_{inv} as being $2/\sqrt{3}$. In Ref. [7] this value is incorrectly stated to be $1/\sqrt{3}$.

ψ	$g_{\psi\bar{\psi}}$	R	$g^{-\frac{3}{2}} e^{K} |\kappa|$
0	$25 \dfrac{\Gamma^5(\frac{4}{5})\Gamma^5(\frac{2}{5})}{\Gamma^5(\frac{1}{5})\Gamma^5(\frac{3}{5})}$	$2\dfrac{\Gamma^{15}(\frac{3}{5})\Gamma^5(\frac{1}{5})}{\Gamma^{15}(\frac{2}{5})\Gamma^5(\frac{4}{5})} - 4$	$\dfrac{\Gamma^{\frac{15}{2}}(\frac{3}{5})\Gamma^{\frac{5}{2}}(\frac{1}{5})}{\Gamma^{\frac{15}{2}}(\frac{2}{5})\Gamma^{\frac{5}{2}}(\frac{4}{5})}$
1	$-a^2 \log r$	$\dfrac{1}{2a^2 r^2 [-\log r]^3}$	$\dfrac{1}{2ar[-\log r]^{\frac{3}{2}}}$
∞	$\dfrac{3}{4|\psi|^2 \log^2 |\psi|}$	$-\dfrac{4}{3}$	$\dfrac{2}{\sqrt{3}}$

Table 4.1: The asymptotic/limiting forms of the metric, the scalar curvature and the invariant Yukawa coupling for $\psi = 0, 1, \infty$. In the middle row a is a positive constant and $r = |\psi - 1|$.

Finally we note that for a one-dimensional manifold that is special Kähler the Ricci scalar is related to the invariant coupling by

$$R + 4 = 2y_{\text{inv}}^2 \, , \tag{4.7}$$

and we present a 3-D plot of the Ricci sclar. A plot of y_{inv} has the same form. The 'bare' value of y_{inv} is $2/\sqrt{3}$. The quantum corrections cause y_{inv} to differ from this constant value and the correction becomes infinite at the conifold.

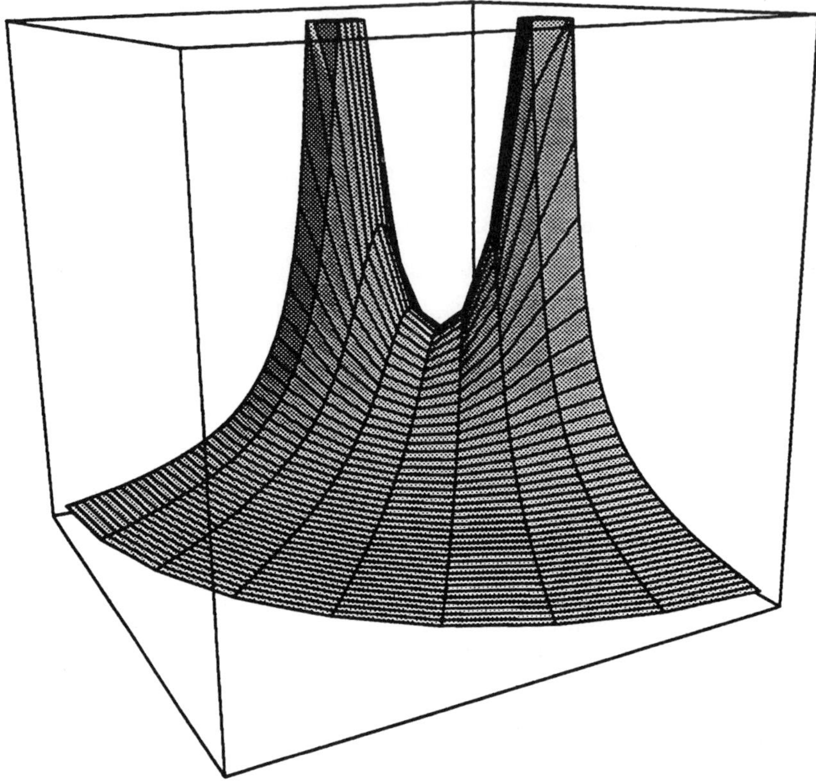

Figure 4.4: A plot of the Ricci scalar, R, against ψ for ψ in the fundamental region. The curvature tends to infinity at $\psi = 1$. As $\psi \to \infty$ the curvature tends asymptotically to the constant value $-\frac{4}{3}$. In virtue of Eq. (4.7) this figure has essentially the same form as a plot of the invariant Yukawa coupling.

5. $\mathbb{P}_4(5)$, The Mirror Map and Quantum Corrections

Recall [20] that prior to receiving quantum corrections the prepotential is given in terms of $b_{11} + 1$ homogeneous coordinates w^j by the expression

$$\mathcal{F}(w) = -\frac{1}{3!}\frac{\kappa_{ABC}\,w^A w^B w^C}{w^0}\ , \qquad A = 1,\ldots,b_{11}\ ,$$

where the

$$\kappa_{ABC} = \int_{\mathcal{M}} e_A \wedge e_B \wedge e_C$$

are the intersection numbers of a basis for $H^2(\mathcal{M},\mathbf{Z})$. For the present case $b_{11} = 1$ and, denoting the generator of $H^2(\mathbb{P}_4(5),\mathbf{Z})$ by e, we have

$$\int_{\mathcal{M}} e \wedge e \wedge e = 5\ . \tag{5.1}$$

Thus, using coordinates (w^1, w^2) rather than (w^0, w^1), we write

$$\mathcal{F} = -\frac{5}{6}\frac{(w^1)^3}{w^2} = -\frac{5}{6}(w^2)^2 t^3\ , \tag{5.2}$$

the latter form being written in terms of an affine coordinate $t \overset{\text{def}}{=} w^1/w^2$. It is simplest to set $w^2 = 1$, after differentiation, then the Kähler potential is given by

$$K = -\log\left\{i\left(\bar{w}^j\frac{\partial\mathcal{F}}{\partial w^j} - w^j\frac{\partial\overline{\mathcal{F}}}{\partial\bar{w}^j}\right)\right\} \tag{5.3}$$

$$= -\log\left(\frac{20}{3}(t_2)^3\right)\ .$$

From this we find that the metric and Ricci-tensor on the parameter space are given by

$$g_{t\bar{t}} = \frac{3}{4(t_2)^2}$$

$$R_{t\bar{t}} = -\frac{3}{2}g_{t\bar{t}}\ .$$

We summarize these results in the form of a table which should be compared with the $\psi = \infty$ row of Table 4.1.

We have seen that the large complex structure limit of the invariant Yukawa coupling and the geometry of $\mathcal{W}$ agree with the bare values of the corresponding quantities for $\mathcal{M}$. We wish to examine the quantum corrections to the bare quantities by comparing corresponding quantities on $\mathcal{W}$ and $\mathcal{M}$.

t	$g_{t\bar{t}}$	R	$g^{-\frac{3}{2}}e^{K}\vert\kappa\vert$
All t	$\dfrac{3}{4t_2^2}$	$-\dfrac{4}{3}$	$-\dfrac{2}{\sqrt{3}}$

Table 5.1:   The metric, the scalar curvature and the invariant Yukawa coupling as functions of t.

A detailed comparison requires the explicit form of the mirror map $t \to \psi(t)$ between the two parameter spaces. In order to do this it turns out that we need to understand the quadratic terms in the prepotential and these in turn involve the loop corrections so we shall first discuss the loop corrections. Perhaps surprisingly we shall see that there is a single loop correction to the bare prepotential.

5.1. The Loop Term

We have of course a nonrenormalization theorem [5], to the effect that there are no sigma model loop corrections to the superpotential or equivalently, to the couplings $\dfrac{\partial^3 \mathcal{F}}{\partial w^a \partial w^b \partial w^c}$. The prepotential is homogeneous of degree two so it is determined by its third derivative up to a quadratic term of the form

$$\Delta\mathcal{F} = \frac{1}{2}w^T \mathcal{Z} w \ ,$$

with $\mathcal{Z}$ a symmetric constant matrix. We can decompose $\mathcal{Z}$ into its real and imaginary parts

$$\mathcal{Z} = \mathcal{X} + i\mathcal{Y} \ .$$

The real and imaginary parts affect the prepotential differently. From (5.3) which may be written in the form

$$e^{-K} = 2w^\dagger(\Im m\mathbb{F})w \ ,$$

with $\mathbb{F}$ the matrix of second derivatives of $\mathcal{F}$, we see that the real part of $\mathcal{Z}$ contributes neither to the Yukawa couplings nor to the metric of the moduli space. Such a term is of no consequence an could be discarded. Alternatively we can absorb it into the bare part, $\mathcal{F}_0$, of the prepotential by extending the range of the indices on the intersection numbers κ_{ABC} to include the value zero so that

$$\mathcal{F}_0 = -\frac{1}{3!}\frac{\kappa_{ABC}w^A w^B w^C}{w^0} + \frac{1}{2}\mathcal{X}_{ab}w^a w^b = -\frac{1}{3!}\frac{\kappa_{abc}w^a w^b w^c}{w^0} \qquad a,b = 0,1,\dots,b_{11} \ .$$

The imaginary part of $\mathcal{Z}$ does affect the metric and since the contribution of $\mathcal{Y}$ to the prepotential

$$\mathcal{F}_{loop} \overset{\text{def}}{=} iw^T \mathcal{Y}w \ ,$$

contains powers of the radius (recall that for the case of $\mathbb{P}_4(5)$, $t = w^1/w^2$ and the imaginary part of t is the square of the radius) it is precisely the loop correction to $\mathcal{F}_0$.

We shall now argue that $\mathcal{F}_{loop}$ contains a four loop correction and that this is the only loop correction to the prepotential[5]. In fact $\mathcal{F}_{loop}$ is closely related to the four loop term found by Grisaru, van de Ven and Zanon [14] in their investigation of the β-function for type $(2,2)$-supersymmetric σ-models. The loop corrections arise because there are loop corrections to the sigma model. When the heavy modes are integrated out of the sigma model, part of the loop term survives into the low energy theory. When we refer to a four loop term, we mean a four loop contribution to the sigma model. This corresponds to a *third* order correction to the prepotential.

The particular case of $\mathbb{P}_4(5)$ corresponds to $b_{11} = 1$ but it is simpler and more general to leave b_{11} arbitrary for the present. The large radius limit is the limit $w^0 \to 0$ for fixed w^A (this is the limit $t \to \infty$ for $\mathbb{P}_4(5)$) and the loop counting is that $1/w^0$ corresponds to one loop and each additional power of w^0 corresponds to an additional loop. Thus the term $iw^T \mathcal{Y}w$ contains, in principle, two-loop, three-loop and four-loop terms. These terms must be constructed from local polynomials in the curvature of $\mathcal{M}$. These terms have to be universal, that is independent of any structure, such as complex structure or the dimension of the space,

---

[5] The discussion presented here grew out of animated conversations with A. Strominger.

that is particular to $\mathcal{M}$. Furthermore the loop contributions must vanish for hyper-Kähler manifolds. These contributions which are clearly very restrictive were investgiated in relation to the counterterms that can arise for the β-functions for type (2,2) supersymmetric σ-models [26]. It is known that there are no curvature polynomials with the requisite properties that could correspond to two or three loops and that there is a single such quantity at four loops

$$S \stackrel{\text{def}}{=} R_{ij}{}^{kl} R_{kl}{}^{mn} R_{mn}{}^{ij} - 2R_i{}^k{}_j{}^l R_k{}^m{}_l{}^n R_m{}^i{}_n{}^j \, ,$$

where R_{ijkl} is the curvature tensor of $\mathcal{M}$. A further remarkable property of S for Calabi–Yau manifolds in six dimensions is that S is proportional to the Euler integrand. It can be shown that

$$S g^{\frac{1}{2}} d^6 x = 12(2\pi)^3 c_3 \, ,$$

where c_3 is the third Chern class.

Consider now the possible structure of the components $\mathcal{Y}_{ab}$. The component $\mathcal{Y}_{00}$ can be of the form

$$\mathcal{Y}_{00} = a \int_{\mathcal{M}} d^6 x g^{\frac{1}{2}} S = 12(2\pi)^3 a\chi \, ,$$

with a a numerical coefficient. The other components of $\mathcal{Y}$ all vanish. The terms $\mathcal{Y}_{AB} w^A w^B$ and $\mathcal{Y}_{0A} w^0 w^A$ are respectively one and two loop terms for which, as already mentioned, suitable curvature polynomials do not exist.

We therefore know the form of $\mathcal{F}_{loop}$ up to a numerical coefficient independent of $\mathcal{M}$. One way to fix the coefficient would be to actually evaluate a four loop graph. It is simpler however to compare with the corrected prepotential for $\mathbb{P}_4(5)$ which we do in Eq. (5.10) below. In this way we find that

$$\mathcal{F}_{loop} = \frac{i\zeta(3)}{2(2\pi)^3}\chi \, . \tag{5.4}$$

For the case of $\mathbb{P}_4(5)$ it is this contribution to the prepotential which produces the subleading term in the metric in Eq. (4.3) and is responsible for the fact that, for large ψ, the Ricci scalar of the moduli space differs from its limiting value by inverse powers of $\log \psi$ as is evident from Figure 4.3.

5.2. The Mirror Map

We find the relation between the coordinates ψ and t by relating the period vector Π to the corresponding vector $\amalg$ for $\mathbb{P}_4(5)$. The idea is that we expect the prepotential $\mathcal{G}$ calculated in the previous section to be the fully corrected form of the bare prepotential $\mathcal{F}_0$ given by (5.2). The complication is that the form of the prepotential depends on the choice of symplectic basis. So we must first find the relation between the symplectic bases for $\mathcal{W}$ and $\mathcal{M}$. We do this by relating Π and $\amalg$ for large radius and large complex structure.

The leading behaviour of $\amalg$ is derived from the leading behaviour of the prepotential $\mathcal{F}$ which we take to be

$$\mathcal{F}_{pert} \stackrel{\mathrm{def}}{=} \mathcal{F}_0 + \mathcal{F}_{loop} = -\frac{5}{6}\frac{(w^1)^3}{w^2} + \frac{1}{2}a(w^1)^2 + bw^1 w^2 + \frac{1}{2}c(w^2)^2 \ .$$

The first term is as in (5.2) and we have allowed for quadratic terms in line with our discussion above. From $\mathcal{F}_{pert}$ we obtain the leading behaviour of the period vector

$$\amalg_{pert} \stackrel{\mathrm{def}}{=} \begin{pmatrix} \frac{\partial \mathcal{F}_{pert}}{\partial w^1} \\ \frac{\partial \mathcal{F}_{pert}}{\partial w^2} \\ w^1 \\ w^2 \end{pmatrix} = w^2 \begin{pmatrix} -\frac{5}{2}t^2 + at + b \\ \frac{5}{6}t^3 + bt + c \\ t \\ 1 \end{pmatrix} \ . \tag{5.5}$$

We have the elementary fact that $\amalg_{pert}$ is cubic in t. On the other hand we have as $\psi \to \infty$.

$$\frac{\Pi}{\mathcal{G}_2} \sim t^3 \begin{pmatrix} 0 \\ 0 \\ 0 \\ -\frac{5}{6} \end{pmatrix} + t^2 \begin{pmatrix} \frac{5}{2} \\ 0 \\ 5 \\ 0 \end{pmatrix} + t \begin{pmatrix} \frac{11}{2} \\ 0 \\ 10 \\ -\frac{25}{12} \end{pmatrix} + \begin{pmatrix} -\frac{25}{12} \\ 1 \\ -\frac{25}{6} \\ 0 \end{pmatrix} + \frac{25i}{\pi^3}\zeta(3) \begin{pmatrix} 0 \\ 0 \\ 0 \\ 1 \end{pmatrix} \ , \tag{5.6}$$

where

$$\mathcal{G}_2 \sim \left(\frac{2\pi i}{5}\right)^3 \ ,$$

with asymptotic equality meaning equality up to terms that are $\mathcal{O}(\psi^{-5} \log^3(5\psi))$. We have also taken

$$t \sim -\frac{5}{2\pi i} \log(5\psi)$$

as before and we again mean equality up to terms that involve ψ^{-5} multiplied by logarithms. Because of the relation between t and ψ the neglected terms are $\mathcal{O}(t^k e^{2\pi i t})$ and with a slight abuse of language we shall refer to such terms as being exponentially small in the large radius ($\Im m t \to \infty$) limit. We allow for a symplectic transformation N and set

$$\amalg_{pert} \sim N \Pi \; . \tag{5.7}$$

Since both (5.5) and (5.6) are cubics we can solve for N

$$N = \begin{pmatrix} -1+2a' & b' & -a' & 0 \\ 2b' & c' & -b' & -1 \\ 2 & 0 & -1 & 0 \\ 0 & 1 & 0 & 0 \end{pmatrix}$$

where

$$a = -\frac{11}{2} + a' \; , \quad b = \frac{25}{12} + b' \; , \quad \text{and} \quad c = -\frac{25i}{\pi}\zeta(3) + c' \; .$$

If we take a', b', $c' \in \mathbb{Z}$ then $N \in \mathrm{Sp}(4;\mathbb{Z})$ for all a', b', c'. The fact that $\amalg_{pert}$ and Π are related by a symmplectic matrix is a consequence of mirror symmetry. The fact that they may be related by an integral symplectic matrix is the observation already made by Aspinwall and Lütken [27] that the mirror symmetry identifies the two lattices

$$\Lambda = H^3(\mathcal{W}, \mathbb{Z}) \; , \quad \text{and} \quad V = \bigoplus_{i=0}^{3} H^{2i}(\mathcal{M}, \mathbb{Z}) \; .$$

We shall here take a', b', c' to vanish, other choices corresponding merely to a further $\mathrm{Sp}(4,\mathbb{Z})$ transformation. So the final form of N is

$$N = \begin{pmatrix} -1 & 0 & 0 & 0 \\ 0 & 0 & 0 & 1 \\ -2 & 0 & -1 & 0 \\ 0 & -1 & 0 & 0 \end{pmatrix} \; .$$

With sufficient prescience we could of course have chosen a basis for Π so that N would have turned out to be the identity.

With N in hand we may write down the quantum version of $\amalg_0$, with the gauge choice $w^2 = \mathcal{G}_2$

$$\amalg = N \Pi \qquad \boxed{w^2 = \mathcal{G}_2} \; .$$

The gauge invariant form of this relation is

$$\amalg = \frac{w^2}{\mathcal{G}_2} N\Pi \ . \tag{5.8}$$

Our aim here is to compare the quantum prepotential $\mathcal{F}$ that derives from $\amalg$ with that derived from $\mathcal{F}_0$. We set

$$\amalg = \begin{pmatrix} \mathcal{F}_1 \\ \mathcal{F}_2 \\ w^1 \\ w^2 \end{pmatrix} \ ,$$

and refer back to (3.23) and (3.25), recalling that the latter expression for Π is in the gauge $\mathcal{G}_2 = \left(\dfrac{2\pi i}{5}\right)^3 \varpi_0$, we find the following expression for t

$$\begin{aligned}
t &= \frac{w^1}{w^2} = \frac{2(\varpi_1 - \varpi_0) + \varpi_2 - \varpi_4}{5\varpi_0} \\
&= -\frac{5}{2\pi i}\left\{ \log(5\psi) - \frac{1}{\varpi_0(\psi)}\sum_{m=1}^{\infty} \frac{(5m)!}{(m!)^5(5\psi)^{5m}}[\Psi(1+5m) - \Psi(1+m)] \right\} \ .
\end{aligned} \tag{5.9}$$

This equation gives the mirror map. We have introduced the coordinate t partly to make contact with previous work but primarily because $2k\pi it$ is the value of the action evaluated on a rational curve of degree k (that is on an instanton of degree k). The coordinate t transforms in a complicated way under the generators $\mathcal{A}$ and $\mathcal{T}$ of modular transformations although under $(\mathcal{T}\mathcal{A})^{-1}$ we have the simple rule

$$(\mathcal{T}\mathcal{A})^{-1} : t \to t + 1 \ .$$

We also have

$$\mathcal{F} = \frac{1}{2}w^a \mathcal{F}_a = (w^2)^2\left\{ -\frac{5}{6}t^3 - \frac{11}{4}t^2 + \frac{25}{12}t - \frac{25i}{2\pi^3}\zeta(3) + \text{ exponentially small} \right\} \ , \tag{5.10}$$

where the last expression follows on inverting (5.9) to give $\psi = \psi(t)$. Note the structure of $\mathcal{F}$. First there are terms that are cubic, quadratic and linear in t that have real coefficients. These correspond to the bare prepotential $\mathcal{F}_0$. Then there is the term independent of t, which we identify as the loop term and which fixes the coefficient in (5.4). Lastly there are exponentially small terms which may be thought of as instanton corrections.

5.3. The Sum Over Instantons

We wish to examine these exponentially small terms. The idea is to examine the difference between $\mathcal{G}$ and $\mathcal{F}_0$. However, as mentioned above, one must also allow for the effect of the symplectic transformation N. It is simpler to consider first the instanton contribution to the Yukawa coupling, which is of interest in its own right, the corresponding contributions to $\mathcal{F}$ can then be obtained by integration. We have

$$\kappa_{ttt} = \left(\frac{w^2}{\mathcal{G}_2}\right)^2 \kappa_{\psi\psi\psi} \left(\frac{d\psi}{dt}\right)^3 , \tag{5.11}$$

where $\kappa_{\psi\psi\psi}$ is given by (4.6) and the prefactor expresses the gauge freedom. If we choose $w^2 = 1$ and recall that $\kappa_{\psi\psi\psi}$ was derived in the gauge $\mathcal{G}_2 = \left(\dfrac{2\pi i}{5}\right)^3 \varpi_0$ we find

$$\kappa_{ttt} = 5 + 2875 e^{2\pi i t} + 4876875 e^{4\pi i t} + \cdots , \qquad \boxed{w^2 = 1} . \tag{5.12}$$

To understand the structure of this sum consider the contribution of a rational curve $\mathcal{L}$ of degree k to the coupling. A rational curve of degree k is a $\mathbb{P}_1$ embedded by equations of degree k. Equivalently it is characterized by the fact that

$$\int_{\mathcal{L}} e = k ,$$

where e is the generator of $H^2(\mathbb{P}_4(5), \mathbb{Z})$. The contribution of $\mathcal{L}$ to the coupling is [7]

$$\exp\left(2\pi i t \int_{\mathcal{L}} e\right) \left(\int_{\mathcal{L}} e\right)^3 = k^3 e^{2\pi i k t} .$$

It is necessary to consider also multiple covers of the rational curves. For example, at degree 2 there are embeddings $\mathbb{P}_1 \hookrightarrow \mathbb{P}_4(5)$ give by

$$(i): \quad (u,v) \longrightarrow (u^2, v^2, uv, 0, 0)$$

$$(ii): \quad (u,v) \longrightarrow (u^2, v^2, 0, 0, 0) .$$

These are given as embeddings $\mathbb{P}_1 \hookrightarrow \mathbb{P}_4$ but for suitable quintics the curves lie in the quintic hypersurface. The image in case (i) is a rational curve of degree 2 while case (ii) is a double cover of the line $(u, v, 0, 0, 0)$ which is a rational curve of degree 1. Note that the two cases are intrinsically different, we cannot by means of a coordinate transformation transform case (ii) to case (i). More generally an m-fold multiple cover of a rational curve of degree k is an embedding $\mathbb{P}_1 \hookrightarrow \mathbb{P}_4(5)$ such that the homogeneous coordinates of the $\mathbb{P}_4$ are polynomials of degree k in quantities (U, V) that are themselves polynomials of degree m in the homogeneous coordinates of the $\mathbb{P}_1$. One difference between the multiple covers and the single covers is that it appears that the single covers have no parameters (it is proved in [28] that the single covers have no parameters for $k \leq 7$) while m-fold covers with $m > 1$ do have parameters. Since the quantities U and V are polynomials of degree m in the coordinates (u, v) of the $\mathbb{P}_1$ they each contain $m + 1$ parameters. Taking into account the three degrees of freedom in a reparametriation

$$(u, v) \longrightarrow (au + bv, cu + dv) , \quad ad - bc = 1 ,$$

of the $\mathbb{P}_1$ and the freedom to multiply U and V by a common scale we find

$$2(m + 1) - 3 - 1 = 2(m - 1)$$

parameters.

We shall assume that an m-fold cover contributes an amount $k^3 e^{2\pi i m k t}$ to the coupling. Since

$$\int_{m\mathcal{L}} e = mk ,$$

this amounts to the assumption that m-fold covers have an associated prefactor $1/m^3$ (otherwise the contribution would be $(mk)^3 e^{2\pi i m k t}$) we believe that this prefactor derives from the zero modes associated with the parameters that the m-fold covers enjoy but we have not performed the computation. The reason for assuming this specific form for the prefactor will become apparent as we proceed. With this understanding we find that the contribution to the coupling of a rational curve $\mathcal{L}$ of degree k together with all its multiple covers is

$$k^3 \sum_{m=1}^{\infty} e^{2\pi i k m t} = \frac{k^3 e^{2\pi i k t}}{1 - e^{2\pi i k t}} .$$

Let n_k be the number of rational curves of degree k, then we have the following expression for the coupling as a sum over instantons

$$k_{ttt} = 5 + \sum_{k=1}^{\infty} \frac{n_k k^3 e^{2\pi i k t}}{1 - e^{2\pi i k t}} = 5 + n_1 e^{2\pi i t} + (2^3 n_2 + n_1) e^{4\pi i t} + \cdots . \qquad (5.13)$$

It is gratifying that we find that $n_1 = 2875$ which is indeed the number of lines [29][6] (rational curves of degree one) and $n_2 = 609\ 250$ which is known to be the number of conics [28] (rational curves of degree two). Clemens has shown [30] that $n_k \neq 0$ for infinitely many k and has conjectured that $n_k \neq 0$ for all k, but it seems that the direct calculation of these numbers becomes difficult beyond $k = 2$ (see also [28]). It is however straightforward to develop the series (5.12) to more terms and to find the n_k by comparison with (5.13). We present the first few n_k in Table 5.2. These numbers provide compelling evidence that our assumption about the form of the prefactor is in fact correct. The evidence is not so much that we obtain in this way the correct values for n_1 and n_2 but rather that the coefficients in (5.12) have remarkable divisibility properties. For example asserting that the second coefficient 4,876,875 is of the form $2^3 n_2 + n_1$ requires that the result of substracting n_1 from the coefficient yields an integer that is divisible by 2^3. Similarly the result of substracting n_1 from the third coefficient must yield an integer divisible by 3^3. These conditions become increasingly intricate for large k. It is

---

[6] This number is misleadingly given as 375 in [6]. This is due to the fact that the calculation is performed for a special class of quintics and no account taken of multiplicity. If the counting is done for a generic quintic then the result is 2875.

therefore remarkable that the n_k calculated in this way turn out to be integers.

k	n_k
1	2875
2	6 09250
3	3172 06375
4	24 24675 30000
5	22930 58888 87625
6	248 24974 21180 22000
7	2 95091 05057 08456 59250
8	3756 32160 93747 66035 50000
9	50 38405 10416 98524 36451 06250
10	70428 81649 78454 68611 34882 49750

Table 5.2:   The numbers of rational curves of degree k for $1 \le k \le 10$.

The values for the n_k shown in the table are paritcular to $\mathbb{P}_4(5)$ however we can abstract from (5.13) a form for the mirror map which we conjecture to be of general validity

$$\mathcal{W}_w = \mathcal{M}_w + \sum_{\mathcal{L} \in \mathcal{M}} \frac{e^{2\pi i \mathcal{L}[w]}}{1 - e^{2\pi i \mathcal{L}[w]}} \mathcal{L}^3 \, , \tag{5.14}$$

where we regard the complex strcuture of $\mathcal{W}$ as being parametrized by the complex Kähler form $w = B + iJ$ of $\mathcal{M}$ and

$$\mathcal{L}[w] = \int_{\mathcal{L}} w \, .$$

Eq. (5.14) embodies the moral of the present work. There is a 'bare manifold' $\mathcal{M}$ and a 'quantum manifold' $\mathcal{W}$ and the quantum manifold is the bare manifold together with its rational curves.

5.4. The Number of Rational Curves of Large Degree

It is immediately apparent from Table 5.2 that the n_k grow very rapidly with k. It is of interest to observe that the distribution of the n_k for large k is governed by the conifold at

$\psi = 1$. Consider again the series (5.12) which gives the Yukawa coupling as a power series in $e^{2\pi i t}$. The radius of convergence of the series is determined by the singularity of the Yukawa coupling and the coupling is singular only when $\psi^5 = 1$ so we conclude that the series (5.12) converges for

$$\Im mt > \Im mt(1) \ .$$

From this it follows that n_k cannot grow faster than $e^{-2\pi i k t(1)}$ for large k. However we can do better and find the asymptotic form of n_k from the singularity of the coupling.

First we need to know the behaviour of $t(\psi)$ as $\psi \to 1$. From (5.9) we find

$$t(\psi) - t(1) \sim -\frac{5^{\frac{3}{2}}}{4\pi^2}\frac{it_2(1)}{\varpi_0(1)}(\psi - 1)\log(\psi - 1) \ .$$

Together with (5.11) this enables us to find the leading behaviour of the singularity of the coupling as $\psi \to 1$

$$\kappa_{ttt} \sim \frac{(2\pi)^3}{5^{\frac{3}{2}}}\frac{\varpi_0^3(1)}{t_2^3(1)}\frac{1}{(\psi - 1)[-\log(\psi - 1)]^3} \ . \tag{5.15}$$

On the other hand the severity of the singularity at $\psi \to 1$ must be dictated by the asymptotic behaviour of the n_k for large k. If we set

$$n_k \sim Bk^{\rho - 3}\log^\sigma ke^{2\pi kt_2(1)} \ ,$$

then we have

$$\begin{aligned}
\kappa_{ttt} &\sim B\sum_{k=1}^{\infty} k^\rho \log^\sigma k\, e^{-2\pi k(t_2(\psi) - t_2(1))} \\
&\sim B\int_0^\infty dk\, k^\rho \log^\sigma k\, e^{-kx} \ , \quad x \overset{\text{def}}{=} 2\pi(t_2(\psi) - t_2(1)) \\
&\sim B\frac{\Gamma(1 + \rho)}{x^{1+\rho}}(-\log x)^\sigma \ .
\end{aligned} \tag{5.16}$$

By comparing (5.16) with (5.15) we see that $\rho = 0$, $\sigma = -2$ and we find also a value for B. Thus we have

$$n_k \sim \left(\frac{2\pi\varpi_0(1)}{t_2(1)}\right)^2 \frac{e^{2\pi kt_2(1)}}{k^3\log^2 k} \ , \qquad \begin{aligned} t_2(1) &= 1.208128077077918638192\ldots \\ \varpi_0(1) &= 1.070725868430155800571\ldots \end{aligned} \tag{5.17}$$

the next order terms being smaller by inverse powers of $\log k$. Note that, as anticipated, the asymptotic form just refers to quantities evaluated at $\psi = 1$.

As a check we have plotted the values of

$$r_k^{(0)} = \frac{n_k}{\nu_k} , \tag{5.18}$$

where ν_k is the expression on the right hand side of (5.17) in Figure 5.1 for k in the range $2 \le k \le 25$. The series $\{r_k^{(0)}\}$ converges very slowly owing to the fact that the subleading terms in the asymptotic expansion fall off as inverse powers of $\log k$. To speed the convergence we apply a variant of a Richardson transformation [31] to the series, the aim being the eliminate the effect of the subleading terms. We set

$$r_k^{(m)} = \frac{r_{k+1}^{(m-1)} \log(k+1) - r_k^{(m-1)} \log k}{\log(\frac{k+1}{k})} ,$$

and we record the values of the series $\{r_k^{(m)}\}$ in Figure 5.1 for $1 \le m \le 3$. We see that the third transform converges rapidly to unity.

5.5. Some Further Remarks on the Modular Group

We saw previously that the modular group Γ acts most simply on a modular parameter γ. A puzzling fact is that γ does not appear to be the quantity that is of most direct physical interest. For example, the coordinate that arises naturally in the instanton sum is the coordinate t which is related to the complex Kähler-form by the relation $B + iJ = te$. The coordinate t is, as we have seen, proportional to the value of the instanton action and has a direct significance $\Im mt = R^2$ where R, in this context, is the radius of the Calabi–Yau manifold. The puzzle is that t and γ do not appear to be related in a simple way, though of course both are functions of ψ. It is perhaps worth recalling here the relations

$$t = -\frac{5}{2\pi i} \left\{ \log \psi + \frac{\displaystyle\sum_{n=1}^{\infty} \frac{\Gamma(n+\frac{1}{5})\Gamma(n+\frac{2}{5})\Gamma(n+\frac{3}{5})\Gamma(n+\frac{4}{5})}{(n!)^4 \psi^{5n}} \left[4\Psi(n+1) - \sum_{r=1}^{4} \Psi(n+\frac{r}{5})\right]}{5 \displaystyle\sum_{n=1}^{\infty} \frac{\Gamma(n+\frac{1}{5})\Gamma(n+\frac{2}{5})\Gamma(n+\frac{3}{5})\Gamma(n+\frac{4}{5})}{(n!)^4 \psi^{5n}}} \right\}$$

$$\frac{\gamma}{2\tan\frac{2\pi}{5}} - \frac{1}{2} = -\frac{5}{2\pi i} \left\{ \log \psi + \frac{\displaystyle\sum_{n=0}^{\infty} \frac{\Gamma(n+\frac{2}{5})\Gamma(n+\frac{3}{5})}{(n!)^2 \psi^{5n}} \left[2\Psi(n+1) - \sum_{r=2}^{3} \Psi(n+\frac{r}{5})\right]}{5 \displaystyle\sum_{n=0}^{\infty} \frac{\Gamma(n+\frac{2}{5})\Gamma(n+\frac{3}{5})}{(n!)^2 \psi^{5n}}} \right\} ,$$

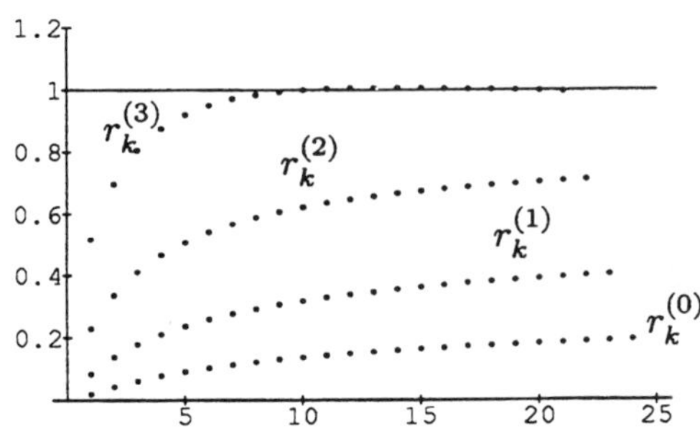

Figure 5.1:   A plot of the series of ratios $\{r_k^{(0)}\}$, for $2 \le k \le 25$, together with its first three Richardson transforms. The third transform converges rapidly to unity.

to show that these functions definitely come from the same stable but are in fact different.

The transformation laws for t under the action of Γ follows from the first of equations (5.9). We observed previously that $t \to t - 1$ under the action of $\mathcal{T}\mathcal{A}$. The transformation of t under the action of $\mathcal{A}$, say, is considerably more complicated. In Figure 5.2 we have plotted the images of the lines arg ψ = constant. The shaded region is the image of the fundamental region and the other four regions are the images of the fundamental region under $\mathcal{A}$, $\mathcal{A}^2$, $\mathcal{A}^3$ and $\mathcal{A}^4$. Since the tips of the latter four regions come down to touch the real axis we see that large R, that is large $\Im mt$, is mapped to small R under the action of $\mathcal{A}$. The precise relation however is not as simple as $R \to 1/R$.

6. A Mechanism for Supersymmetry Breaking

The purpose of this section is to present a speculative mechanism which breaks supersymmetry for the low energy theory. The proposal is logically independent of the mirror symmetry that is the focus of the present article however we include it here because it arises naturally as part of the discussion of the conifold at $\psi = 1$.

In Ref. [17], [18] it has been observed that the nodes of a conifold can be resolved in

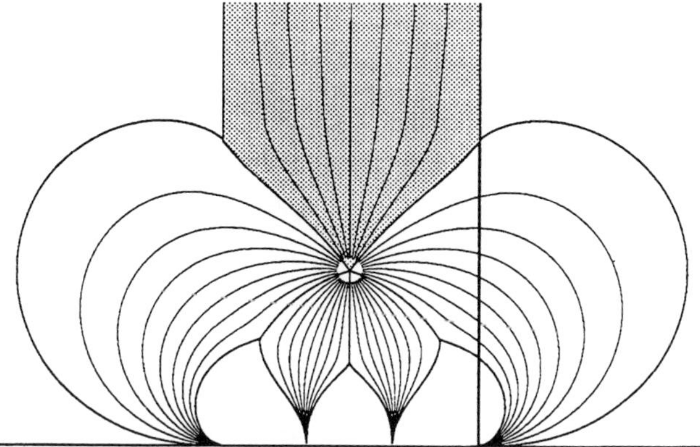

Figure 5.2: The images of the lines arg ψ = constant in the t-plane. The shaded region is the image of the fundamental region and the other four regions are the images of the shaded region under the action of $\mathcal{A}$, $\mathcal{A}^2$, $\mathcal{A}^3$ and $\mathcal{A}^4$. Although it is not clear at this scale, the tangents to the boundaries coincide at the branch points so the deficit angles are zero as are the angles at the tips of the images. The image of the point $\psi = 0$ is $t = -\dfrac{1}{2} + \dfrac{4i}{5}\sin\dfrac{2\pi}{5}$ and the four cusps touch the real axis at $t = 0,\ -\dfrac{1}{3},\ -\dfrac{2}{3},\ -1$.

different ways. One is by smoothing whereby the node is replaced by an S^3 as in Figure 2.1. Another way is to replace the node by an S^2 as indicated by Figure 6.1.

The construction that achieves this is entirely local in nature. However in replacing the node by an S^2 we interfere with the cohomology group H^2 and it need no longer be the case that there remain any positive (1,1)-forms. If this is so then the resulting manifold $\check{\mathcal{M}}$ is a Moishezon manifold but is not Kähler. The case to hand is of just this type and this follows from the fact that the conifold has just one node. Let $\mathcal{C}$ be the resolving $S^2 = \mathbb{P}_1$ then the essential point is that $\mathcal{C}$ is a boundary[7]. We can see this by considering the sequence of operations depicted in Figure 6.1. In passing from $\mathcal{M}$ to $\check{\mathcal{M}}$ a puncture is made in B_2 and the puncture is blown up

---

[7]We are indebted to T. Hübsch for explaining this point to us.

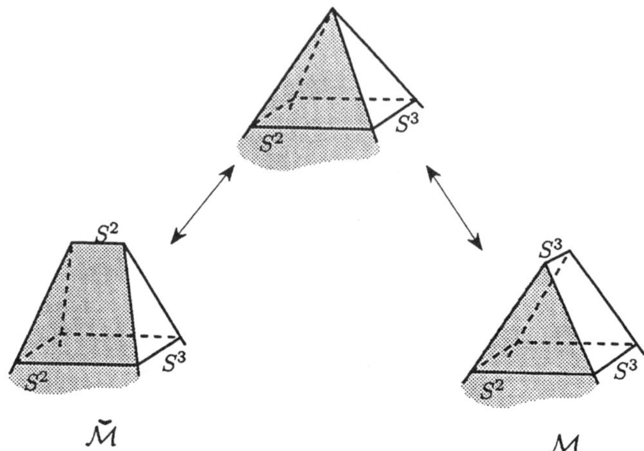

Figure 6.1: Local neighborhoods of a *node* in a conifold, its *small resolution* in $\tilde{\mathcal{M}}$, and its *deformation* in $\mathcal{M}$. The conifold is singular while both $\tilde{\mathcal{M}}$ and $\mathcal{M}$ are smooth.

into $\mathcal{C}$. Thus B_2 now has a boundary $\partial B_2 = \mathcal{C}$. Let also $J = ig_{\mu\bar{\nu}}dx^\mu \wedge dx^{\bar{\nu}}$ be the putative Kähler form then we have

$$\int_{\mathcal{C}} J = \int_{\partial B_2} J = \int_{B_2} dJ \ .$$

The first integral is the area of $\mathcal{C}$ which is strictly positive. Thus dJ cannot vanish identically and hence $\tilde{\mathcal{M}}$ cannot be Kähler.

It was observed in [32] that such a compactification of string theory has certain attractive features. The holonomy group of $\tilde{\mathcal{M}}$ is presumably $O(6)$ so embedding the spin connection in the gauge group breaks $E_8 \times E_8$ to $E_8 \times O(10)$ and we can have broken supersymmetry while retaining vanishing cosmological constant. This proposal is cast in the geometrical language of Kaluza-Klein theory, because it is geometrical in nature and we do not yet know how to formulate these ideas in terms of conformal field theory. It does however show a way of resolving what was regarded as a difficult problem in the days of Kaluza-Klein theory which was how to introduce a length scale, different from the compactification scale, which could correspond to low energy supersymmetry breaking. Here we have a second length scale arising as the radius of the resolving $\mathbb{P}_1$. In some appropriate sense supersymmetry is only slightly broken if this parameter is small, though we have no argument as to why the scale of the resolving $\mathbb{P}_1$ should be vastly different from the compactification scale.

A. A Second Look at the Homology of $\mathcal{W}$

The function of this appendix is to make a more detailed study of the homology of $\mathcal{W}$ and to close the gap left in §3 by showing that the cycles A^2 and B_2 do indeed intersect in a point. We also make some further observations concerning the monodromy of the cycles and find cycles Q_j corresponding to the periods ϖ_j.

We begin by defining three-chains V_j, $j = 0,\ldots,4$

$$V_j(\psi) = \{x_k | x_5 = 1,\; x_1, x_2, x_3 \text{ real and positive, the branch of}$$
$$x_4 \text{ being chosen such that arg } x_4 \to \pi + \frac{2\pi j}{5} \text{ as } \psi \to 0\} \;.$$

The five V_j are chains rather than cycles however a little thought shows that in virtue of the identities (2.2) they have a common boundary, that is ∂V_j is independent of j. So the difference of any two V_j's is a cycle. We find x_4 by solving the quintic which we write in the form

$$x_4^5 - 5\psi x_1 x_2 x_3 x_4 + \Delta = 0\;, \quad \Delta \stackrel{\text{def}}{=} 1 + x_1^5 + x_2^5 + x_3^5\;.$$

Set also

$$x_4 = \Delta^{\frac{1}{5}}\eta\;, \quad u = \frac{x_1 x_2 x_3}{\Delta^{\frac{4}{5}}}\;,$$

then we have

$$\eta^5 - 5\psi u \eta + 1 = 0\;. \tag{A.1}$$

This equation has of course five roots for given (x_1, x_2, x_3). For ψ sufficiently small these can be found by rewriting (A.1) in the form

$$\eta_j = -\alpha^j (1 - 5\psi u \eta_j)^{\frac{1}{5}}\;, \tag{A.2}$$

and iterating the equation. It is easy to show that u is bounded, in fact, u varies in the range

$$0 \leq u \leq 4^{-\frac{4}{5}} \,, \tag{A.3}$$

so for ψ sufficiently small the root η_j is unambiguously defined by the iteration. It is clear that for small ψ the V_j have no points in common apart from their boundaries.

We wish next to enquire how the V_j vary with ψ. In order to accomplish this we first consider the behaviour of the roots of (A.1). It is evident that (A.1) cannot have a purely imaginary solution for η if ψ is real since then the first two terms would be purely imaginary and the third term real. Next we observe that (A.1) has a double root if in addition η satisfies the equation

$$\eta^4 = \psi u \,.$$

It follows that (A.1) has a double root if and only if

$$(\psi u)^5 = 4^{-4} \,, \tag{A.4}$$

and that a double root satisfies

$$\eta^5 = \frac{1}{4} \,. \tag{A.5}$$

One immediately sees that (A.1) cannot have a triple root or two double roots for any value of ψ. As a consequence of (A.3) and (A.4) η will have a double root for some value of (x_1, x_2, x_3) if and only if ψ^5 is real and $|\psi| \geq 1$. It follows that we can unambiguously extend the definition of the V_j to the ψ-plane cut as in Figure 2.3. Consider now the behaviour of the roots as ψ runs from 0 to ∞ through real values. For $\psi = 0$ the five roots are, as we have already seen, $\eta_j = -\alpha^j$. Since there are no purely imaginary roots for any real ψ it follows that three of the roots have negative real part and two have positive real part. Moreover since there cannot be a double root that is real and negative, in virtue of (A.5), the three roots with negative real part will always consist of a real root and a conjugate pair of complex roots.

On the other hand the two roots with positive real part will consist of a conjugate pair for

$$\psi < \frac{4^{-\frac{4}{5}}}{u} \,,$$

and will be distinct and real for

$$\psi > \frac{4^{-\frac{4}{5}}}{u} \ ,$$

with a double root precisely when

$$\psi = \frac{4^{-\frac{4}{5}}}{u} \ .$$

The situation for large ψ was discussed in §3. One of the roots approaches 0, while the other four recede to infinity along trajectories that are asymptotic to the four semi-axes. The two roots with positive real part therefore approach 0 and ∞ respectively while the three roots with negative real part all become infinite. A plot of the trajectories is presented in Figure A.1. These considerations will shortly allow us to compute $A^2 \cap B^2$.

We now look at the chains $V_j(\psi)$ for $\psi > 1$ which we define as the limit of $V_j(\psi)$ with $\Im m\psi$ positive. It follows from the foregoing that V_0, V_1, V_4 and $V_2 \cup V_3$ are disjoint and that V_2 intersects V_3. Let X be the subset of $V_2 \cup V_3$ for which

$$\frac{4^{-\frac{4}{5}}}{\psi} \leq u \ . \tag{A.6}$$

X consists of two three-chains corresponding to the two positive solutions of (A.1) that have positive real part with their boundaries identified because there is a double root precisely when equality holds in (A.6). A little reflection should convince the reader that the two three-chains are topologically three balls and that the boundaries of these three balls are identified with opposite orientation so that X is an S^3. In fact X is the cycle A^2 of §3.

We are now in a position to compute $A^2 \cap B_2$. The definition (3.1) of A_2 restricts all coordiantes to be real and positive. The definition (3.4) of B_2 requires that $|x_1| = |x_2| = |x_3| = \delta$ so in fact

$$x_1 = x_2 = x_3 = \delta \ .$$

Because δ in the definition of B^2 is chosen to exclude multiple values of x_4 and in fact forces strict inequality in (A.6) there are exactly two points which satisfy all these restrictions, corresponding to the two positive real roots for η. However, as we have already seen, exactly one of these is on the branch for which $x_4 \rightarrow 0$ as $\psi \rightarrow \infty$. Consequently the intersection of A^2 and B_2 consists of a single point.

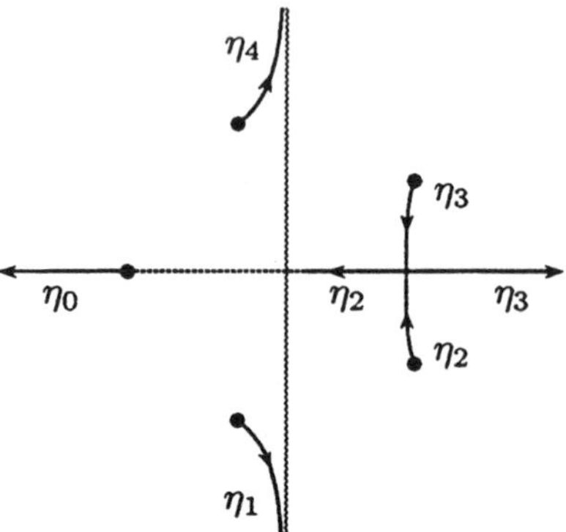

Figure A.1: The trajectories of the roots of (A.1) as ψ increases through real values. The roots η_2 and η_3 become real and move as shown if ψ has an infinitesimal positive imaginary part.

The last point we shall address in this part of the discussion is the monodromy of the cycles

$$V_{ij} \stackrel{\text{def}}{=} V_i - V_j \; ,$$

with respect to a loop circling the point $\psi = 1$. From the discussion above, it follows that as we go around such a loop, the chains V_j remain unchanged except for $j = 2, 3$ and that the chains V_2 and V_3 exchange the balls corresponding to Eq. (A.6). This can be expressed as adding X to V_2 and substracting X from V_3 (there is an orientation of X implicit in this choice). The monodromy of the cycles can now be readily computed. We leave this as an exercise for the diligent reader.

A.1. Cycles Corresponding to the Periods ϖ_j

As a final topic in this appendix we wish to find cycles Q_j corresponding to the periods ϖ_j. The Q_j together with the relation (3.23) serve to define the cycles A^1 and B_1 which have hitherto been defined only implicitly. Of course we know that the period corresponding to

$B_2(\psi)$ is proportional to $\varpi_0(\psi)$ so cycles corresponding to the ϖ_j are proportional to $B_2(\alpha^j\psi)$ with the continuation performed along paths that go from ψ to $\alpha^j\psi$ without crossing the cuts. We shall here give an alternative definition of the Q_j and relate these cycles to the V_j discussed above.

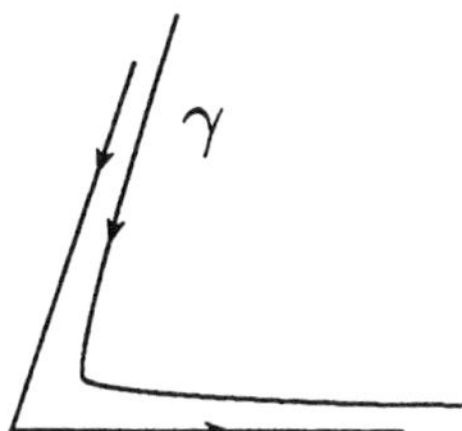

Figure A.2: Two choices for the cycle γ_i. One composed of straight lines, the other smooth.

Let γ_i, $i = 1,2,3$, be the one-cycles consisting of the union of the two half-lines arg $x_i = 2\pi/5$ and arg $x_i = 0$. The reason for choosing the cycle as we have is that u is bounded on the cycle. For other cycles, such as taking the x_i to be real, this is no longer true so that the term ψu in (A.2) is not necessarily small for small ψ. We define the cycle Q_k by taking $(x_1, x_2, x_3) \in \gamma_1 \times \gamma_2 \times \gamma_3$ and x_4 to be given by the $(k+3)$-rd branch of the quintic.

Solving (A.2) by iteration we find

$$\eta_k^{(0)} = -\alpha^k$$

$$\eta_k^{(1)} = -\alpha^k(1 + \alpha^k\psi u)$$

$$\eta_k^{(2)} = -\alpha^k(1 + \alpha^k\psi u - (\alpha^k\psi u)^2)$$

etc..

The important point is that

$$\eta_k(\psi) = \alpha^k\eta_0(\alpha^k\psi) \ . \tag{A.7}$$

Let $q_k(\psi)$ be the period evaluated on Q_k

$$q_k(\psi) = \psi \int_{\gamma_1 \times \gamma_2 \times \gamma_3} \frac{dx_1\, dx_2\, dx_3}{\Delta^{\frac{4}{5}}(\eta_{k+3}^4 - \psi x_1 x_2 x_3 \Delta^{-\frac{4}{5}})} \ . \tag{A.8}$$

In virtue of (A.7) we see that

$$q_k(\psi) = q_0(\alpha^k \psi) \, ,$$

so it is sufficient to study q_0. The factor $(\eta^4 - \psi x_1 x_2 x_3 \Delta^{-\frac{4}{5}})^{-1}$ can be expanded as a power series in ψ:

$$(\eta_3^4 - \psi x_1 x_2 x_3 \Delta^{-\frac{4}{5}})^{-1} = \alpha^3 \sum_{n=0}^{\infty} c_{n+1} \left(\frac{5 \alpha^3 \psi x_1 x_2 x_3}{\Delta^{\frac{4}{5}}} \right)^n \, .$$

Substituting this expression into (A.8) we have

$$q_0(\psi) = \frac{1}{5} \sum_{m=1}^{\infty} c_m I_m \alpha^{3m} (5\psi)^m \, , \tag{A.9}$$

with

$$I_m = \int_{\gamma_1 \times \gamma_2 \times \gamma_3} \frac{dx_1 dx_2 dx_3 (x_1 x_2 x_3)^{m-1}}{\Delta^{\frac{4m}{5}}} \, .$$

The integrals I_m can be evaluted in closed form. First observe that

$$\int_{\gamma_1} \frac{dx_1 x_1^{m-1}}{\Delta^{\frac{4}{5}}} = (1 - \alpha^m) \int_0^{\infty} \frac{dx_1 x_1^{m-1}}{\Delta^{\frac{4}{5}}} \, ,$$

so we have

$$I_m = (1 - \alpha^m)^3 \int_0^{\infty} \int_0^{\infty} \int_0^{\infty} \frac{dx_1 dx_2 dx_3 (x_1 x_2 x_3)^{m-1}}{\Delta^{\frac{4m}{5}}} \, . \tag{A.10}$$

On introducing new variables $y_i = x_i^{\frac{5}{2}}$ and then going over to polar coordinates we find that the integral factorizes into three integrals that are easily evaluated in terms of B-functions. The result is:

$$I_m = (-1)^{m+1} \left(\frac{2\pi i}{5} \right)^3 \alpha^{-m} \frac{\Gamma(\frac{m}{5})}{\Gamma(\frac{4m}{5}) \Gamma^3(1 - \frac{m}{5})} \, , \tag{A.11}$$

where in writing this last relation we have used the fact that $\alpha^{\frac{1}{2}} = -\alpha^3$. Substituting (A.11) into (A.9) we find

$$q_0(\psi) = \left(\frac{2\pi i}{5} \right)^3 \frac{1}{5} \sum_{m=1}^{\infty} (-1)^{m+1} c_m \alpha^{2m} \frac{\Gamma(\frac{m}{5})}{\Gamma(\frac{4m}{5}) \Gamma^3(1 - \frac{m}{5})} (5\psi)^m \, .$$

When the coefficients c_m are calculated we find

$$c_m = (-1)^{m+1} \frac{\Gamma(\frac{4m}{5})}{\Gamma(m) \Gamma(1 - \frac{m}{5})} \, ,$$

yielding

$$q_0(\psi) = \left(\frac{2\pi i}{5}\right)^3 \frac{1}{5} \sum_{m=1}^{\infty} \frac{\alpha^{2m}\Gamma(\frac{m}{5})(5\psi)^m}{\Gamma(m)\Gamma^4(1-\frac{m}{5})} \ .$$

We recongnize that, in virtue of (3.15), the right hand side of this relation is porportional to ϖ_0. In fact

$$q_0(\psi) = -\left(\frac{2\pi i}{5}\right)^3 \varpi_0(\psi) \ .$$

Finally we observe from (A.10) that the Q-cycles are related to the V-chains by

$$Q_j = (1 - \mathcal{A})^3 V_{j+3} \ ,$$

with $\mathcal{A}$, as previously, the operation that replaces ψ by $\alpha\psi$. Thus we have

$$Q_j = V_{j-2} - 3V_{j-1} + 3V_j - V_{j+1} \ .$$

B. Further Properties of the Periods

We record here some further results pertaining to the periods $\varpi_j(\psi)$. Recall that $\varpi_j(\psi) = \varpi(\alpha^j \psi)$ so for $|\psi| < 1$ we have, from (3.15),

$$\varpi_j(\psi) = -\frac{1}{5} \sum_{m=1}^{\infty} \frac{\alpha^{2m}\Gamma(m/5)(5\alpha^j\psi)^m}{\Gamma(m)\Gamma^4(1-m/5)} \ , \quad |\psi| < 1 \ . \tag{B.1}$$

Our main purpose here is to obtain explicit expressions for the $\varpi_j(\psi)$ valid throughout the fundamental region. It is simplest to begin by discussin the basis (3.13). It is perhaps notationally simpler to regard the hypergeometric function that appears on the right hand side of (3.13) as a $_5F_4$ which has $a_5 = c_4 = 1$. We choose the basis

$$\tilde{\varphi}_k(\psi) = \frac{\Gamma^5(k/5)}{\Gamma(k)}(5\psi)^k \, _5F_4\left(\frac{k}{5}, \frac{k}{5}, \frac{k}{5}, \frac{k}{5}, 1; \frac{k+1}{5}, \frac{k+2}{5}, \frac{k+3}{5}, \frac{k+4}{5}; \psi^5\right)$$
$$= \sum_{n=0}^{\infty} \frac{\Gamma^5(n+k/5)(5\psi)^{5n+k}}{\Gamma(5n+k)} \ , \quad |\psi| < 1$$

the last equality following in virtue of the multiplication formula (3.12). To analytically continue these functions we write them as integrals

$$\widetilde{\varpi}_k(\psi) = -\int_C \frac{ds}{(e^{2\pi i s} - 1)} \frac{\Gamma^5(s + k/5)(5\psi)^{5s+k}}{\Gamma(5s + k)} \ , \qquad 0 \le \arg \psi \le \frac{2\pi}{5} \ .$$

For $|\psi| > 1$ the contour can be closed to the left. Note that there are no poles when $s = -N$, $N = 0, 1, \ldots$ for then $5s + k = k - 5N$ and the factor $1/\Gamma(5s + k)$ renders the integrand finite. There are however fourth order poles when $s = -N - k/5$. These are not of fifth order, again because of the $\Gamma(5s + k)$ in the denominator. Extracting the residues involves the expansion of ψ^{5s} about $s = -N - k/5$. This produces the $\log \psi$, $\log^2 \psi$ and $\log^3 \psi$ terms. The ϖ_j are given in terms of the $\widetilde{\varpi}_k$ by the relation

$$\varpi_j(\psi) = -\frac{1}{80\pi^4} \sum_{k=1}^{4} a^{jk}(\alpha^k - 1)^4 \widetilde{\varpi}_k(\psi) \ .$$

The upshot is that the ϖ_j have series expansions of the form

$$\varpi_j(\psi) = \sum_{r=0}^{3} \log^r(5\psi) \sum_{n=0}^{\infty} b_{jrn} \frac{(5n)!}{(n!)^5 (5\psi)^{5n}} \ , \qquad |\psi| > 1 \ .$$

With the coefficients b_{jrn} given by somewhat lengthy expressions. Defining

$$S_{jm} = \sum_{k=1}^{5} \alpha^{k(j+1)}(\alpha^k - 1)^m \ , \quad \text{and} \quad \Phi(z) = \Psi(1 + z) - \Psi(1 + 5z)$$

we have

$$b_{j0m} = -\frac{1}{6(2\pi i)^3}\{6(2\pi i)^3(S_{j0} - 1)$$
$$+ 6(2\pi i)^2[2\pi i + 5\Phi(m)]S_{j1}$$
$$+ 3(2\pi i)[2(2\pi i)^2 + 5(2\pi i)\Phi(m) - 5\Phi'(m) + 25\Phi^2(m)]S_{j2}$$
$$+ 5[\Phi''(m) + 5(2\pi i)^2\Phi(m) - 15\Phi(m)\Phi'(m) + 25\Phi^3(m)]S_{j3}\}$$

$$b_{j1m} = -\frac{5}{2(2\pi i)^3}\{2(2\pi i)^2 S_{j1}$$
$$+ (2\pi i)[2\pi i + 10\Phi(m)]S_{j2}$$
$$+ \frac{5}{3}[(2\pi i)^2 + 15\Phi^2(m) - 3\Phi'(m)]S_{j3}\}$$

$$b_{j2m} = -\frac{25}{2(2\pi i)^3}\{2\pi i S_{j2} + 5\Phi(m)S_{j3}\}$$

$$b_{j3m} = -\frac{125}{6(2\pi i)^3}S_{j3} \ .$$

Finally we record the series expansion of z^2 about $\psi = 1$

$$z^2(\psi) = \frac{4\pi^2}{5^{3/2}} \sum_{m=0}^{\infty} a_m(1 - \psi^5)^{m+1} \ , \quad |\psi^5 - 1| < 1 \ ,$$

where the coefficients satisfy the recurrence relation

$$0 = 625m^2(m^2 - 1)a_m - 125m(m - 1)(20m^2 - 40m + 23)a_{m-1}$$
$$+ 125(m - 1)(30m^3 - 150m^2 + 261m - 157)a_{m-2}$$
$$- (2500m^4 - 22500m^3 + 76625m^2 - 116875m + 67226)a_{m-3}$$
$$+ 625(m - 3)^4 a_{m-4}$$

with the initial values

$$a_0 = -\frac{1}{5} \ , \quad a_1 = -\frac{3}{50} \ .$$

Acknowledgements

It is pleasure to acknowledge instructive discussions with Paul Aspinwall, Willy Fischler, Brian Greene, Vadim Kaplunovsky, Sheldon Katz, Andy Lütken, Fernando Quevedo, Graham Ross and Andy Strominger. We are indebted also to Vadim Kaplunovsky and Karen Uhlenbeck for making computer resources available.

References

[1] P. Candelas, M. Lynker and R. Schimmrigk, Nucl. Phys. **B341** (1990) 383.

[2] B.R. Greene and M.R. Plesser, *(2,2) and (2,0) Superconformal Orbifolds*, Harvard University Report HUTP-89/B241. *Duality in Calabi–Yau Moduli Space*, Harvard University Report HUTP-89/A043.

[3] P. Aspinwall, A. Lütken and G.G. Ross, Phys. Lett. **241B** (1990) 373.

[4] A. Strominger and E. Witten, Commun. Math. Phys. **101** (1985) 341.

[5] E. Witten, Nucl. Phys. **B268** (1986) 79,

E. Martinec, Phys. Lett. **171B** (1986) 189,

M. Dine and N. Seiberg, Phys. Rev. Lett. **57** (1986) 2625.

[6] M. Dine, N. Seiberg, X.G. Wen and E. Witten, Nucl. Phys. **B278** (1987) 769, Nucl. Phys. **B289** (1987) 319.

[7] J. Distler and B.R. Greene, Nucl. Phys. **B309** (1988) 295.

[8] L. Dixon and D. Gepner, unpublished.

[9] W. Lerche, C. Vafa and N.P. Warner, Nucl. Phys. **B324** (1989) 427.

[10] P. Candelas, Nucl. Phys. **B298** (1988) 458.

[11] A. Giveon, E. Rabinovici and G. Veneziano, Nucl. Phys. **B322** (1989) 167,

V.P. Nair, A. Shapere, A. Strominger and F. Wilczek, Nucl. Phys. **B287** (1987) 402,

A. Shapere and F. Wilczek, Nucl. Phys. **B320** (1989) 669.

[12] S. Ferrara, D. Lüst, A. Shapere and S. Theisen, Phys. Lett. **225B** (1989) 363,

S. Ferrara, D. Lüst and S. Theisen, Phys. Lett. **233B** (1989) 147,

W. Lerche, D. Lüst and N.P. Warner, Phys. Lett. **B231B** (1989) 417,

S. Ferrara, D. Lüst and S. Theisen, Phys. Lett. **242B** (1990) 39,

B. de Wit and A. Van Proeyen, *Symmetries of Dual-Quaternionic Manifolds*, University of Utrecht Report THU-90/18, Cern Report CERN-TH.5865/90.

[13] For a reviw see J. Kollar, Bull. Am. Math. Soc. **17** (1987) 211.

[14] M.T. Grisaru, A. van de Ven and D. Zanon, Phys. Lett. **173B** (1986) 423, Nucl. Phys. **B277** (1986) 388, Nucl. Phys. **B277** (1986) 409.

[15] M. Lynker and R. Schimmrigk, *Landau–Ginzburg Theories as Orbifolds*, University of Texas Report UTTG-22-90, Santa Barbara Institute for Theoretical Physics Report NSF-ITP-90-88.

[16] P. Candelas, P.S. Green, M. Lynker and R. Schimmrigk, *Calabi–Yau Manifolds in Weighted* $\mathbb{P}_4$, University of Texas Report, in preparation.

[17] P. Candelas, P. Green and T. Hüsch, Nucl. Phys. **B330** (1990) 49.

[18] P. Candelas and X.C. de la Ossa, Nucl. Phys. **B342** (1990) 246.

[19] R. Bryant and P. Griffiths, Porgress in Mathematics **36** pp. 77-102, (Birkhäuser, Boston, 1983).

[20] P. Candelas and X.C. de la Ossa, *Moduli Space of Calabi–Yau Manifolds*, University of Texas Report UTTG-07-90.

[21] A. Strominger, *Special Geometry*, University of California at Santa Barbara Report, UCSBTH-89-61.

[22] A. Erdélyi, F. Oberhettinger, W. Magnus and F.G. Tricomi, *Higher Transcendental Functions*, McGraw-Hill 1953.

[23] V.I. Arnold, S.M. Gusein-Zade and A.N. Varchenko, *Singularities of Differentiable Maps*, vol. II, Monographs in Mathematics **83**, Birkhäuser 1988.

[24] L.J. Slater, *Generalized Hypergeometric Functions*, Cambridge 1966.

[25] D. Gepner, Phys. Lett. **199B** (1987) 380.

[26] C.N. Pope, M.F. Sohnius and K.S. Stelle, Nucl. Phys. **B283** (1987) 192,
M.D. Freeman and C.N. Pope, Phys. Lett. **174B** (1986) 48,
P. Candelas, M.D. Freeman, C.N. Pope, M.F. Sohnius and K.S. Stelle, Phys. Lett. **177B** (1986) 341,
M.D. Freeman, C.N. Pope, M.F. Sohnius and K.S. Stelle, Phys. Lett. **178B** (1986) 199.

[27] P. Aspinwall and A. Lütken, *Geometry of Mirror Manifolds*, Oxford University Report, April 1990, *Quantum Algebraic Geometry of Superstring Compactifications*, Oxford University Report, September 1990.

[28] S. Katz, Compositio Math. **60** (1986) 151.

[29] J. Harris, Duke Math. Jour. **46** (1979) 685.

[30] H. Clemens, *Some Results on Abel-Jacobi Mappings in Topics in Transcendental Algebraic Geometry*, Princeton University Press 1984.

[31] C.M. Bender and S.A. Orzag, *Adanced Mathematics for Scientists and Engineers*, McGraw Hill, 1978.

[32] P. Candelas, G. Horowitz, A. Strominger and E. Witten, Nucl. Phys. **B258** (1985) 46.

AMS/IP Studies in Advanced Mathematics
Volume **9**, 1998

Topological Mirrors and Quantum Rings

Cumrun Vafa[1]

Lyman Laboratory of Physics

Harvard University

Cambridge, Ma 02138, USA

Abstract

Aspects of duality and mirror symmetry in string theory are discussed. We emphasize, through examples, the importance of loop spaces of a deeper understanding of the geometrical origin of dualities in string theory. Moreover we show that mirror symmetry can be reformulated in very simple terms as the statement of equivalence of two classes of topological theories: Topological sigma models and topological Landau-Ginzburg models. Some suggestions are made for generalization of the notion of mirror symmetry.

11/91

---

[1]Talk presented at MSRI conference on Mirror Symmetry, May, 1991, Berkeley.

© 1998 American Mathematical Society
and International Press

1. Introduction

One of the most fascinating aspects of string theory is the way it modifies our intuition of classical geometry. It modifies it in ways which in some sense makes the classical geometry *more symmetrical*, and thus, in a sense simpler. This is probably most manifest in the principle of duality in string theory, which states that two classically inequivalent geometries (target spaces for strings) can nevertheless be identical from the string point of view. The aim of this paper is to develop this notion emphasizing the basic physical reasons for believing in its universal existence. My presentation is written with the mathematically oriented reader in mind and even though I will not be fully rigorous I hope that the main ideas are more or less clear to mathematicians.

I will first discuss some general aspects of Hilbert space of strings propagating in a target space in a geometrical way and discuss the notion of duality in this set up (section 2). Then I give some simple examples of this duality for bosonic strings (section 3). In section 4, I will discuss aspects of fermionic (super-) string vacua highlighting aspects which are relevant for mirror symmetries. As we will see an important ingredient in this setup is the notion of *quantum cohomology ring* of Kahler manifolds which is a deformation of the ordinary cohomology ring. In section 5 the relation between singularity theory and solutions of superstrings is discussed. This turns out to be a convenient bridge between target space interpretation and abstract conformal field theory definition of string theory. In section 6 the topological formulation of mirror symmetry is discussed. This turns out to be a very effective language to describe mirror symmetry. In this setup, mirror symmetry is stated as the equivalence of two seemingly inequivalent topological theories. This topological formulation has the advantage of simplifying the conformal theory to a much simpler theory which is the relevant piece needed for the discussion of mirror symmetry. Finally in section 7 I discuss some puzzles for mirror symmetry and their potential resolutions. I also discuss some potential generalizations of mirror symmetries and some possible connections with quantum groups and Donaldson theory.

2. String Hilbert space

In this section we discuss the basic structure of string vacua which involves the Hilbert space and operatorial formulatiion of the theory (this aspect is discussed much more extensively in the talks of Friedan in this conference; for a mathematical introduction see [1]). Consider a closed string (one dimensional parametrized circle) sitting in a Riemannian manifold M. The space of all such configurations is given by the (parametrized) loop space of M which we denote by $\mathcal{L}M$. The geometrical questions that arise in string theory basically correspond to probing the geometry of $\mathcal{L}M$. The Hilbert space of *bosonic strings* is an 'appropriate' category of function space on $\mathcal{L}M$, which we denote by

$$\mathcal{H}_{bosonic} = \Phi(\mathcal{L}M)$$

with norm inherited from the metric on M. The Hilbert space of *fermionic or superstrings* is the space of semi-infinite forms on $\mathcal{L}M$:

$$\mathcal{H}_{fermionic} = \Lambda^{\infty}(\mathcal{L}M)$$

In addition to this Hilbert space, there is a more or less canonical one to one correspondence between the states $|v\rangle$ in the Hilbert space and some 'special' operators 0_v acting on the Hilbert space. Roughly speaking, these operators are characterized by the fact that they are 'invariant' under reparametrizations of the string and that when they act on a special state $|0\rangle$ (the vacuum state) in the Hilbert space, they give the corresponding state $(O_v|0\rangle = |v\rangle)$. These form a complete operator product algebra, in the sense that the product of any two of these operator is another such operator. Choosing a basis, we have

$$O_i O_j = \sum_k C_{ij}^k O_k$$

where the sum over k is generically an infinite sum.

A convenient method of computing C_{ij}^k is as follows: In string theory to find the amplitude of how a number of loops $l_i \in \mathcal{L}M$ ends up changing to the loops $\bar{l}_j \in \mathcal{L}M$ we have to sume over all interpolating surfaces Σ immersed in M, $f(\Sigma) \subset M$ whose boundary is

$$\partial f((\Sigma)) = \bigoplus l_i - \bigoplus \tilde{l}_j$$

weighed by $exp(-E)$ where E is the energy functional of the surface immersed in M (a natural extension of this applies to fermionic strings). We can choose a 'basis' for our Hilbert space of delta functions corresponding to fixed loops in the manifold. The above prescripton then gives a way to compute the amplitude that two of these basis elements ends up with the thrid one. This can be extended to the full Hilbert space by multilinearity of the amplitude. The amplitude thus computed for the two string state $|i\rangle$ and $|j\rangle$ to end up with the third one $|k\rangle$ can be obtained by integrating the 'wave function' of these states against the basic amplitude with the delta functions. The resulting answer is in fact the same as C_{ij}^k.

There are consistency conditions that Hilbert space and these coefficients need to satisfy for a consistent theory (following from the associativity of the operator products and modular invariance of string amplitudes). Once we are given such a structure, we can forget about M altogether and talk bout the 'string vacuum', meaning this abstract Hilbert space with some canonical set of operators satisfying some 'nice' operator product properties. Let us denote such a structure by S and call it a *string vacuum*. Then two string vacua are equivalent, or isomorphic, if there is an isomorphism between the corresponding Hilbert spaces and the operators. Now it may happen that strings on two different manifolds M_1 and M_2 give rise to isomorphic string vacua

$$M_1 \neq M_2 \quad but \quad S(M_1) = S(M_2) \ .$$

In other words *the map from manifolds to string vacua may be many to one.* In such a case we call the manifolds M_1 and M_2 dual or mirror pairs. Actually the choice of the terminology is unfortunate, as it may happen that more than two manifolds may give rise to the same string vacuum. One could also ask the reverse question: Does every string vacuum come from a manifold, i.e., is this map onto? The answer seems to be no (see for example [2]).

The existence of mirror symmetry is thus simply the statement of the existence of different geometrical ways to realize a string vacuum. We can use any representation we please. in such case, if we try to study some aspects of the string vacuum we can choose any realization and may thus end up equating a 'hard' geometrical computation in one representation to an 'easy' one in another realization. In this lies the power of mirror symmetry transforming a hard problem to an easy one. In the next section we give some examples of mirror pairs in the context of bosonic strings.

3. Examples of Bosonic Mirrors

In this section we consider examples of mirror manifolds which lead to the same string vacuum for the bosonic strings. We will give two classes of examples: In one class the mirror Riemannian manifolds are tpologically the same but geometrically distinct, and in the second class the mirror manifolds are even topologically distinct.

Let M_1 be the d dimensional torous indentified (as a Riemannian manifold) with

$$M_1 = \frac{E^d}{\Gamma}$$

where Γ is a d dimensional discrete lattice group acting by isometry on flat Euclidean space E^d. Let us consider the Hilbert space of strings on M_1 which is related to the function space on $\mathcal{L} M_1$. First note that $\mathcal{L} M_1$ naturally splits to infinitely many components, corresponding to each element of Γ which can be identified with $H_1(M_1, Z)$. Moreover, the function space on each component splits to the functions of the center of strings which is isomorphic to ordinary function space on M_1, and functions of oscillations of loops (which is universal and independent of Γ). The function space on M_1 is canonically isomorphic to Γ^*, the dual lattice to Γ suing Fourier transform. So the dependence of Hilbert space of strings on Γ, appears as a choice of loop component (an element of Γ) and the Fourier component of functions of center of string (an element of Γ^*), i.e., the dependence comes through a choice of element of

$$\Gamma + \Gamma^*$$

This implies that if we consider the second manifold M_1

$$M_2 = \frac{E^d}{\Gamma^*}$$

Then the Hilbert spaces of strings based on M_1 and M_2 are isomorphic, both depending on the *self dual* lattice $\Gamma + \Gamma^*$. This turns out to extend to the full string vacuum structure, i.e., to the operators and their products. So M_1 and M_2 are mirror pairs. In physical terms this implies that there is no physical experiment one can do in string theory to distinguish strings on M_1 from strings on M_2. This means in particular that the notion of 'length' is not a universally invariant way to decide if two manifolds are different as far as strings are concerned. This simple example illustrates the basic structure of duality or mirror symmetry in bosonic strings. This in fact was the first example of mirror symmetry discovered in string theory [3]. The rest of the examples are just extensions of this to more intricate cases.

For our second class of example we consider a simple laced compact Lie group G. Let H denote its Cartan torus. Consider an element $g \in G$ of finite order which belongs to the normalizer of H (i.e., it acts as a Weyl transformation on H). This means that

$$H \to g\,H\,g^{-1}$$

Let us denote the cyclic group generated by this transformation Λ_1 (we take g to act non-trivially on H). Choose an element $h \in H$ conjugate to $g \in G$. Consider the action

$$H \to h\,H$$

and denote the group action generated by this cyclic group Λ_2. Consider taking the quotients of H by these two different group actions:

$$M_1 = \frac{H}{\Lambda_1} \qquad M_2 = \frac{H}{\Lambda_2}$$

These two spaces are completely different. In fact M_1 is not even a manifold, but an orbifold, as g acts by fixed points on H, but M_2 is simply another torus, as h simply generates translations on H. It turns out that (bosonic) strings propagating on M_1 and M_2 are equivalent [4]. It is somewhat surprising that M_1 which is not even a manifold behaves very much like the smooth manifold M_2 as far as strings are concerned. This means, in the mathematical sense (as is also seen in examples for superstrings [4][5]) that loop space of an orbifold is a far better behaved object than the orbifold itself and in a sense provides a kind of universal space for resolution of orbifold singularity.

It should also be clear from the above examples that we can construct examples where three (or more) inequivalent Riemannian manifolds lead to the same string vacuum.

4. Superstring Vacua and Quantum Cohomology Rings

Most of our discussion up to now has been on bosonic strings. This is the case in which the Hilbert space is roughly speaking the function space on the loop space of manifold. However fermionic string is the physically (and mathematically) more interesting case. This is the case corresponding to the Hilbert space of semi-infinite forms on the loop space. In most application one considers target spaces which are Kahler manifolds. In this case the Hilbert space and the operators acting on it naturally admit $Z \oplus Z$ grading, corrsponding to the (holomorphic, anti-holomorphic) degree of the diferential forms. Let $\mathcal{O}$ denote the space of physical operators. Then we have the decomposition according to the degrees of the forms:

$$\mathcal{O} = \bigoplus_{p,q \in Z} \mathcal{O}_{p,q}$$

Naturally under operator products the degrees add, as expected. Noted that since we are dealing with semi-infinite differential forms, the degree of operators runs from $-\infty$ to $+\infty$. This is an important difference with respect to the differential forms on the ordinary manifolds where the degree of differential forms is positive. As we shall see later this is one of the main reasons for the prediction of mirror symmetry in the fermionic strings. There is an anti-unitary involution which implies that $O_{p,q}$ is the conjugate of $O_{-p,-q}$. The existence of this anti-unitary involution is the statement of CPT invariance of the theory. In the language of forms, since we are dealing with semi-infinite forms, it is roughly the statement that operation of 'adding' and 'subtracting' forms are conjugate operations. This turns out to be an important piece of physics in the story of mirror symmetry.

Since the manifold M is naturally embedded in $\mathcal{L} M$, one expects that at least the differential forms on M are related to a subset of those on $\mathcal{L} M$ and in particular the cohomology ring of M should correspond to some closed operator algebra (modulo addition of cohomologically trivial elements) of operators acting on the fermionic Hilbert space. Let d denote the complex dimensions of M. Then we expect that there exist a special set of operators $A_\alpha \in \mathcal{O}_{p,q}$ with $0 \leq p, q \leq d$, such that the operator algebra of A_α correspond to the cohomology ring of M. This expectation turns out to be correct and we denote this subsector of the operators by $H^{*,*}$. In fact more is true [6]: There is a natural way to define the product of these operators which yields a closed truncated operator algebra when restricted to this special finite subspace of operators which becomes finite and related to the cohomology ring[2]. There is one important subtlety however, Unlike the ordinary cohomology ring, the ring we get *depends* on the Kahler class of the metric on M. Only in the limit where we rescale the metric $g \to \lambda g$ and let $\lambda \to \infty$ do the ring of A_α's become exactly the cohomology ring of M. The deviation from the classical result is due to instanton correction [7] (an explicit exact result for instanton correction on Z orbifold is discussed in [8]). So string theory deforms the cohomology ring. A nice description of this deformation is as follows [9]. In order to describe this it is more convenient to go to the dual basis (i.e., homology). Let A^α denote the dual basis. Each α can be represented by a cycle in M. In order to specify the ring, it is sufficient to give the trilinear pairing between cycles. The ordinary ring is obtained by defining this pairing to be

$$\langle A^\alpha A^\beta A^\gamma \rangle = \#(C^\alpha \cap C^\beta \cap C^\gamma)$$

i.e., the number of common intersection points of the three cycles (and defining it to be zero if the common intersection has dimension bigger than zero). To define the ring we obtain in string theory we have to consider the space of holomorphic maps from CP^1 to the manifold M (rational curves in M), with the restriction that three fixed points on CP^1 get mapped to points in C^α, C^β and C^γ respectively. Again if the dimension of moduli of such maps is positive they do not contribute to the cohomology ring. The isolated ones contribute weighed by the instanton action. Let us denote an element of the space of such holomorphic maps by $h^{\alpha\beta\gamma}$. Let U denote the image of the sphere under h. Let k denote the Kahler form on M. Then the

---

[2]This is unlike the ordinary cohomology ring of manifold, in that the actual product of harmonic representatives does not form a closed operator algebra.

definition of the defomred ring (which is commutative and associative as shown in [9]) is

$$\langle A^\alpha A^\beta A^\gamma \rangle = \sum_{h^{\alpha\beta\gamma}} \#(C^\alpha \cap U) \cdot \#(C^\beta \cap U) \cdot \#(C^\gamma \cap U) \exp - \int h^{*\alpha\beta\gamma}(k) \qquad (4.1)$$

Note that in the limit $k \to \infty$ only the constant holomorphic maps survive in this sum and that gives back the ordinary definition of intersection between cycles. So in this way we have a *quantum deformed* cohomology ring. To actually derive (4.1) in the context of string theory (and define it properly for multiple covers of holomorphic maps)[3] is achieved by showing the topological nature of computation (and showing that on the cylinder it can be rephrased as a computation in a topological sigma model [9] which is discussed briefly in section 6). Without going to much detail let me at least indicate why its form is reasonable from what we have discussed up to this point. As we have discussed before to compute the algebra of operators in string theory we have to consider maps of a sphere with three discs cut out, to the manifold with three fixed boundary circles mapped to specific loops on the manifold. For constant loops or loops which are 'close' to being constant, we can take the limit in which the discs shrink to points, and map a specific point on CP^{-1} to a particular point on the manifold. Now the string loop amplitude computation tells us that we have to sum over all such loops weighed with e^{-E}, which in this case is nothing but the exponential of the pull back of the Kahler form on the map, as it appears in (4.1). The factors in front of exponential simply counts how many inequivalent ways a fixed rational curve could map to the three cycles (which is accomplished by an $SL(2,C)$ transformation of CP^1 to move the three points on the sphere). The fact that we sum over only holomorphic maps in (4.1) and get an exact answer and its precise definition can be best understood in the topological description of sigma models [9].

It is quite natural to speculate that this defomred ring may be the actual cohomology ring on a properly defined loop space. One way this may be realized is to consider the space of holomorphic maps from the disc to the manifold. The map from the boundary of the disc to the manifold induced from such maps may be viewed as a 'modified' loop space. In this loop space the points of the manifold will be represented more than once in the loop space; in fact

---

[3] Recent progress from this viewpoint has been made in [10].

if we look for the space of constant loops which was previously isomorphic to the manifold, that would be the same as looking for holomorphic maps which take the boundary of the disc to a point, which is basically a holomorphic map from the sphere to the manifold. So in this case the manifold and all the holomorphic curves in it are representing the original manifold in this loop space. In this set up it is likely to expect that there exists a fixed point formula for the cohomology elements (corresponding to the circle action on the loop) which reduces the computation of cohomology elements to the fixed point subspace which consists of the manifold and the holomorphic curves in it. This would then (presumably) give rise to the cohomology ring defined in (4.1) with $k = 0$. We can then expect to get the deformed ring by twisting the cohomology ring, which allows us to weigh the different fixed points (i.e., different holomorphic maps) differently, and thus obtain the formula (4.1) with $k \neq 0$. This line of thought is worth pursuing further and may lead to a better geometrical understanding of the loop space itself.

As an example, if one considers strings on CP^{-1}, if we denote by x the standard $(1,1)$ cohomology element, the classical cohomology ring is generated by x with

$$x^2 = 0$$

Let $\beta = \exp - \int k$ integrated over the nontrivial 2-cycle. Then the quantum deformed cohomology ring can be computed from its definition given above and is generated by x but the relation is deformed to [9]

$$x^2 = \beta$$

This can be generalized to CP^n [11] with the result that the quantum cohomology ring is defined by

$$x^{n+1} = \beta$$

We will discuss the conjectured generalized of this to the Grassmanians in the next section (see also [11]).

Note that in the above examples the deformed or quantum cohomology ring does not respect the grading of differential forms (in physics terminology we say that the instantons have destroyed chiral fermion number conservation), but the amount of violation of grading can be understood. The point is that the (formal) dimension of moduli space $\mathcal{M}$ of holomorphic maps h is given by

$$\dim \mathcal{M} = d + c_1(h)$$

where d is the dimension of manifold and $c_1(h)$ denotes the evaluation of the first chern class on the image of h. By the definition of quantum cohomology ring we see that the sum of dimensions of cohomology elements will have to be $d + c_1(h)$ in order to get a non-vanishing result, which means that we have a violation by $c_1(h)$. This explains the cohomology ring structure for CP^n discussed above (where $c_1 = n + 1$ for the fundamental cycle). Note the fundamental role played by Kahler manifolds where $c_1 = 0$, i.e., the Calabi-Yau manifolds. In this case there is no violation of the grading, and we indeed get a quantum cohomology ring which respects the cohomology grading. For Calabi-Yau manifolds of dimensions one and two (torus and $K3$), there are generically no holomorphic maps (this is due to the fact that if there were any there would be a three dimensional family of them by Mobious transofrmations, and so this would be in contradiction with the above formal dimension). So the first case of interest in terms of the deformation with the above formal dimension). So the first case of interest in terms of the deformation of cohomology rings is the case of Calabi-Yau 3-fold, which has also been the case of most interest for string theory[4].

To obtain a 'static' solution to superstring theory, it turns out that the target Kahler manifold M should admit a Ricc-flat metric, i.e., by Yau's theorem it should be a Calabi-Yau manifold[5]. In such a case the dimensions of $H^{d,0}(M)$ is one, and thus we have in our theory an operator corresponding to this element in $H^{d,0}$. This operator induces an isomorphism on

---

[4]For manifolds which have $c_1 < 0$, by which I mean there are some two cycles where c_1 evaluates to a negative number, the underlying theory is not very well behaved (i.e., it is not asymptotically free) and it seems that similarly the quantum cohomology ring is somewhat ill defined (in the Landau-Ginzburg description to be mentioned in section 5 it corresponds to perturbing the action by non-renormalizable terms with charge greater than 1). So quantum cohomology rings make better sense for $c_1 \geq 0$. However it would be interesting to see, and there is some indication [12] that maybe the mirror map acts on the space of *all* Kahler manifolds (possibly non-compact) by flipping the sing of c_1, which in particular sends a Calabi-Yau manifold to another Calabi-Yau manifold.

[5]Physically we should not ignore other manifolds as is commonly done, since one can use them to construct interesting non-static solutions of string theory, of the type relevant for cosmology (see for example [13]).

the space of operators by multiplication [6]. This is known as the spectral flow and gives the isomorphism

$$O_{p,q} \sim O_{p+d,q}$$

The fact that this is an isomorphism is related to the existence of the conjugate (or inverse) operator. In other words By conjugation *there must also exist conjuge operators* in $H^{-d,0}$. This operator induces a correspondence between operators:

$$O_{p,q} \sim O_{p-d,q}$$

(similar statements of course hold for conjugate sectors of the Hilbert space and amounts to shifting the anti-holomorphic degree by $-d$). This isomorphism in particular applies to the special operators operators $H^{p,q}$ with $0 \leq p,q \leq d$ which represent cohomology of M and thus suggest that there are also 'special' operators which we denote by $H^{p-d,q}$, by shifting the holomorphic degree by $-d$. These special operators have the following properties which follows by the above isomorphism:

$$\dim H^{0,0} = \dim H^{-d,0} = \dim H^{0,d} = \dim H^{-d,d} = 1$$

$$\dim H^{-p,q} = \dim H^{d-p,q} = h^{d-p,q}$$

where $h^{*,*}$ denote the hodge numbers M. It looks as if the operators in $H^{-p,q}$ describe the cohomology of a d-dimensional manifold which has the same hodge diamond as M except that it is flipped. In fact from the structure of string vacuum [6] it follows that there is a closed operator ring among these states which is additive in terms of their $Z \oplus Z$ grading just as was the case for the operators $H^{p,q}$ with $0 \leq p,q \leq d$. Note that the correspondence between cohomology elements of $H^{-p,q}$ and $H^{d-p,q}$ do not respect the ring structure and is thus not an isomorphisms of these rings. So we learn that for *for any Calabi-Yau manifold we find not one but two rings-only one of which is related to the deformed cohomology ring of the manifold.* This second ring we call the *complex ring* of the manifold as it will turn out to (generically) characterize the complex structure of the Calabi-Yau manifold[6].

---

[6]The complex ring can be viewed geometrically as the ring generated by wedging $H^q(\Lambda^p\Theta)$ where Θ represents the holomorphic tangent bundle [14]. That their dimension is related to that of $H^{d-p,q}$ can be easily infered from the existence of a holomorphic d form for the Calabi-Yau case.

So far we have described the Hilbert space and operators corresponding to strings imbedded in a Calabi-Yau manifold M. But usually we are given not a Calabi-Yau manifold, but the string vacuum itself, i.e., a Hilbert space and a set of operators acting on it. Note that in the isomorphism class of string vacua, if we just relabel the labels of $O_{p,q}$ by $O_{-p,q}$ we have not changed the string vacuum, and we obtain an isomorphic vacuum. This involution of one of the gradings simply exchanges the two rings that we discussed above. In this abstract setting how do we decide which of these two rings are 'preferred' in the sense that it corresponds to the deformation of the cohomology ring of a manifold? Since these two rings are absolutely on the same footing as far as the string vacuum is concerned, i.e., that there is an isomorphic string vacuum which relabels the sign of one of the gradings, the only way to restore the impartiality is *to postulate that for every Calabi-Yau manifold M there is another manifold $\tilde{M}$, such that the string vacuum on either M or $\tilde{M}$ gives rise to both cohomology rings.* This in particular means that

$$h^{p,q}(\tilde{M}) = h^{d-p,q}(M) \tag{4.2}$$

This is the basic idea of mirror symmetry [15] [6]. Note that this idea applies to a Calabi-Yau manifold of any dimension (not just three as is mostly applied to). Also note that the dimension of complex deformations of M which is equal to $h^{1,d-1}(M)$ is equal to the dimension of Kahler deformations of $\tilde{M}$ and vice versa. So under this mirror symmetry the shape and size of the manifolds get exchanged. Since the quantum cohomology ring encodes the information about the Kahler class in it, and under mirror symmetry shape and size get exchanged, this explains why the second ring, the complex ring, is characterizing the complex structure of the manifold.

Let us consider the simplest examples of mirror pairs: As we have discussed before for bosonic strings, strings propagating on a torus and the dual torus give identical vacua and form mirror pairs. It turns out that these are in fact also the simplest examples of mirror vacua for fermionic strings. Let us explain this briefly in the context of simplest complex torus, a one dimensional complex torus which is geometrically the product of two circles with radii R_1 and R_2. Then the complex structure τ of the torus and its volume $-i\rho$ are given by

$$\tau = i\frac{R_1}{R_2} \qquad \rho = iR_1 R_2$$

Now we apply the duality described in section 2 for bosonic strings in the case of target space being a torus. This duality works equally well for bosonic and fermionic strings. Let us apply that to the second circle of this example sending $R_2 \to 1/R_2$ and we thus end up exchanging $\rho \leftrightarrow \tau$. This is an example of the general phenomena described above namely that the moduli controlling the shape and the size of the Calabi-Yau manifolds are exchanged under such a duality[7]. This is the simplest example of mirror symmetry. It is worth emphasizing that the other beautiful examples that have been found are highly non-trivial to describe geometrically [16] [17] [18] and have far more out reaching consequences. Nevertheless the basic idea remains the same, and fits very naturally into the general framework of duality just as we saw for the bosonic strings.

5. Catastrophes and Superstring Vacua

In this section we describe a link between string vacua and catastrophe theory. The origin of this direction of study of strings was motivated by trying to ignore geometry of target space and classify all string vacua directly (as had been emphasized by Friedan). So far we have mostly described string vacua arising from strings propagating in some target space. However, there are other useful ways to describe string vacua which may or may not be related to such a picture. The main idea is to note that the string amplitude was defined as a sum over all interpolating Riemann surfaces weighed by energy functional $exp(-E)$. Here $E = \int |Dx|^2$ where x denotes the map which defines an immersion of the Riemann surface into the target space (with appropriate addition of fermionic terms in the case of superstrings). The basic idea to generalize this is to think of E as a functional of some fields (functions) defined on the Riemann surface. This defines a quantum field theory in two dimensions. There are many interesting examples of such field theories, but we will mention the one most relevant for superstring vacua which is the case of Landau-Ginzburg theories. Without going to too much detail it turns

---

[7]This duality extends to the full moduli of the torus not just to the case that it is geometrically the product of two circles. In the more general case the size also is a complex modulus due to the appearance of the anti-symmetric tensor fields which effectively complexifies the Kahler cone.

out that in this case the field theory is characterized by a single holomorphic function $W(x_i)$ where x_i are superfields. It was found [19] that quasi-homogeneous W's which have an isolated critical point at $x_i = 0$ give rise to a nice class of (super conformal) theories. In this way the classification of quasihomogeneous singularities became very relevant for the classification of string vacua. Moreover, it was found [20] that if the index of the singularity[8] is integral and equal to the number of variables x_i minus 2, they are related to string vacua propagating on the Calabi-Yau 'manifold' defined by (the possibly singular variety) $W(x_i) = 0$ in weighted projective space with a very particular Kahler metric. This clarified the geometrical meaning of the important discovery of Gepner [21] in this construction of string vacua. Note that the complex structure of the Calabi Yau is fixed by $W = 0$, but the Kahler structure of Calabi-Yau is only implicitly specified by W (through its quantum symmetries) [22] [20]. As an example if we take

$$W = x^4 + y^4 + z^2 + a\,x^2 y^2$$

Setting $W = 0$ in weighted projective two space, we get a one dimensional torus whose moduli is fixed by a. The volume of the torus is implicitly fixed (by the existence of quantum Z_4 symmetry) which teaches us that the volume of this torus is 1 for all a (and the anti-symmetric field vanishes) [22]. So in this way the study of strings propagating on Calabi-Yau manifolds can be very effectively studied using this picture, and this has become an important tool in the recent discovery of interesting class of exmaples of mirror symmetric pairs of string vacua.

For strings on Calabi-Yau manifolds, as we discussed before we automatically get two rings, only one of which is the cohomology ring of the manifold. What is the other, the complex ring, geometrically? Well, a *subring* of this second ring can be described geometrically, when the Calabi-Yau theory is represented by a variety defined by $W = 0$ in a weighted projective space. In this case if we consider the (integral dimension) ring of the singularity defined by

$$\mathcal{R} = \frac{C[x_i]}{dW} \tag{5.1}$$

---

[8]For a quasihomogenous function the index is defined as follows: By assigning degree one to a quasi-homogeneous W we can obtain fractional weights q_i of variables x_i. The index of W is simply $\sum(1 - 2q_i)$.

they generate a subring of $H^{-p,q}$ discussed before (where p corresponds to the degree of the ring element). It would be interesting to see if one can extend this picture to the full ring for all $H^{-p,q}$ (and not just the diagonal elements). Note that this ring certainly does depend on the complex moduli of Calabi-Yau manifold (as that changes as we change W). This is consistent with the mirror picture, namely the mirror ring depends on Kahler moduli (as the quantum deformed cohomology ring does depend on Kahler moduli).

So can we describe the quantum deformed cohomology ring of some manifolds using singularity ring for some W? The answer to this question should be in the affirmative if the mirror picture is valid. After all the mirror map changes $H^{-p,p} \to H^{p,p}$ and so maps (part of) the singularity ring to the diagonal elements of the deformed cohomology ring. The computation of cohomology ring for Calabi-Yau manifolds is in general rather difficult. So in this way we map a difficult problem (computation of deformed cohomology ring) to a simple problem (computation of the ring of a singularity) once we know the right transformation.

The quantum cohomology rings are easy to compute in some cases, as we mentioned before. For example for CP^n we mentioned that the deformed cohomology ring is

$$x^{n+1} = \beta$$

This of course can be written in the 'mirror' picture by the ring of W according to (5.1):

$$W(x) = \frac{x^{n+2}}{n+2} - \beta x$$

This can also be generaized to Grassmanians[9]. As discussed before for the case where $c_1 \neq 0$ we expect to violate the grading of the ring, which means the corresponding W would not be quasi-homogeneous. For Calabi-Yau manifolds as we mentioned before the grading of the ring is respected by the deformation, so if the deformed ring is that of a singularity ring the corresponding W will again be quasi-homogeneous.

---

[9]The cohomology ring of Grassmanian $U(n+k)/U(n) \times U(k)$ can be written as the singularity ring [6] generated by a single potential $W(x_i) = \sum z_i^{n+k+1}/n+k+1$ where x_i are symmetric polynomials of degree i in z_j (with no monomial appearing more than once) and i runs from 1 to n. The x_i correspond to the Chern classes of the n-dimensional tautological vector bundle on the Grassmanians. The quantum deformation of this ring is naturally conjectured to be $W \to W - \beta x_1$. The motivation for this comes from the fact that $c_1 = n + k$.

6. Topological Mirrors

So far we have talked about mirror symmetry in the following sense: We have strings propagating on two manifolds M_1 and M_2, which lead, as described before, to two Hilbert spaces each equipped with an infinite set of operators acting on them. Then if these two structures, or vacua, are isomorphic, we call M_1 and M_2 mirror pairs. Establishing this isomorphism at the level of Hilbert spaces is in general a complicated task. It would have been nice if there were a simple criterion to establish their equivalence. This question is also the same as asking how do we find a simple way to classify string vacua.

Classifying string vacua (and in particular static solutions which correspond to conformal field theories in two dimensions) has been investigated intensively in the past seven years. We are unfortunately still far from a complete classification. However for the fermionic vacua, an interesting class of vacua have, as discussed before, a simple description in terms of quasi-homogeneous signularities. In fact it is believed that for any quasi-homogeneous function W there is a unique string vacuum. In other words it is believed that the information about W is enough to reconstruct the full Hilbert space of strings and operators acting on it. More generally, whether or not the theory comes from a quasi-homogeneous singularity, it is believed that essentially given the chiral rings in the theory one has enough information to reconstruct the full theory. Applied to the special case of strings propagating on manifolds this may sound a little strange: We seem to be saying that given the cohomology ring of a manifold, we can find the manifold, which is certainly false. However it is for the special case of Calabi-Yau manifolds that we are considering this and in such cases just specifying the hodge numbers may go a long way in determining the manifold itself. Moreover we have *two* rings the *quantum cohomology ring* and the *complex ring*, which fix the Kahler class and the complex structure of the manifold respectively. Thus from Yau's proof of Calabi's conjecture which shows that knowing the Kahler class uniquely fixes the Ricci flat metric we can reconstruct the metric on the manifold by the information encoded in these rings.

Having said all these, it becomes clear that the phenomena of mirror symmetry can be formulated more compactly by stating that the two rings we get for one manifold are exchanged in the mirror manifold. In other words we can forget about the rest of the structure of string vacua and Hilbert spaces and the full set of operators acting on them and concentrate simply on this finite dimensional subset of special operators. In fact this concept can be formalized. Consider strings propagating on a Kahler manifold. It turns out there is a *twisted* or *topological* version of this theory [9] [23] which can be obtained by a simpole modification of the definition of the theory (by shifting the spin of fermions) which has the effect that the only physical operators we obtain are the ones corresponding to the cohomology classes and that they form the quantum cohomology ring of the manifold. If in addition the manifold in question is a Calabi-Yau manifold this twisting can be done in two inequivalent ways (by shifting the spins of fermions chirally, which is allowed for Calabi-Yau manifolds because of absence of sigma model anomalies since $c_1(M) = 0$), one of which gives the quantum cohomology ring and the other gives the complex ring, which (except for the diagonal elements) has a less clear geometrical meaning. In this way we can get both rings depending on which twist we choose. However, it is clear that in this topological description the ordinary cohomology ring has a more 'natural' origin, and it seems to be 'preferred'. However, there is another way to describe (fermionic) string vacua and that is via a Landau-Ginzburg theory. In this case we can also twist the theory and obtain a topological version [24] whose only (physical) operators correspond to the singularity ring of W. Again, if W is quasi-homogeneous, this can be done in two different ways, one of which corresponds to the singularity ring which when W describes a Calabi-Yau manifold correspond to its complex ring, and the other which has a less clear geometrical meaning (as it appears in the twisted sectors) correspond to the deformed cohomology ring. So we see that again we have two rings, but the complex ring is 'preferred'.

The notion of mirror symmetry can be simply translated to the *quivalence of a topological sigma model with a topological Landau-Ginzburg model*, where the 'preferred' ring of the sigma model (the quantum cohomology ring or Kahler ring) gets mapped to the 'preferred' ring of the Landau-Ginzburg model (the singularity ring or complex ring). Stated in this way this mirror symmetry is more general than Calabi-Yau manifolds, as the W may or may not correspond to a Calabi-Yau manifold (even if it is quasi-homogeneous (see next section)). Also W may not be quasi-homogeneous as the example of the Grassmannians mentioned before illustrates (i.e., it goes beyond conformal theories) but nevertheless we have a mirror symmetry in the sense defined above.

7. Some Puzzles and Conclusion

It would be nice to be able to state the mirror symmetry in full generality. In geometrical terms, in the sense that strings on manifold M_1 behave the same way as strings on manifold M_2 this would be rather difficult to do. It is difficult even to fix precisely which category of geometrical objects we are considering. If we fix the category to be that of Calabi-Yau manifolds this would be false because there are examples of Calabi-Yau manifolds which are rigid (i.e., do not admit complex deformations) therefore their mirror would not admit Kahler deformations (i.e., $h^{1,1} = 0$), which means that the mirror would not even be a Kahler manifold! So in this sense we have lost the mirror. However in the sense of equivalence of two topological theories, i.e., equivalence of a topological theory based on a sigma model and tht on a Landau-Ginzburg model this way still be possible. In fact now we will give an example where this is indeed what happens. Consider a three-fold Calabi-Yau manifold defined by taking the product of three two dimensional tori, with Z_3 symmetry, and modding out by a $Z_3 \times Z_3$ symmetry generated by the elements $(\omega, \omega^{-1}, 1))$, $(1, \omega, \omega^{-1})$ acting on the three tori, where ω denotes the Z_3 action. It is possible to resolve the fixed point singularities and obtain a smooth Calabi-Yau manifold. This manifold is rigid, in that it does not admit any complex deformations $h^{1,2} = 0$. The dimension of Kahler deformations is $h^{1,1} = 84$. What is the mirror for this manifold? The answer turns out to be easy in this case: It is the landau-Ginzburg theory defined by

$$W = \sum_{i=1,\dots,9} x_i^3 + \sum_{i \neq j \neq k} a_{ijk}\, x_i x_j x_k$$

The way we know this is that at $a_{ijk} = 0$ we can explicitly construct the Landau-Ginzburg theory and compare it explicitly with the geometrical description which also turns out to be exactly solvable (before blowing up the singularities) and one finds that (with the metric and the antisymmetric field of tori corresponding to the point of enhanced Z_3 symmetry) they agree. Moreover one can map the fields $x_i x_j x_k$ to the Kahler classes of the manifold. So in this way the 84 Kahler deformations of the manifold (which includes the blow up modes) will get mapped to the deformation of W which are captured through varying a_{ijk} above. This description of mirror symmetry is enough to capture the counting of instantons on the original manifold by studying variations of Hodge structure characterized by W [14] so for the purposes of 'simplifying' the instanton counting it works as well. So in a sense we do not really need a geometrical mirror; or if we insist we can say that the geometrical mirror in this case is a 7 fold defined by $W = 0$ in CP^8. But this description is only valid as far as we are relating the variation of its Hodge structure with the deformed cohomology ring of the original Calabi-Yau manifold[10]. This example reinforces another view of mirror symmetry, namely, the abstract property of the rings that may arise in conformal theory is the same whether or not they come from the cohomology ring or the complex ring. So somewhat the lesson is to *forget* about the underlying manifold altogether and concentrate on abstract properties of the rings, and the classification of the kinds of rings that can appear. This is very much the question of classification of variation of Hodge structures [25]. This is in fact the point of view advocated by Cecotti [14]. In this setup the existence of mirror symmetry is probably related to the 'scarcity' of inequivalent types of variations of hodge structure (with some given topological variants).

The same idea of mirror picture applies even to the general case of manifolds with $c_1 \neq 0$, for example the Grassmanians, where the 'mirror symmetry' allows us to compute exactly instanton corrections to the analog of 'Weil-Petersson' metric for such manifolds [26]. This follows from the structure of special geometry which exists even off criticality. This reinforces the picture that *we should not restrict our attention to Calabi-Yau manifolds if we are to have a deeper understanding of mirror symmetry.*

---

[10]It would be interesting to see if turning on (possible singular) dilation fields and torsion on this 7-fold, gives a sigma model which is equivalent to the three fold Calabi-Yau we started with. As is well known turning on dilation field shifts the effective dimension (central charge) of the theory. In such a picture the freezing of Kahler degrees of freedom would be related to solving dilation equations of motion.

The notion of 'quantum' cohomology ring might remind one of seemingly unrelated subject of 'quantum' groups. As is well known these groups have representation ring which is a 'quantum deformation' of the classifical representation ring of the group. The deformations being parametrized by a parameter k which is the level of quantum group, and as $k \to \infty$ we recover the classical representation ring. Indeed this k seems to pay a very similar role to the role kahler class k plays in quantum cohomology ring in the infinite limit of which one recovers the classical cohomology ring. It turns out these two different 'quantum rings' are not as unrelated as might seem at first sight! In particular it has been shown [27] that for special class of such theories the fusion ring (representation ring) of quantum groups get mapped to the chiral ring of a Landau-Ginzburg theory (see also [28] [11]). For example, if one considers $SU(n)$ quantum group with level $k = 1$, its representation ring is isomorphic to the quantum cohomology ring with level $k = 1$, its representation ring is isomorphic to the quantum cohomology ring of CP^{n-1} (discussed before). It would be interesting to see whether or not all the rings of quantum groups can be interpreted as the quantum cohomology ring of some manifold (for example which manifold has the quantum cohomology ring related to Chebychev polynomial?). This connection has become even more intriguing with the discovery [26] that precisely these Landau-Ginzburg theories seem integrable field theories in the sense that they have an integrable classical equation describing the generalized special geometry) some further evidence for their integrability has been found in [29]). It was further conjecture in [26] that whenever the ring of a supersymmetric theory corresponds to that of a RCFT (rational conformal field theories), i.e. a solution to quantum group representation ring, the corresponding field theory is integrable. These connections we believe are very important to understand better for a more abstract understanding of 'mirror symmetry' and 'quantum rings'.

We have learned that mirror symmetry is the statement of equivalence of two topological theories, one which is difficult to compute and the other thing happens is easy. It is natural to continue this line of thinking and suggest that the same thing happens for other topological theories. In particular Donaldson theory which captures some invariants for differentiable manifolds in four dimensions, has a topological field theory description [3]. It is in general very difficult to compute Donaldson invariants, just as it is in general difficult to compute the number of rational (holomorphic) curves in a manifold. But we have seen in the latter case that there is a simpler topological theory which is the Landau-Ginzburg description. It is tempting to conjecture that there is a similar thing going to happen in four dimensions [31], namely that there must be a topological mirror theory, far simpler than Donaldson theory, which via an appropriate mirror map allows us to effectively compute Donaldson invariants. It remains to be seen if this conjecture is valid.

It is a pleasure to thank D. Cecotti for many discussions which has greatly influenced my thinking on this subject. I would like to thank D. Kazhdan for a careful reading of this manuscript and for making suggestions for its improvement. I also am thankful to I. Singer and S.-T. Yau for encouraging me to participate in this conference. This work was supported in part by the Packard Foundation and NSF grants PHY-89-57162 and PHY-87-14654.

References

[1] G. Segal, *Conformal Field Theory*, Oxford preprint; and lecture at the IAMP Congress, Swansea, July, 1988.

[2] K.S. Narain, M.H. Sarmadi and C. Vafa, Nucl. Phys. **B288** (1987) 551;
J. Harvey, G. Moore and C. Vafa, Nucl. Phys **B304** (1988) 269.

[3] K. Kikkawa and M. Yamasaki, Phys. Lett. **B149** (1984) 357;
N.Sakai and I. Senda, Prog. Theor. Phys. **75** (1984) 692.

[4] L. Dixon, J. Harvey, C. Vafa and E. Witten, Nucl. Phys. **B274** (1986) 285;
J. Lepowski, Proc. of the Nat. Acad. of Sci. **82** (1985) 8295.

[5] S.S. Roan, Int. Jour. of Math., **v.1** (1990) 211.

[6] W. Lerche, C. Vafa and N. Warner, Nucl. Phys. **B324** (1989) 427.

[7] M. Dine, N. Seiberg, X.G. Wen and E. Witten, Nucl. Phys. **B278** (1987) 769; **B289** (1987) 319.

[8] S. Hamidi and C. Vafa, Nucl. Phys. **B279** (1987) 465;
L. Dixon, D. Friedan, E. Martinec and S. Shenker, Nucl. Phys. **B282** (1987) 13.

[9] E. Witten, Comm. Math. Phys. **118** (1988) 411; Nucl. Phys. **B340** (1990) 281.

[10] P.S. Aspinwall and D.R. Morrison, *Topological Field Theory and Rational Curves*, preprint, OUTP-91-32p. DUK-M-91-12.

[11] K. Intriligator, *Fusion Residues*, Harvard preprint, HUTP-91/A041.

[12] This idea arose in discussions with G. Horowitz.

[13] A. Tseytlin and C. Vafa, preprint HUTP-91/A049; JHU-TIPAC-910028.

[14] S. Cecotti, Int. J. Mod. Phys. **A6** (1991) 1749;
S. Cecotti, Nucl. Phys. **B355** (1991) 755.

15] L. Dixon, unpublished.

[16] B.R. Greene and M.R. Plesser, Nucl. Phys. **B338** (1990) 15.

[17] P. Candelas, X.C. de la Ossa, P.S. Green and L. Parkes, Nucl. Phys. **B359** (1991) 21; Phys. Lett. **258B** (1991) 118;
P. Candelas, M. Lynker, and R. Schimmrigk, Nucl Phys. **B341** (1990) 383.

[18] P.S. Aspinwall, C.A. Lutken, and G.G. Ross, Phys. Lett. **241B** (1990) 373;

P.S. Aspinwall and C.A. Lutken, Nucl. Phys. **B355** (1991) 482.

[19] C. Vafa and N. Warner, Phys. Lett. **B218** (1989) 51;

E. Martinec, Phys. Lett. **B217** (1989) 431.

[20] B.R. Greene, C. Vafa and N. Warner, Nucl. Phys. **B324** (1989) 371;

E. Martinec, *Criticality, Catastrophe and Compactifications*, V.G. Knizhnik memorial volume, 1989.

[21] D. Gepner, Phys. Lett. **199B** (1987) 380; Nucl. Phys. **B296** (1987) 380.

[22] C. Vafa, Mod. Phys. Lett. **A4** (1989) 1615.

AMS/IP Studies in Advanced Mathematics
Volume **9**, 1998

Mirror Manifolds and Topological Field Theory

Edward Witten*

School of Natural Sciences

Institute for Advanced Study

Olden Lane

Princeton, N.J. 08540

Abstract

These notes are devoted to sketching how some of the standard facts relevant to mirror symmetry and its applications can be naturally understood in the context of topological field theory. If X is a Calabi–Yau manifold, the usual nonlinear sigma model governing maps of a Riemann surface Σ to X can be twisted in two ways to give topological field theories, which I call the A model and the B model. Mirror symmetry relates the A model (of one Calabi–Yau manifold) to the B model (of its mirror). The correlation functions of the A and B models can be computed, respectively, by counting rational curves and by calculating periods of differential forms. This can be proved as a consequence of a reduction to weak coupling (as in §3-4 of these notes) or by a sort of fixed point theorem for the Feynman path integral (see §5). The correlation functions of the twisted models coincide, as explained in §6, with certain matrix elements of the

---

*Research supported in part by NSF Grant 91-06210.

© 1998 American Mathematical Society
and International Press

physical, untwisted model – namely those that determine the superpotential. The conventional moduli spaces of sigma models can be thickened, in the context of topological field theory, to extended moduli spaces, indicated in §7, which are probably the natural framework for understanding the still mysterious "mirror map" between moduli spaces.

1. Introduction

The purpose of these notes is to explain aspects of the mirror manifold problem that can be naturally understood in the context of topological field theories. (For other aspects of the problem, and detailed references, the reader should consult other articles in this volume.) Most of what I will say can be found in the existing literature, but to isolate the facts most relevant to mirror symmetry may be useful. The points that I want to explain are as follows.

First, we will consider the standard supersymmetric nonlinear sigma model in two dimensions, governing maps of a Riemann surface Σ to a target space which for our purposes will be a Kahler manifold X of $c_1 = 0$. We will see that there are two different topological field theories that can be made by twisting the standard sigma model. There are not standard names for these topological field theories; I will call them the A theory and the B theory, or $A(X)$ and $B(X)$ when I want to specify the choice of X. The A theory has been studied in detail in [18], where many facts sketched below are explained; the B theory has been studied less intensively. (The A and B theories were discussed qualitatively, in a general context of $N = 2$ superconformal field theories, by Vafa in this volume [16].)

Unlike the ordinary supersymmetric nonlinear sigma model, the twisted models are "soluble" in the sense that the problem of computing all the physical observables can be reduced to classical questions in geometry. This is done by a sort of fixed point theorem in field space, which gives the following results: the correlation functions of the A theory are determined by counting holomorphic maps $\Sigma \to X$ obeying various conditions; the correlation functions of the B theory can be computed by calculating periods of classical differential forms. (More generally, "counting" rational curves must be replaced by computing the Euler class of the vector bundle of antighost zero modes, as I have explained elsewhere [19, §3.3]; this has been implemented in the mirror manifold problem by Aspinwall and Morrison [1].)

For the most direct physical applications one is not interested in the twisted A or B theories, but in the original, "physical" sigma model. However, there is one very important case in which physical observables coincide with observables of the A and B theories. This happens for the Yukawa couplings, which are certain quantities that one wishes to compute for Σ of genus zero.[1] In particular, the $\overline{27}^3$ and 27^3 Yukawa couplings of superstring models coincide with certain observables of the A and B theories, respectively. These Yukawa couplings are therefore determined by the fixed point theorem mentioned in the last paragraph; from this recover results that were originally found by more detailed arguments, such as the fact that the 27^3 Yukawa couplings have no quantum corrections, and the instanton sum formula [7] for the $\overline{27}^3$ Yukawa couplings.

In §2 we recall the definition of the standard sigma model. The twisted A and B models are described in §3 and §4, along with an explanation of their main properties, including the reduction to instanton moduli spaces and to constant maps in the A and B models respectively. This latter reduction is reinterpreted as a sort of fixed point theorem in §5. In §6, I explain why certain observables of the standard physical sigma model coincide with observables of the twisted models. This occurs whenever the canonical bundle of the Riemann surface becomes trivial after deleting the points at which fermion vertex operators have been inserted – in practice, mainly for amplitudes in genus zero with precisely two fermions. In §7 – the only part of these notes containing some novelty – I look a little more closely at the A and B models and describe the full families of topological field theories of which they are part. I strongly suspect that many properties of mirror symmetry that are not now well understood, including the structure of the mirror map between the parameter spaces, can be better understood in looking at the full topological families.

Since "elliptic genera" (see [11]) arise in the same supersymmetric nonlinear sigma models

---

[1]The restriction to genus zero appears in two ways: the superpotential that determines the Yukawa couplings has no higher genus corrections because of nonrenormalization theorems [7]; alternatively, in relating the superpotential to an observable of the twisted model, we will require in §6 that the canonical bundle of Σ with two points deleted be trivial, which is of course true only in genus zero.

that are the basis for the study of mirror manifolds, I will also along the way make a few observations about them. In particular we will note the easy fact that if X, Y are a mirror pair, they have the same elliptic genus. This is trivial in complex dimension three (since the elliptic genus of a three dimensional Calabi–Yau manifold is zero), but becomes interesting in higher dimension.

Perhaps I should emphasize that if X and Y are a mirror pair, then mirror symmetry relates *all* observables of the sigma model on X to corresponding observables on Y – and not just the few observables that can be related naturally to the twisted models and thus to topological field theory. The literature on mirror symmetry has focussed on these particular observables, which of course are also the ones we will by studying here, because of their phenomenological importance, and because, since the B model is soluble classically, the relation given by mirror symmetry between observables of the $A(X)$ model and observables of the $B(Y)$ model is particularly useful.

2. Preliminaries

To begin with, we recall the standard supersymmetric nonlinear sigma model in two dimensions.[2] It governs maps $\Phi : \Sigma \to X$, with Σ being a Riemann surface and X a Riemannian manifold of metric g. If we pick local coordinates z, $\bar{z}$ on Σ and ϕ^I on X, then Φ can be described locally via functions $\phi^I(z, \bar{z})$. Let K and $\overline{K}$ be the canonical and anti-canonical line bundles of Σ (the bundles of one forms of types $(1, 0)$ and $(0, 1)$, respectively), and let $K^{1/2}$ and $\overline{K}^{1/2}$ be square roots of these. Let TX be the complexified tangent bundle of X. The fermi fields of the model are ψ_+^I, a section of $K^{1/2} \otimes \Phi^*(TX)$, and ψ_-^I, a section of $\overline{K}^{1/2} \otimes \Phi^*(TX)$.

---

[2]This discussion will be at the classical level, and we will not worry about the anomalies that arise and spoil some assertions if the target space is not a Calabi–Yau manifold.

The Lagrangian is[3]

$$L = 2t \int d^2 z \Big(\frac{1}{2} g_{IJ}(\Phi) \partial_z \phi^I \partial_{\bar{z}} \phi^J + \frac{i}{2} g_{IJ} \psi_-^I D_z \psi_-^J$$
$$+ \frac{i}{2} g_{IJ} \psi_+^I D_{\bar{z}} \psi_+^J + \frac{1}{4} R_{IJKL} \psi_+^I \psi_+^J \psi_-^K \psi_-^L \Big) . \tag{2.1}$$

Here t is a coupling constant, and R_{IJKL} is the Riemann tensor of X. $D_{\bar{z}}$ is the $\bar{\partial}$ operator on $K^{1/2} \otimes \Phi^*(TX)$ constructed using the pullback of the Levi-Civita connection on TX. In formulas (using a local holomorphic trivialization of $K^{1/2}$),

$$D_{\bar{z}} \psi_+^I = \frac{\partial}{\partial \bar{z}} \psi_+^I + \frac{\partial \phi^J}{\partial \bar{z}} \Gamma_{JK}^I \psi_+^K , \tag{2.2}$$

with Γ_{JK}^I the affine connection of X. Similarly D_z is the ∂ operator on $\overline{K}^{1/2} \otimes \Phi^*(TX)$.

The supersymmetries of the model are generated by infinitesimal transformations

$$\delta \Phi^I = i\epsilon_- \psi_+^I + i\epsilon_+ \psi_-^I$$
$$\delta \psi_+^I = -\epsilon_- \partial_z \phi^I - i\epsilon_+ \psi_-^K \Gamma_{KM}^I \psi_+^M \tag{2.3}$$
$$\delta \psi_-^I = -\epsilon_+ \partial_{\bar{z}} \phi^I - i\epsilon_- \psi_+^K \Gamma_{KM}^I \psi_-^M$$

where ϵ_- is a holomorphic section of $K^{-1/2}$, and ϵ_+ is an antiholomorphic section of $\overline{K}^{-1/2}$. The formulas for the Lagrangian (2.1) and the transformation laws (2.3) have a natural interpretation upon formulating the model in superspace, that is in terms of maps of a super-Riemann surface to X. I will not discuss this here; for references see Rocek's lecture in this volume.

As a small digression, let me note that usually, in discussing superconformal symmetry, one considers ϵ's that are defined locally (say in a neighborhood of a circle $C \subset \Sigma$ if one is studying the Hilbert spaces obtained by quantization on C). One might wonder what happens if we can find global ϵ's. On a Riemann surface of genus $g > 1$, this will not occur as there are no holomorphic sections of $K^{-1/2}$. In genus one K is trivial; if we pick $K^{1/2}$ to be trivial, then there is a one dimensional space of global ϵ_-'s. On the other hand, if we pick $\overline{K}^{1/2}$ to be a non-trivial line bundle (of order two) then globally ϵ_+ must vanish. With the techniques that

---

[3] Here $d^2 z$ is the measure $- i dz \wedge d\bar{z}$. Thus if a and b are one forms, $\int a \wedge b = i \int d^2 z (a_z b_{\bar{z}} - a_{\bar{z}} b_z)$. The Hodge $\star$ operator is defined by $\star dz = i dz$, $\star d\bar{z} = -i d\bar{z}$.

we will use below, one can readily use the symmetry generated by ϵ_- to prove that the partition function of the sigma model, on such a genus one surface, is independent of the metric of X and so is a topological invariant in the target space; it is in fact the elliptic genus of X. (See [11] for an introduction to elliptic genera, and my lecture in that volume for an explanation of the field theoretic approach to them.) As the existence of a holomorphic section of $K^{-1/2}$ was essential here, this construction will not generalize to genus $g > 1$.

The twisted models that I will explain below and which will be the basis for whatever I have to say about mirror manifolds are the closest analogs of the usual supersymmetric sigma model for which global fermionic symmetries exist regardless of the genus. In particular the closest analog of the elliptic genus for $g > 1$ involves the half-twisted model introduced at the end of this section.

Kahler Manifolds and the Twisted Models

We now wish to describe the additional structure – $N = 2$ supersymmetry, to be precise – that arises if X is a Kahler manifold. In this case, local complex coordinates on X will be denoted as ϕ^i; their complex conjugates are $\phi^{\bar{i}} = \overline{\phi^i}$. ($\phi^I$ will still denote local real coordinates, say the real and imaginary parts of the ϕ^i.) The complexified tangent bundle TX of X has a decomposition as $TX = T^{1,0}X \oplus T^{0,1}X$. The projections of ψ_+ in $K^{1/2} \otimes \Phi^*(T^{1,0}X)$ and $K^{1/2} \otimes \Phi^*(T^{0,1}X)$, respectively, will be denoted as ψ_+^i and $\psi_+^{\bar{i}}$. Likewise the projections of ψ_- in $\overline{K}^{1/2} \otimes \Phi^*(T^{1,0}X)$ and $\overline{K}^{1/2} \otimes \Phi^*(T^{0,1}X)$ will be denoted as ψ_-^i and $\psi_-^{\bar{i}}$, respectively. The Lagrangian can be written

$$L = 2t \int_\Sigma d^2z \left(\frac{1}{2} g_{IJ} \partial_z \phi^I \partial_{\bar{z}} \phi^J + i\psi_-^{\bar{i}} D_z \psi_-^i g_{\bar{i}i} + i\psi_+^{\bar{i}} D_{\bar{z}} \psi_+^i g_{\bar{i}i} + R_{i\bar{i}j\bar{j}} \psi_+^i \psi_+^{\bar{i}} \psi_-^j \psi_-^{\bar{j}} \right) . \quad (2.4)$$

As for the fermionic symmetries, these are twice as numerous as before because of the Kahler structure; this is analogous to the decomposition of the exterior derivative on a complex manifold as $d = \overline{\partial} + \partial$. I will write the following formulas out in detail because we will need various specializations of them in describing the twisted models. In terms of infinitesimal fermionic parameters α_-, $\tilde{\alpha}_-$ (which are holomorphic sections of $K^{-1/2}$) and α_+, $\tilde{\alpha}_+$ (antiholomorphic sections of $\overline{K}^{-1/2}$), the transformation laws are

$$\delta\phi^i = i\alpha_-\psi^i_+ + i\alpha_+\psi^i_-$$

$$\delta\phi^{\bar{i}} = i\tilde{\alpha}_-\psi^{\bar{i}}_+ + i\tilde{\alpha}_+\psi^{\bar{i}}_-$$

$$\delta\psi^i_+ = -\tilde{\alpha}_-\partial_z\phi^i - i\alpha_+\psi^j_-\Gamma^i_{jm}\psi^m_+$$

$$\delta\psi^{\bar{i}}_+ = -\alpha_-\partial_z\phi^{\bar{i}} - i\tilde{\alpha}_+\psi^{\bar{j}}_-\Gamma^{\bar{i}}_{\bar{j}\bar{m}}\psi^{\bar{m}}_+ \qquad (2.5)$$

$$\delta\psi^i_- = -\tilde{\alpha}_+\partial_{\bar{z}}\phi^i - i\alpha_-\psi^j_+\Gamma^i_{jm}\psi^m_-$$

$$\delta\psi^{\bar{i}}_- = -\alpha_+\partial_{\bar{z}}\phi^{\bar{i}} - i\tilde{\alpha}_-\psi^{\bar{j}}_+\Gamma^{\bar{i}}_{\bar{j}\bar{m}}\psi^{\bar{m}}_+ \ .$$

Now, the twisted models are constructed as follows:

(1) Instead of taking ψ^i_+ and $\psi^{\bar{i}}_+$ to be section of $K^{1/2} \otimes \Phi^*(T^{1,0}X)$ and $K^{1/2} \otimes \Phi^*(T^{0,1}X)$, respectively, we take them to be sections of $\Phi^*(T^{1,0}X)$ and $K \otimes \Phi^*(T^{0,1}X)$, respectively. I will call this a $+$ twist. The terms in the Lagrangian containing ψ_+ are unchanged (except that $D_{\bar{z}}$ must now be interpreted as the $\bar{\partial}$ operator of the appropriate bundle). Alternatively, we can make what I will call a $-$ twist, taking ψ^i_+ and $\psi^{\bar{i}}_+$ to be sections of $K \otimes \Phi^*(T^{1,0}X)$ and $\Phi^*(T^{0,1}X)$, respectively.

(2) Similarly, we twist ψ_- by either a $+$ twist, taking ψ^i_- to be a section of $\Phi^*(T^{1,0}X)$ and $\psi^{\bar{i}}_-$ to be a section of $\overline{K} \otimes \Phi^*(T^{0,1}X)$, or a $-$ twist, taking ψ^i_- to be a section of $\overline{K} \otimes \Phi^*(T^{1,0}X)$ and $\psi^{\bar{i}}_-$ to be a section of $\Phi^*(T^{0,1}X)$. Again the Lagrangian is unchanged (D_z is now interpreted as the ∂ operator of the appropriate bundle).

if one is twisting only ψ_+ or only ψ_-, it does not much matter if one makes a $+$ twist or a $-$ twist. The two choices differ by a reversal of the complex structure of X, and we are interested in considering all possible complex structures anyway. However, when we twist both ψ_+ and ψ_-, there are two essentially different theories that can be constructed. By making a $+$ twist of ψ_+ and a $-$ twist of ψ_-, we make what I will call the A theory. By making $-$ twists of both ψ_+ and ψ_- we make what I will call the B theory. When I want to make the dependence on X explicit, I will call these theories $A(X)$ and $B(X)$.

Locally the twisting does nothing at all, since locally K and $\overline{K}$ are trivial anyway. In particular, in the twisted models, the transformation laws (2.5) are still valid, but globally the parameters α, $\bar{\alpha}$, etc., must be interpreted as sections of different line bundles. For instance, in the A model, α_- and $\tilde{\alpha}_+$ are functions while α_+ and $\tilde{\alpha}_-$ are sections of $\overline{K}^{-1}$ and K^{-1}. One can therefore canonically pick α_- and $\tilde{\alpha}_+$ to be constants (and the others to vanish); this gives canonical global fermionic symmetries of the A model that are responsible for its simplicity. Similarly the B model has two canonical global fermionic symmetries. These global fermionic symmetries are nilpotent and behave as BRST-like symmetries.

There is an obvious variant that is possible, and that is to twist only ψ_+ or only ψ_-, leaving the order untwisted. I will call this half-twisted model. The half-twisted model has only one canonical fermionic symmetry, just the situation which for $g = 1$ usually leads to the elliptic genus. The half-twisted model would appear to be the most reasonable framework for generalizing the elliptic genus to $g > 1$. It also is of phenomenological interest in the following sense. In superstring compactifications on Calabi–Yau manifolds of $\dim_{\mathbb{C}} X = 3$, the A model is suitable for computing the $\overline{27}^3$ Yukawa couplings and the B model for computing the 27^3 Yukawa couplings. To get a full understanding of the low energy theory, one also needs the $1 \cdot 27 \cdot \overline{27}$ and 1^3 Yukawa couplings; these are not naturally studied in either the A or B model but involve BRST invariant observables of the half-twisted model. The analysis below of the A model can be applied to the half-twisted model to show that correlations of BRST observables reduce to a sum over tree level computations in instanton fields.

Mirror symmetry (since it reverses one of the $U(1)$ quantum numbers in the $N = 2$ superconformal algebra) can be taken to exchange the $+$ and $-$ twists of ψ_+ while leaving ψ_- alone. As a result, mirror symmetry exchanges A and B models (which differ by the choice of twist of ψ_+, for fixed twist of ψ_-) and maps a half-twisted model to another half twisted model (since in these models one makes no twist of ψ_- anywhere).

Since the elliptic genus of a complex manifold is the same as the partition function of the half-twisted (or untwisted) model on a Riemann surface of genus one, mirror pairs have the same elliptic genus.

We now turn to a detailed description of the A and B models.

3. The A Model

In the A model, we regard ψ_+^i and $\psi_-^{\bar{i}}$ as sections of $\Phi^*(T^{1,0}X)$ and $\Phi^*(T^{0,1}X)$, respectively. It is convenient to combine them into a section χ of $\Phi^*(TX)$ (so henceforth $\chi^i = \psi_+^i$, and $\chi^{\bar{i}} = \psi_-^{\bar{i}}$). As for $\psi_+^{\bar{i}}$, it is in the A model a $(1,0)$ form on Σ with values in $\Phi^*(T^{0,1}X)$; we will denote it as $\psi_z^{\bar{i}}$. On the other hand, ψ_-^i is now a $(0,1)$ form with values in $\Phi^*(T^{1,0}X)$, and will be denoted as $\psi_{\bar{z}}^i$.

The topological transformation laws are found from (2.5) by setting $\alpha_+ = \tilde{\alpha}_- = 0$ and setting α_- and $\tilde{\alpha}_+$ to constants, which we will call α and $\tilde{\alpha}$. The result is

$$
\begin{aligned}
\delta\phi^i &= i\alpha\chi^i \\
\delta\phi^{\bar{i}} &= i\tilde{\alpha}\chi^{\bar{i}} \\
\delta\chi^i &= \delta\chi^{\bar{i}} = 0 \\
\delta\psi_z^{\bar{i}} &= -\alpha\partial_z\phi^{\bar{i}} - i\tilde{\alpha}\chi^{\bar{j}}\Gamma^{\bar{i}}_{\bar{j}\bar{m}}\psi_z^{\bar{m}} \\
\delta\psi_{\bar{z}}^i &= -\tilde{\alpha}\partial_{\bar{z}}\phi^i - i\alpha\chi^j\Gamma^i_{jm}\psi_{\bar{z}}^m \; .
\end{aligned}
\tag{3.1}
$$

The supersymmetry algebra of the original model collapses for these topological transformation laws to $\delta^2 = 0$, which holds modulo the equations of motion. (By including auxiliary fields, as in [18], one can get $\delta^2 = 0$ off-shell.)

Henceforth, we will generally for simplicity set $\alpha = \tilde{\alpha}$. (The additional structure that we will overlook is related to the Hodge decomposition of the cohomology of the moduli space of holomorphic maps of Σ to X.) In this case, the first two lines of (3.1) combine to $\delta\Phi^I = i\alpha\chi^I$. Also, we will sometimes express the transformation laws in terms of the BRST operator Q, such that $\delta W = -i\alpha\{Q, W\}$ for any field W. Of course $Q^2 = 0$.

In terms of these variables, the Lagrangian is simply

$$
L = 2t \int_\Sigma d^2z \left(\frac{1}{2} g_{IJ}\partial_z\phi^I\partial_{\bar{z}}\phi^J + i\psi_z^{\bar{i}}D_{\bar{z}}\chi^i g_{\bar{i}i} + i\psi_{\bar{z}}^i D_z\chi^{\bar{i}}g_{\bar{i}i} + R_{i\bar{i}j\bar{j}}\psi_{\bar{z}}^i\psi_z^{\bar{i}}\chi^j\chi^{\bar{j}} \right) \; .
\tag{3.2}
$$

It is now a key fact that this can be written modulo terms that vanish by the ψ equation of motion as

$$
L = it \int_\Sigma d^z z\{Q, V\} + t \int_\Sigma \Phi^*(K)
\tag{3.3}
$$

where

$$V = g_{i\bar{j}}\left(\psi_z^{\bar{i}}\partial_{\bar{z}}\phi^j + \partial_z\phi^{\bar{i}}\psi_{\bar{z}}^j\right) ,\qquad(3.4)$$

while

$$\int_\Sigma \Phi^*(K) = \int_\Sigma d^2z\left(\partial_z\phi^i\partial_{\bar{z}}\phi^{\bar{j}}g_{i\bar{j}} - \partial_{\bar{z}}\phi^i\partial_z\phi^{\bar{j}}g_{i\bar{j}}\right)\qquad(3.5)$$

is the integral of the pullback of the Kähler form $K = -ig_{i\bar{j}}dz^i dz^{\bar{j}}$. Thus $\int \Phi^*(K)$ depends only on the cohomology class of K and the homotopy class of the map Φ. If, for instance, $H^2(X,\mathbb{Z}) \simeq \mathbb{Z}$, and the metric g is normalized so that the periods of K are integer multiples of 2π, then

$$\int_\Sigma \Phi^*(K) = 2\pi n ,\qquad(3.6)$$

where n is an integer, the instanton number or degree. We will adopt this terminology for simplicity; this involves no essential distortion.

Instead of saying that (3.3) is true modulo the ψ equations of motion, we could modify the BRST transformation law of ψ (by adding terms that vanish on shell) to make (3.3) hold exactly. We will not spell out the requisite additional terms in the transformation law, which do not affect the analysis below, since the operators $\mathcal{O}_a$ that we will consider are independent of ψ.

Reduction to Weak Coupling

From equation (3.3), we can give a quick explanation of one of the key properties of the model, which is the reduction to weak coupling. Suppose that we wish to calculate the path integral for fields of degree n. With insertions of some BRST invariant operators $\mathcal{O}_a$ (the details of which we will discuss presently), one wishes to compute

$$\left\langle \prod_a \mathcal{O}_a \right\rangle_n = e^{-2\pi nt}\int_{B_n} D\phi D_\chi D_\psi e^{-it\{Q,\int V\}} \cdot \prod_a \mathcal{O}_a .\qquad(3.7)$$

Here B_n is the component of the field space for maps of degree n, and $\langle\ \rangle_n$ is the degree n contribution to the expectation value. We have made use of (3.6) to pull out an explicit factor $e^{-2\pi nt}$ which will turn out to contain the entire t dependence of $\langle\ \rangle_n$.

Standard arguments using the Q invariance and the fact that $Q^2 = 0$ show that $\langle \{Q, W\} \rangle_n = 0$ for any W. It is therefore also true that, as long as $\{Q, \mathcal{O}_a\} = 0$ for all a, (3.7) is invariant under $\mathcal{O}_a \to \mathcal{O}_a + \{Q, S_a\}$ for any S_a. Thus, the $\mathcal{O}_a$ should be considered as representatives of BRST cohomology classes.

Likewise, and this is the key point, (3.7) is independent of t (as long as Re $t > 0$ so that the path integral converges) except for the explicit factor of $e^{-2\pi n t}$ that has been pulled out. In fact, differentiating the order t dependent factor $\exp(-it\{Q, \int V\})$ with respect to t just brings down irrelevant factors of the form $\{Q, \ldots\}$. Therefore, the path integral in (3.7) can be computed by taking the limit of large Re t. This is the conventional weak coupling limit (for maps of degree n).

Looking back at the original form of the Lagrangian (2.4), or for that matter at the form of V, one sees that for given n, the bosonic part of L is minimized for holomorphic maps of Σ to X, that is maps obeying

$$\partial_{\bar{z}}\phi^i = \partial_z \phi^{\bar{i}} = 0 \ . \tag{3.8}$$

The weak coupling limit therefore involves a reduction to the moduli space $\mathcal{M}_n$ of holomorphic maps of degree n. The entire path integral, for maps of degree n, reduces to an integral over $\mathcal{M}_n$ weighted by one loop determinants of the non-zero modes. (A possibly, $\langle \ldots \rangle_n$ vanishes for $n < 0$, as there are no holomorphic maps of negative degree.

We can also now explain why the model is a topological field theory, in the sense that correlation functions $\langle \prod_a \mathcal{O}_a \rangle$ are independent of the complex structure of Σ and X, and depend only on the cohomology class of the Kahler form K. This is certainly true of $\int_\Sigma \Phi^*(K)$. For the rest, all dependence of the Lagrangian on the complex structure of Σ or X is buried in the definition of V, which appears in the path integral only in the form $\{Q, V\}$; varying the path integral with respect to the complex structure of Σ or X will therefore bring only irrelevant factors of the form $\{Q, \ldots\}$.

The Ghost Number Anomaly

The Lagrangian (3.2) has the classical level a "ghost number" conservation law, with χ

having ghost number 1, ψ having ghost number -1, and ϕ having ghost number 0. The BRST operator Q has ghost number 1.

At the quantum level, the ghost number is not really a symmetry because of the anomaly associated with the index or Riemann-Roch theorem. Let a_n be the number of χ zero modes, that is the dimension of the space of solutions of the equations $D_{\bar{z}}\chi^i = D_z\chi^{\bar{i}} = 0$. Similarly, let b_n be the number of ψ zero modes, solutions of $D_{\bar{z}}\psi^i_z = D_z\psi^i_{\bar{z}} = 0$.[4] The index theorem gives a simple formula for the difference $w_n = a_n - b_n$. In particular, w_n is a topological invariant. For instance, if X is a Calabi–Yau manifold of complex dimension d, and Σ has genus g, then $w_n = 2d(1 - g)$, independent of n. The expression $\langle\prod \mathcal{O}_a\rangle_n$ will vanish unless the sum of the ghost numbers of the $\mathcal{O}_a$ is equal to w_n.

An essential fact is that the equation for a χ zero mode is precisely the linearization of the instanton equation (3.8), and consequently the space of χ zero modes is precisely $T\mathcal{M}_n$, the tangent space to $\mathcal{M}_n$. In particular, if $\mathcal{M}_n$ is a smooth manifold, then a_n is its (real) dimension and in particular is a constant.

The number $w_n = a_n - b_n$ is often called the "virtual dimension" of $\mathcal{M}_n$. The reason for this terminology is that in a sufficiently generic situation (which maybe unattainable in complex geometry) one would expect that if $w_n > 0$ (the only situation that will be interest) then $b_n = 0$ and hence $w_n = a_n$.

Somewhat more generally, as long as $\mathcal{M}_n$ is smooth so that a_n is a constant, b_n will also be constant. Hence the space V of ψ zero modes will vary as the fibers of a vector bundle $\mathcal{V}$ over $\mathcal{M}_n$. At singularities of $\mathcal{M}_n$, a_n and b_n may jump.

Observables of the A Model

To prepare for actual calculations, we need a preliminary discussion of the observables of the A model.

---

[4]Though not indicated in our notation, a_n and b_n may depend on the particular map Φ considered; we return to this presently.

The BRST cohomology of the A model, in the space of local operators, can be represented by operators that are functions of ϕ and χ only.[5] They have the following simple construction.

Let $W = W_{I_1 I_2 \ldots I_n}(\phi) d\phi^{I_1} d\phi^{I_2} \ldots d\phi^{I_n}$ be an n-form on X. We can define a corresponding local operator

$$\mathcal{O}_W(P) = W_{I_1 I_2 \ldots I_n} \chi^{I_1} \cdots \chi^{I_n}(P) . \tag{3.9}$$

The ghost number of $\mathcal{O}_W$ is n. A simple calculation shows that

$$\{Q, \mathcal{O}_W\} = -\mathcal{O}_{dW} , \tag{3.10}$$

with d the exterior derivative on W. Therefore, taking $W \to \mathcal{O}_W$ gives a natural map from he de Rham cohomology of X to the BRST cohomology of the quantum field theory $A(X)$. If one restricts oneself to local operators (a more general class is considered in §7), this map is an isomorphism.

Particularly convenient are the following representatives of the cohomology. Let H be a submanifold of X (or more generally any homology cycle). The "Poincaré dual" of H is a cohomology class that counts intersections with H. It can be represented by a differential form $W(H)$ that has delta function support on H. Hopefully it will cause no confusion if we refer to $\mathcal{O}_{W(H)}$ as $\mathcal{O}_H$. The ghost number of $\mathcal{O}_H$ is the codimension of H.

Evaluation of the Path Integral

Now let us carry out the evaluation of the path integral. We pick some homology cycles H_a, $a = 1 \ldots s$, of codimensions q_a. We also pick points $P_a \in \Sigma$. We want to compute the quantity

$$\langle \mathcal{O}_{H_1}(P_1) \ldots \mathcal{O}_{H_s}(P_s) \rangle_n = e^{-2\pi n t} \int_{B_n} D\phi D_\chi D_\psi e^{it \int \{Q, V\}} \cdot \prod \mathcal{O}_{H_a}(P_a) . \tag{3.11}$$

This quantity will vanish unless w_n, the virtual dimension of moduli space, is equal to $\sum_a q_a$.

---

[5]This happy circumstance prevents complications related to the fact that (3.3) only holds on shell or after modifying the ψ transformation laws.

The path integral in (3.11) reduces, upon using the independence of t and taking $\operatorname{Re} t \to \infty$, to an integral over the moduli space $\mathcal{M}_n$ of instantons. Moreover, as we have picked $\mathcal{O}_{H_a}(P_a)$ to have delta function support for instantons Φ such that

$$\Phi(P_a) \in H_a \ , \tag{3.12}$$

the path integral actually reduces to an integral over the moduli space $\widetilde{\mathcal{M}}_n$ of instantons obeying (3.12).

In a "generic" situation, the dimension a_n of $\mathcal{M}_n$ coincides with the virtual dimension w_n. Moreover, requiring $\Phi(P_a) \in H_a$ involves imposing q_a conditions. Hence "generically" the dimension of $\widetilde{\mathcal{M}}_n$ should be $w_n - \sum_a q_a = 0$. In such a case, $\widetilde{\mathcal{M}}_n$ will consist of a finite set of points. Let $\#\widetilde{\mathcal{M}}_n$ be the number of such points. In determining the contribution of any of those points to the path integral, we can take $\operatorname{Re} t \to \infty$. The computation reduces to evaluation of a ratio of boson and fermion determinants; this ratio is however simply equal to $+1$, because of the BRST symmetry which ensures cancellation between bose and fermi modes.[6]

In a generic situation, we therefore have

$$\langle \prod_{a=1}^{s} \mathcal{O}_{H_a}(P_a) \rangle_n = e^{-2\pi n t} \cdot \#\mathcal{M}_n \ . \tag{3.13}$$

Summing over n we get "generically"

$$\langle \prod_{a}^{s} \mathcal{O}_{H_a}(P_a) \rangle = \sum_{n=0}^{\infty} e^{-2\pi n t} \cdot \#\mathcal{M}_n \ . \tag{3.14}$$

In complex geometry, life is not always "generic" and $\widetilde{\mathcal{M}}_n$ may well have components of positive dimension. Suppose $\widetilde{\mathcal{M}}_n$ has real dimension s (or focus on a component of that dimension). If so, by virtue of the Riemann-Boch theorem, the space V of ψ zero modes is s dimensional and varies as the fibers of a vector bundle $\mathcal{V}$ of (real) dimension s over $\widetilde{\mathcal{M}}_n$.[7] It can

---

[6] In general such a BRST symmetry ensures that the ratio of fermion determinants, in expanding around a BRST fixed point, is $+1$ or -1. In the present case, the ratio is $+1$ since boson determinants are always positive, and the fermion determinant of the $A(X)$ model is also positive since the χ^i, ψ^i_z determinant is the complex conjugate of the $\chi^{\bar{i}}$, $\psi^{\bar{i}}_{\bar{z}}$ determinant.

[7] The ψ zero modes were discussed in the subsection on the ghost number anomaly; however, the equation for such zero modes must now be corrected to permit poles at P_a tangent to H_a. A general extension of index theory to such situations, with the properties essential here, has been given by Gromov and Shubin [10].

be argued on rather general grounds that the generalization of counting the number of points in $\widetilde{\mathcal{M}}_n$ is the evaluation of the Euler class $\chi(\mathcal{V})$ of the bundle $\mathcal{V}$:

$$\#\widetilde{\mathcal{M}}_n \to \int_{\widetilde{\mathcal{M}}_n} \chi(\mathcal{V}) \,. \tag{3.15}$$

I refer §3.3 of [19] for an explanation of this key point (and a detailed field theoretic calculation showing explicitly how $\chi(\mathcal{V})$ arises in a representative example); also, see [2] for some of the background.

The generalization of (3.14) is therefore

$$\langle \prod_{a=1}^{s} \mathcal{O}_{H_a}(P_a) \rangle = \sum_{n=0}^{\infty} e^{-2\pi n t} \cdot \int_{\widetilde{\mathcal{M}}_n} \chi(\mathcal{V}) \,. \tag{3.16}$$

In a real life situation, involving multiple covers of an isolated rational curve in a three dimensional Calabi–Yau manifold, the Euler class of $\mathcal{V}$ has been evaluated by Aspinwall and Morrison [1], who as a result were able to justify a formula that had been guessed empirically by Candelas et al. [5].

Notice that in deriving (3.14) and (3.16) we have not assumed that Σ has genus zero. The restriction to genus zero will arise only when (in §6) we explain the relation of these correlation functions of the twisted model to "physical" correlation functions of the untwisted model.

Although it may be impossible in complex geometry to achieve a "generic" situation (in which the actual dimension of $\mathcal{M}$ coincides with its virtual dimension), this can always be done by perturbing the complex structure of X to a generic non-integrable almost complex structure. (This is allowed in the A model [18].) The importance of (3.16) comes from the fact that it is generally impractical to do calculations based on generic non-integrable deformations. The generic non-integrable deformations are sometimes useful for theoretical arguments; see the end of §7. See [10] for the theory of almost holomorphic curves in almost complex manifolds.

4. The B Model

Now we will consider the B model in a similar spirit.

In the B model, $\psi_\pm^{\bar{i}}$ are sections of $\Phi^*(T^{0,1}X)$, while ψ_+^i is a section of $K \otimes \Phi^*(T^{1,0}X)$ and ψ_-^i is a section of $\overline{K} \otimes \Phi^*(T^{1,0}X)$. It is convenient to set

$$\eta^{\bar{i}} = \psi_+^{\bar{i}} + \psi_-^{\bar{i}}$$
$$\theta_i = g_{i\bar{i}}(\psi_+^{\bar{i}} - \psi_-^{\bar{i}}) \ . \tag{4.1}$$

Also, we combine $\psi_\pm^i$ into a one form ρ with values in $\Phi^*(T^{1,0}X)$; thus, the $(1,0)$ part of ρ is $\rho_z^i = \psi_+^i$, and the $(0,1)$ part of ρ is $\rho_{\bar{z}}^i = \psi_-^i$.

As for the supersymmetry transformations, we now set $\alpha_\pm = 0$, and set $\tilde{\alpha}_\pm$ to constants; in fact, for simplicity we will just set $\tilde{\alpha}_+ = \tilde{\alpha}_- = \alpha$. The transformation laws are then

$$\delta\phi^i = 0$$
$$\delta\phi^{\bar{i}} = i\alpha\eta^{\bar{i}}$$
$$\delta\eta^{\bar{i}} = \delta\theta_i = 0 \tag{4.2}$$
$$\delta\rho^i = -\alpha d\phi^i \ .$$

The BRST operator is again defined by $\delta(\dots) = -i\alpha\{Q,\dots\}$, and obeys $Q^2 = 0$ modulo the equations of motion.

The Lagrangian is

$$L = t \int_\Sigma d^2t \left(g_{IJ}\partial_z\phi^I\partial_{\bar{z}}\phi^J + i\eta^{\bar{i}}(D_z\rho_{\bar{z}}^i + D_{\bar{z}}\rho_z^i)g_{i\bar{i}} \right. \tag{4.3}$$
$$\left. + i\theta_i(D_{\bar{z}}\rho_z^i - D_z\rho_{\bar{z}}^i + R_{i\bar{i}j\bar{j}}\rho_z^i\rho_{\bar{z}}^j\eta^{\bar{i}}\theta_k g^{k\bar{j}}) \right) \ .$$

This can be rewritten

$$L = it \int \{Q,V\} + tW \tag{4.4}$$

where

$$V = g_{i\bar{j}}\left(\rho_z^i\partial_{\bar{z}}\phi^{\bar{j}} + \rho_{\bar{z}}^i\partial_z\phi^{\bar{j}}\right) \tag{4.5}$$

and

$$W = \int_\Sigma \left(-\theta_i D\rho^i - \frac{i}{2}R_{i\bar{i}j\bar{j}}\rho^i \wedge \rho^j \eta^{\bar{i}}\theta_k g^{k\bar{j}} \right) \ . \tag{4.6}$$

Here D is the exterior derivative on Σ (extended to act on forms with values in $\Phi^*(T^{1,0}X)$ by using the pullback of the Levi-Civita connection of X), and $\wedge$ is the wedge product of forms.

We can now see that the B theory is a topological field theory, in the sense that it is independent of the complex structure of Σ and the Kahler metric of X.[8] Under a change of complex structure of Σ or Kahler metric of X, and Lagrangian only change by irrelevant terms of the form $\{Q,\dots\}$. This is obviously true for the $\{Q,V\}$ term on the right hand side of (4.4). As for W, it is entirely independent of the complex structure of Σ, since it is written in terms of differential forms. It is less obvious, but true, that under change of Kahler metric of X, W changes by $\{Q,\dots\}$. These observations are "mirror" to our earlier result that the A theory is independent of the complex structure of Σ and X but depends on the Kahler class of the metric of X.

Similarly, the B theory is independent of the coupling constant t (except for a trivial factor which will appear shortly) as long as $\mathrm{Re}\, t > 0$ so that the path integral converges. Under a change of t, the $t\{Q,V\}$ term changes by $\{Q,\dots\}$. As for the t in tW, this can be removed by redefining $\theta \to \theta/t$ (since V is independent of θ and W is homogeneous of degree one). Hence, the theory is independent of t except for factors that come from the θ dependence of the observables. If $\mathcal{O}_a$ are BRST invariant operators that are homogeneous in t of degree k_a, then the t dependence of $\langle \prod_a \mathcal{O}_a \rangle$ is a factor of $t^{-\sum_a k_a}$, which arises from the rescaling of θ to remove the t from tW. This trivial t dependence of the B theory should be contrasted with the complicated t dependence of the A theory, coming from the instanton sum.

Because the t dependence of the B theory is trivial and known, all calculations can be performed in the limit of large $\mathrm{Re}\, t$, that is, in the ordinary weak coupling limit. In this limit, one expands around minima of the bosonic part of the Lagrangian; these are just the constant maps $\Phi : \Sigma \to X$. The space of such constant maps is a copy of X, so the path integral reduces to an integral over X. We will make this more explicit after identifying the observables.

This is to be contrasted with the A theory, in which one has to integrate over moduli spaces of holomorphic curves. The difference arises because in the A theory, the t dependence becomes standard only after removing a factor of $t \int \Phi^*(K)$, and after doing this, the rest of

---

[8]The theory definitely depends on the complex structure of X, which enters in the BRST transformation laws.

the bosonic part of the action is zero for arbitrary holomorphic curves, not just constant maps. In §5, I will give an alternative and perhaps more fundamental explanation of why calculations in the B theory reduce to integrals over X while in the A theory they reduce to integrals over instanton moduli space.

Anomalies

The fermion determinant of the A model is real and positive (as the χ^i, $\psi_{\bar{z}}^{\bar{i}}$ determinant is the complex conjugate of the $\chi^{\bar{i}}$, $\psi_{\bar{z}}^i$ determinant). In particular, there is no problem in defining this determinant as a function, and the A model, even before taking BRST cohomology, makes at least some sense as a quantum field theory (perhaps with a cutoff, and not conformally invariant) for any complex manifold X, not necessarily Calabi–Yau. (In fact, in [18], the A model was defined for general almost complex manifolds.) The fact that the eventual recipe (3.16) for computing correlation functions does not use the Calabi–Yau condition is related to this.

The B model is very different. Because the zero forms $\eta^{\bar{i}}$, $g^{\bar{i}i}\theta_i$ are sections of $T^{0,1}X$ and the one forms ρ^i are sections of $T^{1,0}X$, the fermion determinant in the B model is complex. The B model does not make any sense as a quantum field theory, even with cutoff, without an anomaly cancellation condition that makes it possible to define the fermion determinants as functions (not just sections of some line bundle). The relevant condition is $c_1(X) = 0$, that is, X should be a Calabi–Yau manifold.[9] Thus, in the B model, the Calabi–Yau condition plays an even more fundamental role than it does in the untwisted model, where it is merely necessary for conformal invariance.

Like the A model, the B model has an important $\mathbb{Z}$ grading by a quantum number that we

---

[9]In two dimensions, anomalies are quadratic in the coupling of fermions to gravitational and gauge fields. To get an anomaly that depends on the twisting (since the untwisted model is not anomalous) and on X (since the twisted model is a non-anomalous free field theory for $X = \mathbb{C}^n$), we must consider a term linear in the gravitational field, that is the spin connection of Σ, and linear in the gauge field, that is the pull-back of the Levi-Civita connection of X. The only invariant linear in the latter is $c_1(X)$, and standard considerations show that $c_1(X)$ is indeed the obstruction to defining the fermion determinant in the B model.

will call the ghost number. The ghost number is 1 for η and θ, -1 for ρ, and zero for ϕ. Q is of degree 1. If X is a Calabi–Yau manifold of complex dimension d, and $\mathcal{O}_a$ are BRST invariant operators of ghost number w_a, then $\langle \mathcal{O}_a \rangle$ vanishes in genus g unless

$$\sum_a w_a = 2d(1-g) \ . \tag{4.7}$$

(There is actually a more refined $\mathbb{Z} \times \mathbb{Z}$ grading, which we have obscured by setting $\tilde{\alpha}_+ = \tilde{\alpha}_-$ and combining $\psi_{\pm}^i$ into ρ.)

The Observables

Now we wish to make the simplest observations about the observables of the B model, analogous to our earlier discussion of the A model.

Instead of the cohomology of X, as in the A model, we consider $(0,p)$ forms on X with values in $\wedge^q T^{1,0} X$, the qth exterior power of the holomorphic tangent bundle of X.[10] Such an object can be written

$$V = d\bar{z}^{i_1} d\bar{z}^{i_2} \ldots d\bar{z}^{i_p} V_{\bar{i}_1 \bar{i}_2 \ldots \bar{i}_p}{}^{j_1 j_2 \ldots j_q} \frac{\partial}{\partial z_{j_1}} \cdots \frac{\partial}{\partial z_{j_q}} \tag{4.8}$$

(V is antisymmetric in the j's as well as in the $\bar{i}$'s.) the sheaf cohomology group $H^p(X, \wedge^q T^{1,0} X)$ consists of solutions of $\bar{\partial} V = 0$ modulo $V \to V + \bar{\partial} S$.

For every V as in (4.8), and $P \in \Sigma$, we can form the quantum field theory operator

$$\mathcal{O}_V = \eta^{\bar{i}_1} \cdots \eta^{\bar{i}_p} V_{\bar{i}_1 \ldots \bar{i}_p}{}^{j_1 \ldots j_q} \psi_{j_1} \ldots \psi_{j_q} \ . \tag{4.9}$$

One finds that

$$\{Q, \mathcal{O}_V\} = -\mathcal{O}_{\bar{\partial} V} \ , \tag{4.10}$$

and consequently $\mathcal{O}_V$ is BRST invariant if $\bar{\partial} V = 0$ and BRST exact if $V = \bar{\partial} S$ for some S. Thus $V \to \mathcal{O}_V$ gives a natural map from $\oplus_{p,q} H^p(X, \wedge^q T^{1,0} X)$ to the BRST cohomology of the

---

[10] In complex geometry, $T^{1,0} X$ might be called simply TX, but we have used that name for the complexification of the real tangent bundle of X.

B model. This is in fact an isomorphism (as long as one considers only local operators; in §7, we will make a slight generalization).

Correlation Functions

Now picking points $P_a \in \Sigma$ and classes V_a in $H^{p_a}(X, \wedge^{q_a} T^{1,0} X)$, we wish to compute

$$\langle \prod_a \mathcal{O}_{V_a}(P_a) \rangle \; . \tag{4.11}$$

We will consider only the case of genus zero. It will be clear that (4.11) vanishes unless

$$\sum_a p_a = \sum_a q_a = d \; . \tag{4.12}$$

(This is related to a more precise grading of the theory, by left- and right-moving ghost numbers, that was alluded to following equation (4.7).)

Taking the large t limit, the calculation reduces as explained earlier to an integral over the space of constant maps $\Phi : \Sigma \to X$. In addition to the bose zero modes – the constant modes of η and θ. The nonzero bose and fermi modes enter only via their one loop determinants; these determinants are independent of the particular constant map $\Phi : \Sigma \to X$ about which one is expanding, and so just go into the definition of the string coupling constant. So one reduces to a computation involving the zero modes only; and correlation functions of the B model will reduce to classical expression.

Once we restrict to the space of zero modes, a function ϕ, η and θ which is of pth order in η and q^{th} order in θ can be interpreted as a $(0, p)$ form on X with values in $\wedge^q T^{1,0} X$. This is course is where $\mathcal{O}$'s came from originally. In multiplying such functions, one automatically antisymmetrizes on the appropriate indices because of fermi statistics. Thus $\prod_a \mathcal{O}_{V_a}$ can be interpreted, using (4.12), as a d form with values in $\wedge^d T^{1,0} X$. The map

$$\otimes_a H^{p_a}(X, \wedge^{q_a} T^{1,0} X) \to H^d(X, \wedge^d T^{1,0} X) \tag{4.13}$$

is the classical wedge product.

What remains is to integrate over X the element of $H^d(X, \wedge^d T^{1,0} X)$ obtained this way. The Calabi–Yau condition is here essential; it ensures that $H^d(X, \wedge^d T^{1,0} X)$ is non-zero and one dimensional. The space of linear forms on this space is thus likewise one dimensional; any such non-zero form gives a method of "integration," unique up to a constant multiple. Of course, the path integral of the B model gives formally a method of evaluating (4.11) and hence of integrating an element of $H^d(X, \wedge^d T^{1,0} X)$; this procedure formally is unique up to a multiplicative constant (a correction to the string coupling constant). We noted in our discussion of anomalies that the B model is anomalous except for Calabi–Yau manifolds.

The restriction to Calabi–Yau manifolds amounts to the fact that what can be integrated naturally are top forms or elements of $H^d(X, \Omega^d X)$. ($\Omega^d X$ is the sheaf of forms of type $(d, 0)$.) In general the relation of $\Omega^d X$ and $\wedge^d T^{1,0} X$ is that they are *inverses*, but in the Calabi–Yau case they are both trivial, and hence isomorphic. Indeed, multiplication by the square of a holomorphic d form gives a map from $\wedge^d T^{1,0} X$ to $\Omega^d X$. Empirically, the choice of a holomorphic d form corresponds to the choice of the string coupling constant, though this relation is still somewhat mysterious.

The Fixed Point Theorem

In the last section, we explained why calculations in the A model reduce to integrals over moduli spaces of holomorphic curves, while calculations in the B model reduce to integrals over spaces of constant maps (and ultimately to classical expressions). I will now (as in [19], §3.1) explain this in an alternative and perhaps more fundamental way, as a sort of fixed point theorem.

Consider an arbitrary quantum field theory, with some function space ε over which one wishes to integrated. Let F be a group of symmetries of the theory. Suppose F acts freely on $\mathcal{E}$. Then one has a fibration $\mathcal{E} \to \mathcal{E}/F$, and by integrating first over the fibers of this fibration, one can reduce the integral over $\mathcal{E}$ to an integral over $\mathcal{E}/F$. Provided one considers only F invariant observables $\mathcal{O}$, the integration over the fibers is particularly simple and just gives a factor of vol(F) (the volume of the group F): '

$$\int_{\mathcal{E}} e^{-L}\mathcal{O} = \mathrm{vol}(F) \cdot \int_{\mathcal{E}/F} e^{-L}\mathcal{O} \ . \tag{5.1}$$

We want to apply this to the case in which F is the $(0|1)$ dimensional supergroup generated by the BRST operator Q. This case has some very special features. The volume of the group F is zero, since for a fermionic variable θ,

$$\int d\theta \cdot 1 = 0 \ . \tag{5.2}$$

Hence (5.1) tells us that if Q acts freely, the expectation value of any Q invariant operator vanishes.

To express this in another way, if F acts freely, then one can introduce a collective coordinate θ for the BRST symmetry. BRST invariance tells us that L and $\mathcal{O}$ are both independent of θ, and since the θ integral of a θ-independent function vanishes, the path integral would vanish.

In general, F does not act freely, but has a fixed point locus $\mathcal{E}_0$, as indicated in figure (1). If so, let $\mathcal{C}$ be a F-invariant neighborhood of $\mathcal{E}_0$ and $\mathcal{E}'$ its complement. Then the path integral restricted to $\mathcal{E}'$ vanishes, by the above reasoning. So the entire contribution to the path integral comes from the integral over $\mathcal{C}$. Here $\mathcal{C}$ can be an arbitrarily small neighborhood, so the result is really a localization formula expressing the path integral as an integral on $\mathcal{E}_0$. The details depend on the structure of Q near $\mathcal{E}_0$. If the vanishing of Q near $\mathcal{E}_0$ is a generic, simple zero, then the fixed point contribution is simply an integral over $\mathcal{E}_0$ weighted by the one loop determinants of the transverse degrees of freedom. This is analogous to, say, the Atiyah-Bott fixed point theorem in topology.

Now let us carry this out in the A and B models. In the A model, the relevant BRST transformation laws read

$$\delta\psi_z^{\bar{i}} = -\alpha\partial_z\phi^{\bar{i}} - i\tilde{\alpha}\chi^{\bar{j}}\Gamma_{\bar{j}\bar{m}}^{\bar{i}}\psi_z^{\overline{m}}$$

$$\delta\psi_{\bar{z}}^{i} = -\tilde{\alpha}\partial_{\bar{z}}\phi^{i} - i\alpha\chi^{j}\Gamma_{jm}^{i}\psi_{\bar{z}}^{m} \tag{5.3}$$

$$\delta\phi^{I} = i\alpha\chi^{I} \ .$$

Requiring $\delta\phi^I = 0$, we get that $\chi^I = 0$ for a BRST fixed point, and setting $\delta\psi = 0$, we see that in addition, a fixed point must have

$$\partial_{\bar{z}}\phi^i = \partial_z\phi^{\bar{i}} = 0 \ . \tag{5.4}$$

This is the equation for a holomorphic curve, and shows the localization of the A model on the space of such curves.

The important part of the BRST transformation law of the B model for our present purposes is

$$\delta\rho^i = -\alpha\, d\phi^i \ . \tag{5.5}$$

Setting $\delta\rho^i = 0$, we see that the condition for a fixed point is $d\phi^i = 0$; that is, $\Phi : \Sigma \to X$ must be a constant map. Thus we recover the localization of the B model on classical, constant configurations.

6. Relation to the "Physical" Model

So far, we have concentrated exclusively on analyzing the twisted A and B theories and their correlation functions. Mirror symmetry is however usually applied to the correlation functions of the untwisted, physical nonlinear sigma models. The purpose of the present section is to explain why certain correlation functions of the twisted models (either A or B; they can be treated together) are equivalent to certain correlation functions of the physical models.

In constructing the twisted models from the physical sigma model, we "twisted" various fields by $K^{1/2}$ and $\overline{K}^{1/2}$ (K being the canonical bundle of a Riemann surface Σ). To state the relation between the physical and twisted models in one sentence, it is simply that the models coincide (with a suitable identification of the observables) whenever K is trivial and we choose $K^{1/2}$ and $\overline{K}^{1/2}$ to be trivial, since in that case the twisting did nothing. Although there are other examples of Riemann surfaces with trivial canonical bundle that might be considered, the important example (for standard applications of mirror symmetry) is the case that Σ is a Riemann surface of genus zero with two points deleted.

Such a surface, of course, can be thought of as a cylinder with a complete, flat metric, say $ds^2 = d\tau^2 + d\sigma^2$, $-\infty < \tau < \infty$, $0 \leq \sigma \leq 2\pi$. In computing path integrals on such a surface, we must pick initial and final quantum states, say $|w\rangle$ and $|w'\rangle$. Considering first the twisted model, we assume that these are Q invariant, and so are representatives of suitable BRST cohomology classes. Picking also points $P_a \in \Sigma$, $a = 1 \ldots s$, and BRST invariant operators $\mathcal{O}_a$, we consider the objects

$$\langle w'| \prod_{a=1}^{s} \mathcal{O}_a(P_a)|w\rangle \tag{6.1}$$

which can be represented by path integrals if we wish.

Of course, (6.1) can be interpreted more symmetrically by compactifying Σ – adding points P and P' and conformally rescaling the metric to bring them to a finite distance. Then the states $|w\rangle$ and $|w'\rangle$ will correspond to BRST invariant operators $\mathcal{O}_w(P)$ and $\mathcal{O}_{w'}(P')$. (In the A and B model, the BRST cohomology is spanned by operators $\mathcal{O}_V$, where V is a de Rham cohomology class or an element of some $H^p(X, \wedge^q T^{1,0}X)$, respectively; so $\mathcal{O}_w$ and $\mathcal{O}_{w'}$ will automatically be operators of this type for some V's.) The matrix element (6.1) is then equivalent to a correlation function

$$\langle \mathcal{O}_{w'}(P')\mathcal{O}_w(P) \prod_{a=1}^{s} \mathcal{O}_a(P_a)\rangle \tag{6.2}$$

on the compactified surface $\widehat{\Sigma}$. This can be evaluate according to the recipes of sections three and four, for the A or B model as the case may be.

Now we go back to the open surface Σ, with its trivial canonical bundle and flat metric, and make the following key observation. As K is trivial, we can pick $K^{1/2}$ to be trivial. If we do so, the twisting by $K^{1/2}$ does nothing. Hence, (6.1) is equivalent to some matrix element in the untwisted model. Of course, the untwisted model has a lot of matrix elements (since the physical states have the multiplicity of a Fock space); we will get only a few of them this way. The operators $\mathcal{O}_a$ will correspond to some bosonic vertex operators of the untwisted model; in fact, as discussed in general terms in [12], if the $\mathcal{O}_a$ are chosen as harmonic representatives of the appropriate BRST cohomology classes, they are standard vertex operators of massless bosons. (In the language of [12, 16], these particular bosonic vertex operators generate the chiral ring – in fact, the ca or cc chiral ring in the case of the A and B model.) Let us call the harmonic representatives B_a.

What about the initial and final states in (6.1)? Equivalence of the twisted and untwisted models depends on choosing $K^{1/2}$ (and $\overline{K}^{1/2}$) to be trivial. This means in the language of the untwisted model that we are working in the Ramond sector; and thus the initial and final states are fermions (and in fact, harmonic representatives of the cohomology classes in question would be ground state fermions of the untwisted model). Let us call these fermi states $|f\rangle$ and $|f'\rangle$. From the point of view of the untwisted model, (6.1) might be written

$$\langle f'| \prod_{a=1}^{s} B_a |f\rangle \ . \tag{6.3}$$

I stress, though, that in going from the twisted to the untwisted model on the flat cylinder Σ, all that we have changed is the notation.

Now, (6.3) is a coupling of two ground state fermions $|f\rangle$ and $|f'\rangle$ to an arbitrary number of ground state bosons B_a (which are all from the same chiral ring). Of the diversity of possible observables of the untwisted model, these are of particular importance as they determine the "superpotential."[11] We have explained how these particular matrix elements can be identified with observables of the twisted model.

Of course, if we wish we can compactify Σ in the context of the untwisted model, adding points P and P' at infinity and conformally scaling the metric to bring them to a finite distance. At this stage the difference between the twisted and untwisted model will come in, as the isomorphism between them depends on a trivialization of K. In the untwisted model, when one projects the points at infinity to a finite distance, the states $|f\rangle$, $|f'\rangle$ will be replaced by fermion vertex operators V_f, $V_{f'}$. Hence, (6.3) has the alternative interpretation as a correlation

---

[11]Actually, the cubic terms in the superpotential come from the case $s = 1$ of the above. To compute higher terms in the superpotential, one must consider the integrated, two form version of the $\mathcal{O}$'s which we will introduce in the next section. The analysis of the relation between twisted and untwisted models on the cylinder is unchanged.

function

$$\langle V_f(P)V_{f'}(P') \prod_{a=1}^{s} B_a \rangle \qquad (6.4)$$

on the closed surface $\widehat{\Sigma}$. Note that in contrast to (6.2), which arose from the analogous compactification in the twisted model, here the "new" vertex operators are of a different type from the old ones. This is possible because the untwisted model has vastly more observables than the twisted model. The fact that the Yukawa couplings are derived from a cubic form (with symmetry between the bose and fermi lines), which is usually regarded as a consequence of space-time supersymmetry, is manifest in the representation (6.2) of the twisted model, because the "fermions" and "bosons" are represented by the same kind of vertex operators.

7. Closer Look at the Observables

In this section we will, finally, take a closer look at the observables of the A and B theories. We will describe a structure – a hierarchy of q form observables for $q = 0, 1, 2$ – which must exist on general grounds. We will then analyze this hierarchy in some detail in the A and B models. In the A model we will obtain a simple answer which moreover has a simple and standard topological description. The analogous calculation in the B model turns out to be far more complicated. We will not pust it through to the end, but we will go far enough to identify the relevant structure, which turns out to be somewhat novel.

In either the A or B model, we described a family of observables, say $\mathcal{O}_V(P)$, where V is a de Rham cohomology class or an element of some $H^p(X, \wedge^q T^{1,0}X)$, in the A or B model, and P is a point in a Riemann surfaces Σ. Correlation function

$$\langle \prod_{a=1}^{s} \mathcal{O}_{V_a}(P_a) \rangle \qquad (7.1)$$

are independent of the P_a, because of the topological invariance of the theory. We will systematically exploit the consequences of this fact. In doing so, we want to think of $\mathcal{O}_V$ as an operator-valued zero form; to emphasize this we write it as $\mathcal{O}^{(0)}$. We fix a particular V and do not always indicate it in the notation.

Topological invariance of the theory – the fact that correlation functions of $\mathcal{O}^{(0)}(P)$ are independent of P – means that $\mathcal{O}^{(0)}$ must be a closed zero form up to BRST commutators,

$$d\mathcal{O}^{(0)} = \{Q, \mathcal{O}^{(1)}\} , \tag{7.2}$$

for some $\mathcal{O}^{(1)}$. This formula, read from right to left, means that the operator-valued one form $\mathcal{O}^{(1)}$ is BRST invariant up to an exact form. Hence, we get new observables in the theory. If C is a circle in Σ (or more generally a one dimensional homology cycle), then

$$U(C) = \oint_C \mathcal{O}^{(1)} \tag{7.3}$$

is a BRST invariant observables.

We can repeat this procedure. Topological invariance means that correlation functions of $U(C)$ must be invariant under small displacements of C; this means that $\mathcal{O}^{(1)}$ must be a closed form up to BRST commutators,

$$d\mathcal{O}^{(1)} = \{Q, \mathcal{O}^{(2)}\} , \tag{7.4}$$

for some $\mathcal{O}^{(2)}$. Also, (7.4) means that the two form $\mathcal{O}^{(2)}$ is BRST invariant up to an exact form, so

$$W = \int_\Sigma \mathcal{O}^{(2)} \tag{7.5}$$

is a new BRST invariant obserable.

In this procedure, if $\mathcal{O}^{(0)}$ has ghost number q, then $\mathcal{O}^{(i)}$ has ghost number $q-i$, for $i = 1, 2$.

Obviously, if X, Y are a mirror pair of Calabi–Yau manifolds, then the mirror symmetry $A(X) \cong B(Y)$ can be applied to the new observables that we have just described. This is likely to be particulary interesting for applications of mirror symmetry with target spaces of complex dimension greater than three.

Now, (7.4) actually leads to the existence of a more general family of topological quantum field theories. If L is the original Lagrangian, and $\mathcal{O}^{(0)}_{V_a}$ are the operator valued zero forms, of ghost number q_a, with which the above procedure begins, then we get a family of topological Lagrangians,

$$L \to L + \sum_a t_a \int_\Sigma \mathcal{O}^{(2)}_{V_a} . \tag{7.6}$$

Let us call this the topological family. As the ghost number $\mathcal{O}^{(2)}_{V_a}$ is $q_a - 2$, Lagrangians in the topological family do not necessarily conserve ghost number (even at the classical level); those that do not are not twistings of standard renormalizable sigma models. In the case of a mirror pair, the whole topological family $A(X)$ is equivalent to the topological family $B(Y)$. So far mirror symmetry has been applied only to the subfamilies of theories that conserve ghost number classically. It is very likely that aspects of mirror symmetry that are not not well understood – like the nature of the mirror map between the moduli spaces – are more transparent in the context of the full topological family.

The A Model

Now we will work out the details of the above for the A model. This is easy enough. If

$$\mathcal{O}^{(0)} = V_{I_1, I_2 \ldots I_n} \chi^{I_1} \chi^{I_2} \cdots \chi^{I_n} , \tag{7.7}$$

for some n form V, then $d\mathcal{O}^{(0)} = \{Q, \mathcal{O}^{(1)}\}$, where

$$\mathcal{O}^{(1)} = -n V_{I_1 I_2 \ldots I_n} d\phi^{I_1} \chi^{I_2} \cdots \chi^{I_n} . \tag{7.8}$$

And $d\mathcal{O}^{(1)} = \{Q, \mathcal{O}^{(2)}\}$, with

$$\mathcal{O}^{(2)} = -\frac{n(n-1)}{2} V_{I_1 I_2 \ldots I_n} d\phi^{I_1} \wedge d\phi^{I_2} \chi^{I_3} \cdots \chi^{I_n} . \tag{7.9}$$

The above field theoretic formulas correspond to the following topological construction. Let $\mathcal{M}$ be the moduli space of holomorphic maps of Σ to X of soem given homotopy type. Thus we have a family of maps $\Phi : \Sigma \to X$ parameterized by $\mathcal{M}$. Alternatively, one can think of this as a single map $\Phi : \Sigma \times \mathcal{M} \to X$. Given now an n dimensional cohomology class $V \in H^*(X)$, we can pull it back to $\Phi^*(V) \in H^*(\Sigma \times \mathcal{M})$. This is an n dimensional cohomology class of $\Sigma \times \mathcal{M}$. To get cohomology classes of $\mathcal{M}$, let γ be an s dimensional submanifold of Σ (for $s = 0, 1,$ or 2 and let $i : \gamma \to \Sigma$ be the inclusion. Then by integration over γ, one gets an $n - s$ dimensional class in the cohomology of $\mathcal{M}$, namely

$$i_*\big(\Phi^*(V)\big) = \int_\gamma \Phi^*(V) . \tag{7.10}$$

For instance, if γ is a point $P \in \Sigma$, then integration over P just means restricting to P, and (7.10) corresponds in the quantum field theory description to our old friend $\mathcal{O}_V^{(0)}(P)$. For γ a one-cycle (say a circle C) or a two-cycle (which must be a multiple of Σ itself), we get the topological counterparts of the objects introduced in (7.3) and (7.5) above.

It is not to hard to verify the precise correspondence between the field theoretic and topological definitions. See [3] for further discussion of some of these matters.

The B Model

To understand the analogous issues in the B model is more difficult. In fact, because the computations involved are rather elaborate, I will first make some qualitative remarks to indicate what must be expected. Then we will just made a few illustrative computations which indicated the form of the general answer.

The operator-valued zero forms $\mathcal{O}^{(0)}$ of the B model are determined by elements of $H^p(X, \wedge^q T^{1,0}X)$ for various p and q. The corresponding two forms $\mathcal{O}^{(2)}$ are possible perturbations of the topological Lagrangian. The case of $p = q = 1$ has particular significance, since $H^1(X, T^{1,0}X)$ is the tangent space to the space of complex structures on X, and the corresponding $\mathcal{O}^{(2)}$'s are just the changes in the Lagrangian required by a change of complex structure. (The explicit calculation showing this would be just analogous to the one we will do presently for perturbations determined by elements of $H^2(X, \wedge^2 T^{1,0}X)$.) So let us discuss what happens when the complex structure of X is changed.

The complex structure of X is determined by the $\bar{\partial}$ operator

$$\bar{\partial} = \sum_i \eta^{\bar{i}} \frac{\partial}{\partial \phi^{\bar{i}}} \; . \tag{7.11}$$

(Mathematically, $\eta^{\bar{i}}$ would usually be written as the $(0,1)$ form $d\phi^{\bar{i}}$.) The transformation laws of the B model (for the fields ϕ, η, θ from which the basic observables are constructed) are just the commutators with $\bar{\partial}$. In (7.11), I have written the $\bar{\partial}$ operator acting on $(0, q)$ forms (that is, functions of ϕ^I and $\eta^{\bar{i}}$), but one can introduce the analogous $\bar{\partial}$ operator for $(0, q)$ forms with values in any holomorphic bundle. In our application, the important holomorphic bundle is $\oplus_q \wedge^q T^{1,0}X$. $(0,q)$ forms with values in this bundle are simply functions of ϕ^I, $\eta^{\bar{i}}$, and θ_j.

If one makes a change in complex structure of X, the $\bar{\partial}$ operator changes. To first order, we get

$$\bar{\partial} \to \eta^{\bar{i}}\left(\frac{\partial}{\partial\phi^{\bar{i}}} + h_{\bar{i}}^{\phantom{\bar{i}}j}\frac{\partial}{\partial\phi^i} - \frac{\partial}{\partial\phi^k}h_{\bar{j}}^{\phantom{\bar{j}}j}\cdot\theta_j\frac{\partial}{\partial\theta_k}\right)\ . \tag{7.12}$$

where $h_{\bar{i}}^{\phantom{\bar{i}}j}$ is a cocycle representing an element of $H^1(X,T^{1,0}X)$. (The perturbed $\bar{\partial}$ operator, acting on functions or $(0,q)$ forms, may be more familiar; it is given by the same expression without the $\theta\cdot\partial/\partial\theta$ terms. This term must be included to give the perturbed $\bar{\partial}$ operator acting on $(0,q)$ forms valued in $\oplus_q \wedge^q T$.) When the complex structure of X is changed, the transformation laws of the $B(X)$ model therefore also change; indeed, taking the commutator with the perturbed $\bar{\partial}$ operator, we find

$$\delta\phi^i = i\alpha\eta^{\bar{i}}h_{\bar{i}}^{\phantom{\bar{i}}j}\ ,$$
$$\delta\theta_j = -i\alpha\eta^{\bar{i}}\partial h_{\bar{i}}^{\phantom{\bar{i}}s}\theta_s\ , \tag{7.13}$$

which replace $\delta\phi^i = \delta\theta_j = 0$ in the uperturbed theory. The non-zero transformation law of θ_j is not so essential in the following sense: it reflects the change in $T^{1,0}X$ (of which θ is a section) under a change in the complex structure of X, and it can be transformed away by rotating the θ^i to a basis appropriate to the new complex structure. The non-zero transformation law of ϕ^i is unavoidable, as it is a basic expression of the change in complex structure.

A Non-Classical Case

So even in a "classical" case, where one is just perturbing the complex structure of X, deformations of the B model require a change in the transformation laws of the basic fields. On must expect this to be true also for other, less classical deformations.

As a typical example, let α be a cocycle representing an element of $H^2(X,\wedge^2 T^{1,0}X)$. The corresponding BRST invariant operator-valued zero form is

$$\mathcal{O}^{(0)} = \alpha_{\bar{i}_1\bar{i}_2}^{\phantom{\bar{i}_1\bar{i}_2}j_1 j_2}\eta^{\bar{i}_1}\eta^{\bar{i}_2}\theta_{j_1}\theta_{j_2}\ . \tag{7.14}$$

We not wish to write $d\mathcal{O}^{(0)} = \{Q, \mathcal{O}^{(1)}\}$, for some $\mathcal{O}^{(1)}$. In constrast to the A model, one finds immediately that (i) this is only true modulo terms that vanish by the equations of motion; (ii) the calculations involved are rather painful. The second point is almost inevitable (in the absence of a powerful computational framework) given the first, and the first point is related, as we will see, to the fact that under perturbation of the Lagrangian, the transformation laws of the fields change.

Eventually one finds that

$$dO^{(0)} = \{Q, O^{(1)}\} + G ,\qquad(7.15)$$

where

$$\mathcal{O}^{(1)} = i\rho^i D_i \alpha_{\bar{i}_1\bar{i}_2}{}^{j_1 j_2} \eta^{\bar{i}_1} \eta^{\bar{i}_2} \theta_{j_1} \theta_{j_2} + 2d\phi^{\bar{i}_1} \alpha_{\bar{i}_1\bar{i}_2}{}^{j_1 j_2} \eta^{\bar{i}_2} \theta_{j_1} \theta_{j_2}$$
$$-2\alpha_{\bar{i}_1\bar{i}_2}{}^{j_1 j_2} \eta^{\bar{i}_1} \eta^{\bar{i}_2} \theta_{j_1} g_{j_2\bar{k}} \star dX^{\bar{k}} .\qquad(7.16)$$

The $\star$ here is the Hodge star operator, and

$$G = 2\alpha_{\bar{i}_1\bar{i}_2}{}^{j_1 j_2} \eta^{\bar{i}_1} \eta^{\bar{i}_2} \theta_{j_1} Z_{j_2} ,\qquad(7.17)$$

where

$$Z_j = D\theta_j - i\rho^m \eta^{\bar{j}} \theta_s R_{j\bar{j}m}{}^s - g_{j\bar{j}} \star D\eta^{\bar{j}}\qquad(7.18)$$

vanishes by the ρ equation of motion.

The next step is to sovle the equations

$$dO^{(1)} = \{Q, O^{(2)}\} + \sum_A \frac{\delta L}{\delta\Phi_A} \cdot \zeta_A .\qquad(7.19)$$

Here Φ_A are all the fields of the theory $(\phi, \eta, \theta, \rho)$. Moreover, $\delta L/\delta\Phi_A$ are the equations of motion of the theory, so any expression that vanishes by the equations of motion is of the form $\sum_A \delta L/\delta\Phi_A \cdot \zeta_A$ for some ζ_A. (7.19) means that in forming the topological family, the generalized Lagrangian

$$\tilde{L} = L + t \int_\Sigma \mathcal{O}^{(2)}\qquad(7.20)$$

is not invariant under the original BRST transformations, but is invariant (to this order; that is, up to terms of order t^2) under

$$\tilde{\delta}\Phi_A = \delta\Phi_A + t\zeta_A \ . \tag{7.21}$$

This shows how the modifications of the transformation laws, which we anticipate from our preliminary discussion of the role of $H^1(X, T^{1,0}X)$, depend upon having non-zero ζ_A.

The computation involved in finding $\mathcal{O}^{(2)}$ in (7.19) (which is guaranteed to exist by the general discussion at the beginning of this section) is very complicated, and would be unilluminating if done without powerful computational methods, such as a superspace formulation. It is much easier to determine the ζ_A. The ζ_A can be determined to eliminate terms in $d\mathcal{O}^{(1)}$ that do not appear in any expression of the general form $\{Q, \mathcal{O}^{(2)}\}$. I will just state the results. First of all, one finds that $\zeta_\eta = 0$. This is in fact inevitable, even without computation, to preserve the fact that $Q^2 = 0$ (when acting on $\phi^{\bar{i}}$). The important novelty, compared to the derivation of equation (7.15), is that ζ_θ and ζ_ϕ are non-zero. In fact,

$$\zeta_{\phi^j} = -2i\alpha_{\bar{i}_1\bar{i}_2}{}^{jk}\eta^{\bar{i}_1}\eta^{\bar{i}_2}\theta_k \tag{7.22}$$

and

$$\zeta_{\theta_i} = iD_i\alpha_{\bar{i}_1\bar{i}_2}{}^{j_1 j_2}\eta^{\bar{i}_1}\eta^{\bar{i}_2}\theta_{j_1}\theta_{j_2} \ . \tag{7.23}$$

Including the terms necessitated by (7.22) and (7.23), the BRST transformation laws of the topological family (or rather, the one parameter subfamily determined by the particular $\mathcal{O}^{(2)}$ considered here) are

$$\begin{aligned}
\delta\phi^{\bar{i}} &= i\alpha\eta^{\bar{i}} \\
\delta\eta^{\bar{i}} &= 0 \\
\zeta_{\phi^j} &= -2it\alpha_{\bar{i}_1\bar{i}_2}{}^{jk}\eta^{\bar{i}_1}\eta^{\bar{i}_2}\theta_k \\
\zeta_{\theta^i} &= itD_i\alpha_{\bar{i}_1\bar{i}_2}{}^{j_1 j_2}\eta^{\bar{i}_1}\eta^{\bar{i}_2}\theta_{j_1}\theta_{j_2} \ .
\end{aligned} \tag{7.24}$$

We sort of perturbed $\overline{\partial}$ operator will generate such transformations? Evidently, we need

$$\overline{D} = \eta^{\bar{i}}\frac{\partial}{\partial\phi^{\bar{i}}} - 2it\eta^{\bar{i}_1}\eta^{\bar{i}_2}\alpha_{\bar{i}_1\bar{i}_2}{}^{j_1 j_2}\theta_{j_1}\frac{\partial}{\partial\phi^{j_2}} + it\eta^{\bar{i}_1}\eta^{\bar{i}_2}D_k\alpha_{\bar{i}_1\bar{i}_2}{}^{j_1 j_2}\theta_{j_1}\theta_{j_2}\frac{\partial}{\partial\theta_k} \ . \tag{7.25}$$

Interpretation

From this sample computation, it is possible to guess the general structure, as we will now indicated. Like the original $\bar{\partial}$ operator, $\overline{D}$ is a first order differential operator, which acts on functions of ϕ, η, θ. As it is the BRST operator of the perturbed model, it obeys $\overline{D}^2 = 0$ (after adding terms of order t^2 and higher, which we have not analyzed).[12] However, because of the terms of third and higher order in fermions, it is definitely not a classical $\bar{\partial}$ operator.

Let $\mathcal{M}$ be the moduli space of complex structures on X, modulo diffeomorphism. $\mathcal{M}$ parametrizes the $B(X)$ models as we constructed them originally in §4. The "topological family" of theories (with Lagrangian $L \to L + \sum_a t_a \int \mathcal{O}_a^{(2)}$) is a more general family of theories parametrized by a thickened moduli space $\mathcal{N}$ which contains $\mathcal{M}$ as a subspace; the tangent space to $\mathcal{N}$ at $\mathcal{M} \subset \mathcal{N}$ is $T\mathcal{N}|_{\mathcal{M}} = \oplus_{p,q=0}^{d} H^p(X, \wedge^q T^{1,0} X)$.

The moduli space $\mathcal{M}$ of complex structures on X is the space of standard $\bar{\partial}$ operators (first order operators whose leading symbol is linear in $\eta^{\bar{i}}$ and independent of θ), which obey $\bar{\partial}^2 = 0$, modulo diffeomorphisms of the ϕ^I. The enriched moduli space $\mathcal{N}$ of topological models is apparently, in view of (7.25), the space of general $\overline{D}$ operators – first order operators on the same space, but with the leading symbol allowed to have a general dependence on ϕ, η, θ – obeying $\overline{D}^2 = 0$, modulo diffeomorphisms of ϕ^I, $\eta^{\bar{i}}$, θ_i. (Thus, $\bar{\partial}$ operators are classified up to ordinary diffeomorphisms of ϕ only, but $\overline{D}$ operators are classified up to diffeomorphisms of the ϕ, η, θ supermanifold). The replacement of classical $\bar{\partial}$ operators by the more general $\overline{D}$ operators has some of hte flavor of string theory, where classical geometry is generalized in a somewhat analogous, but far more drastic, way.

I will call $\mathcal{N}$ the extended moduli space. The analogous extended moduli space in the A model is just $\oplus_{n=0}^{2d} H^n(X, \mathbb{C})$. The study of the extended moduli space is probably the proper framework for understanding the mirror map between the A and B models, and so is potentially rewarding.

---

[12] $\overline{D}$ is a symmetry of the perturbed model, so $\overline{D}^2$ is likewise a symmetry, which moreover is bosonic and of ghost number two. The perturbed or unperturbed model has no such symmetries, so it must be that $\overline{D}^2 = 0$.

The Mirror Map

The weakest link in existing studies of the consequences of mirror symmetry is the construction of the mirror map between the mirror moduli spaces. Given, in other words, a mirror pair X, Y, one would like to identify the mirror map between the $A(X)$ moduli space and the $B(Y)$ moduli space. Understanding this map is essential for extracting the consequences of mirror symmetry. The discussion in [5] involved an elegant and very successful but not fully understood ansatz.

To construct the mirror map, it suffices to identify some sufficiently rich class of observables that can be computed both in the $A(X)$ model and in the $B(Y)$ model. I do not know how to do this, but will make a few comments. Since the $B(Y)$ model reduces to classical algebraic geometry, "everything" is computable in the $B(Y)$ model. The $A(X)$ model is another story; obserables in the $A(X)$ model depend on complicated instanton sums and so in general are not really calculable (except perhaps via mirror symmetry which requires already knowing the mirror map).

The association $\mathcal{O}^{(0)} \to \mathcal{O}^{(2)}$ that we have described above gives a natural identification between the tangent space to the extended moduli space (which is the space of $\mathcal{O}^{(2)}$'s) and the BRST cohomology classes of local operators (the $\mathcal{O}^{(0)}$'s). The two point function in genus zero

$$\eta_{ab} = \langle \mathcal{O}_a^{(0)} \mathcal{O}_b^{(0)} \rangle \tag{7.26}$$

therefore defines a metric on the extended moduli space.[13] (It is essential here to work with the extended moduli space $\mathcal{N}$; the induced metric on the ordinary moduli space $\mathcal{M} \subset \mathcal{N}$ is typically degenerate or even zero.) Moreover, it can be shown that this metric is flat. For topological field theories (like the A and B models for Calabi–Yau manifolds) that arise by twisting conformal field theories, this was shown in [6]. The main ingredients in the argument are the $\mathbb{C}^*$ action on a genus zero surface with two marked points, and the fact that for the

---

[13]I should stress that this "topological" metric in no way coincides with the Zamolodchikov metric on the moduli space – which is much more difficult to study and involves what has been called "topological antitopological fusion" [17].

particular BRST invariant local operator $\mathcal{O}^{(0)} = 1$, the corresponding two form is $\mathcal{O}^{(2)} = 0$. For the A model, I will give another proof of flatness presently (by showing that the metric has no instanton corrections); this proof uses the $\mathbb{C}^*$ action in a different way, and does not require the Calabi–Yau condition.

For the $A(X)$ model, the extended moduli spaces is $\sum_{n=0}^{2d} H^2(X, \mathbb{C})$, and the flat metric is simply the metric on this vector space given by Poincaré duality; there are no instanton corrections, as we will show shortly. For topological field theories obtained by twisting Landau-Ginzbery models, the metric was computed by Vafa [15] up to a conformal factor; that the metric that so arises is flat (after a proper choice of the conformal factor, which has not yet been analyzed in the quantum field theory) is part of the rather deep study of singularities by K. Saito [14], as was explained by Blok and Varchenko [4]. For the $B(Y)$ model, the extended moduli space was roughly described above, but the metric is not yet understood. It seems to be very hard to find any obserables of the $A(X)$ model except the metric (and the related "exponential map" noted below) that are effectively computable, without instanton corrections. So understanding the metric on the extended moduli space of the $B(Y)$ model would appear to be a promising way to understand the mirror map. (If the metrix were known for both $A(X)$ and $B(Y)$, the mirror map would be uniquely determined up to an isometry; isometries depend on finitely many constants, which one might determine by matching a few coefficients at infinity.)

Vanishing of Instanton Corrections to the Metric

It remains to show that instantons do not contribute to the metric of the $A(X)$ model. The proof is essentially dimension counting, taking account of the $\mathbb{C}^*$ action on a genus zero surface with two marked points. Let X be a complex manifold, not necessarily Calabi–Yau, and Σ a Riemann surface of genus zero. Consider a component $\mathcal{U}$ of the moduli space of holomorphic maps $\Phi : \Sigma \to X$ of virtual dimension w. let H and H' be submanifolds (or homology cycles) of codimension q and q' with $q + q' = w$, and let P and P' be two points in Σ. Let $\mathcal{U}'$ be the subspace of $\mathcal{U}$ parameterizing Φ's with $\Phi(P) \in H$ and $\Phi(P') \in H'$. If $\mathcal{U}'$ is empty, the contribution of this homotopy class to $\langle \mathcal{O}_H^{(0)} \mathcal{O}_{H'}^{(0)} \rangle$ is zero. As the virtual dimension of $\mathcal{U}'$ is zero, it appears superficially that $\mathcal{U}'$ need not vanish generically.

However, let us take account of the $\mathbf{C}^*$ action on Σ fixing P and P'. Let $\widetilde{\mathcal{U}} = \mathcal{U}'/\mathbf{C}^*$. $\mathcal{U}'$ must be empty if $\widetilde{\mathcal{U}}$ is. One can think of $\widetilde{\mathcal{U}}$ as the moduli space of rational curves $C \in X$ that interest both H and H' (without specifying any parametrization of C or map from Σ to C). The virtual dimension of $\widetilde{\mathcal{U}}$ is -2. Hence, $\widetilde{\mathcal{U}}$ is "generically" empty, and will really the empty at worst after making a generic nonintegrable deformation of the almost complex structure of X. So the instanton corrections to the metric vanish.

As a very concrete example of this counting, let X be a particular Calabi–Yau manifold, a quintic hypersurface in P^4. The virtual dimension of $\mathcal{U}$ is six (regardless of the homotopy class of the map considered). Let H, H' be two homology cycles in X the sum of whose codimensions is six. So one of them, say H, has codimension at least three. With this particular X, even for generic integrable complex structures, it is believed that the number of rational curves of given positive degree in X is finite; if so, the union Z of these curves has codimension four. Since $4 + 3 > 6$, H can be perturbed slightly so as not to intersect Z; hence $\mathcal{U}'$ is empty, and the instanton contribution to the metric vanishes.

The Exponential Map

The description of the topological family by a family of Lagrangians

$$\widetilde{L} = L + \sum_a t_a \int_\Sigma \mathcal{O}_a^{(2)} \tag{7.27}$$

shows that once we pick a base Lagrangian L, corresponding to a base point $P \in \mathcal{N}$, there is a natural linear structure on $\mathcal{N}$. This might be described as an exponential map from the tangent space to $\mathcal{N}$ at P to $\mathcal{N}$. In the case of the A model, this structure corresponds to the linear structure on $\oplus_n H^n(X, \mathbf{C})$. In the case of the B model, it is not presently understood. Understanding the exponential map should be roughly similar to understanding the metric on the moduli space. It is puzzling, in particular, that in the A model, the linear structure on the moduli space does not seem to depend on the choice of base point, while in the B model such as dependence seems almost inevitable.

                                  EDWARD WITTEN

References

[1] P.S. Aspinwall and D.R. Morrison, *Topological Field Theory and Rational Curves*, Oxford and Duke preprint OUTP-91-32P, DUK-M-91-12.

[2] M.F. Atiyah and L. Jeffrey, *Topological Lagrangians and Cohomology*, J. Geom. Phys. **7** (1990) 119.

[3] L. Baulieu and I.M. Singer, *The Topological Sigma Model*, Commun. Math. Phys. **125** (1989) 227.

[4] B. Blok and A. Varchenko, *Topological Conformal Field Theories and the Flat Coordinates*, IASSNS 91-5.

[5] P. Candelas, P. Green, L. Parke and X. de la Ossa, *A Pair of Calabi–Yau Manifolds as An Exactly Soluble Superconformal Field Theory*, Nucl. Phys. **B359** (1991) 21.

[6] R. Dijkgraaf, E. Verlinde and H. Verlinde, *Topological Strings in $D < 1$*, IASSNS-HEP-90/71.

[7] M. Dine, N. Seiberg, X.-G. Wen and E. Witten, *Nonperturbative Effects on the String World Sheet I, II*, Nucl. Phys. **B278** (1986) 769, **289** (1987) 319.

[8] B. Greene and R. Plesser, *Duality in Calabi–Yau Moduli Space*, Nucl. Phys. **B338** (1990) 15.

[9] M. Gromov, *Pseudoholomorphic Curves in Symplectic Manifolds*, Invent. Math. **82** (1985) 307.

[10] M. Gromov and M.A. Shubin, *The Riemann-Roch Theorem for General Elliptic Operators*, preprint.

[11] P.S. Landweber, ed., *Elliptic Curves and Modular Forms in Algebraic Topology*, Lecture Notes in Mathematics **1326** (Springer-Verlag, 1988).

[12] W. Lerche, C. Vafa and N.P. Warner, *Chiral Rings in $N = 2$ Superconformal Theories*, Nucl. Phys. **B324** (1989) 427.

[13] D.R. Morrison, *Mirror Symmetry and Rational Curves on Quintic Threefolds: a Guide for Mathematicians*, Duke preprint DUK-M-91-01.

[14] K. Saito, *Period Mapping Associated to a Primitive Form* Publ. RIMS, Kyoto Univ. **19**

(1983) 1231.

[15] C. Vafa. *Topological Landau-Ginzbery Models*, Harvard preprint, (1990).

[16] C. Vafa. *Topological Mirrors and Quantum Rings*, contribution to this volume.

[17] S. Cecotti and C. Vafa, *Topological Anti-Topological Fusion*, Harvard University preprint (1991).

[18] E. Witten, *Topological Sigma Models*, Commun. Math. Phys. **118** (1988) 411.

[19] E. Witten, *The N Matrix Model and Gauged WZW Models*, IAS preprint, to appear in Nucl. Phys. B.

FIGURE CAPTIONS

1. The space $\mathcal{E}$ of all fields, the locus $\mathcal{E}_0$ of fixed points of the BRST symmetry, and a tubular neighborhood of the latter.

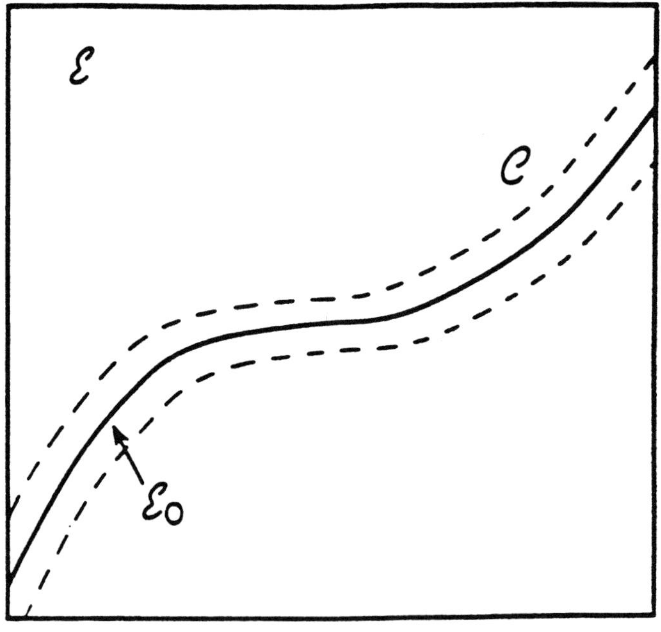

AMS/IP Studies in Advanced Mathematics
Volume **9**, 1998

Rational Curves and Classification of Algebraic Varieties

Yujiro Kawamata

Department of Mathematics

Faculty of Science

University of Tokyo

Hongo Tokyo 113 Japan

In this note we shall review some results on the classification of complex algebraic varieties of dimension 3 or higher, with the emphasis on the existence of rational curves.

A complete reduced irreducible curve is called *rational* if its normalization is isomorphic to $\mathbf{P}^1$. Since the canonical divisor of $\mathbf{P}^1$ is negative, the existence of rational curves is related to the negativity of the canonical divisor. We start with the following theorem which is proved by using a deformation theory in positive characteristic.

Theorem 1. *([MM]) Let X be a nonsingular complex projective variety of dimension n, H an ample divisor, C a non-rational curve on X, and $x \in C$ a point. Assume that $(K_X.C) < 0$. Then there exists a rational curve C' such that $x \in C'$ and*

$$(H.C') \leqq 2n(H.C)/(-K_X.C) \, .$$

Even in the case in which $(K_X.C) = 0$, we may have some kind of negativity of $-K$ along the tangent direction of C if it is negative along the normal direction. We can formulate this

© 1998 American Mathematical Society
and International Press

fact by using the "log" formalism. We consider the pair (X, B) consisting of a variety X and an effective $\mathbf{Q}$-divisor B, a divisor with rational coefficients. We replace the canonical divisor K_X by the "log canonical divisor" $K_X + B$. For example, if $(K_X.C) = 0$ and $(B.C) < 0$, then $((K_X + B).C) < 0$, and by using the following Theorems 2 and 3, we can deduce the existence of rational curves on X. Here B may be singular along C so that there is no "normal direction" in the ordinary sense. These theorems are also applicable to the case where X has singularities and the canonical divisor has no direct differential geometric meaning.

Let us fix our notation. Let X be a normal complex algebraic variety. A subvariety of codimension 1 is called a *prime divisor*. A formal finite sum $B = \sum_i b_i B_i$ with mutually distinct prime divisors B_i is called a *Weil divisor* (resp. *$\mathbf{Q}$-divisor*) if $b_i \in \mathbf{Z}$ (resp. $\mathbf{Q}$). Two Weil divisors B and B' are said to be *linearly equivalent*, denoted by $B \sim_{\text{lin}} B'$, if there exists a rational function g on X such that $B = B' + \text{div}(g)$. A Weil divisor is called a *Cartier divisor* if it is expressed locally as a divisor of a rational function. B is called *$\mathbf{Q}$-Cartier* if there is a positive integer m such that mB is Cartier. X is called *$\mathbf{Q}$-factorial* if every prime divisor is $\mathbf{Q}$-Cartier. For a curve C (which is always meant to be irreducible and reduced) and a $\mathbf{Q}$-Cartier divisor B, the *intersection number* $(B.C) \in \mathbf{Q}$ is defined by $m(B.C) = \deg_{C'} \nu^*(mB)$ for the normalization $\nu : C' \to C$. B is called *nef* if $(B.C) \geq 0$ for all curves C. A *1-cycle* $C = \sum_i c_i C_i$ is a formal finite sum of curves with integer coefficients. We can extend the definition of intersection numbers to 1-cycles by linearity. The *numerical equivalences of $\mathbf{Q}$-Cartier divisors and 1-cycles* are defined by

$$B \sim_{\text{num}} B' \quad \text{iff} \quad (B.C) = (B'.C) \quad \text{for all} \quad C$$
$$C \sim_{\text{num}} C' \quad \text{iff} \quad (B.C) = (B.C') \quad \text{for all} \quad B .$$

We define dual vector spaces

$$N^1(X) = (\{\mathbf{Q} - -\text{Cartier divisors}\} \sim_{\text{num}} \otimes_{\mathbf{Q}} \mathbf{R}$$
$$N_1(X) = (\{\text{1-cycles}\} / \sim_{\text{num}}) \otimes_{\mathbf{z}} \mathbf{R} .$$

Their dimension, which is known to be finite, is called the *Picard number* and denoted by $\rho(X)$.

Let (X, B) be a pair of a normal variety and a $\mathbf{Q}$-divisor $B = \sum_i b_i B_i$ with mutually distinct prime divisors B_i such that $0 < b_i < 1$. (X, B) is said to have only *log terminal singularities* if the following conditions are satisfied:

(1) $K_X + B$ is $\mathbf{Q}$-Cartier,

(2) there is a projective birational morphism $\mu : Y \to X$ from a nonsingular variety Y with a normal crossing divisor $F = \sum_j F_j$ such that $Y \setminus F \xrightarrow{\approx} X \setminus (\mathrm{Supp}\, B \cup \mathrm{Sing} X)$ and

$$K_Y = \mu^*(K_X + B) + \sum_j a_j F_j \quad \text{with} \quad a_j > -1 \quad \text{for all} \quad j \, .$$

If $B = 0$ and $a_j > 0$ (resp. ≥ 0) for all j, then X is said to have only *terminal* (resp. *canonical*) singularities. There is also a notion of *weak log terminal singularities* where the coefficients b_i are allowed to be equal to 1 (cf. [KMM]). A typical example of a weak log terminal pair (X, B) is the case in which X is nonsingular and B is a reduced divisor with only normal crossing.

We list up some properties of terminal, canonical or log terminal singularities.

(1) If X is terminal, then X is canonical; if X is canonical, then $(X, 0)$ is log terminal; if (X, B) is log terminal and B is $\mathbf{Q}$-Cartier, then $(X, 0)$ is log terminal.

(2) If (X, B) is log terminal and D is an effective $\mathbf{Q}$-Cartier divisor, then $(X, B + \varepsilon D)$ is long terminal for sufficiently small positive rational number ε.

(3) If $(X, 0)$ is log terminal and K_X is Cartier, then X is canonical.

(4) Let $X = \mathbf{C}^n / \Gamma$ be a quotient by a finite group $\Gamma \subset GL(n, \mathbf{C})$ such that the fixed point set of any $\gamma \in \Gamma$ with $\gamma \neq I$ has codimension greater than 1. Then $(X, 0)$ is log terminal; X is canonical if and only if $\Gamma \subset SL(n,)$; it is difficult to determine when X is terminal.

(5) Let $\mu : Y \to X$ be a resolution of singularities. If $K_Y = \mu^* K_X$, then X is canonical. Moreover, if the exceptional locus of μ has codimension greater than 1, then X is terminal.

(6) In case $\dim X = 2$, X is terminal (resp. X is canonical, $(X, 0)$ is log terminal) if and only if X has only nonsingular (resp. rational double, quotient singular) points.

(7) In case $\dim X = 3$, terminal singularities are isolated. If K_X is Cartier, then they are hypersurface singularities of type "cDV" ([R]); in the case in which K_X is not Cartier, they are classified as quotients of cDV singularities ([M2]).

Theorem 2 (Cone Contraction Theorem). ([K2]) *Let (X, B) be a pair of a normal complex projective variety and a $\mathbf{Q}$-divisor which has only log terminal singularities, and H an ample divisor on X. Assume that $K_X + B$ is not nef. Then*

$$\alpha = \max\{t \in \mathbf{R} \,|\, H + t(K_X + B) \quad \text{is nef}\}$$

is a rational number. Moreover, there exists a surjective morphism $\varphi : X \to Y$ with connected fibers to a normal projective variety Y such that any curve C on X is contracted to a point on Y if and only if

$$((H + \alpha(K_X + B)).C) = 0 \; .$$

Since H is φ-ample, so is $-(K_X + B)$, i.e., the log canonical divisor is relatively negative for φ. If we start with a general H, then any two curves C_1 and C_2 which are contractible by φ are numerically proportional; there exists a positive rational number q such that $C_1 \sim_{\text{num}} qC_2$. In this case φ is called an *elementary contraction*. The numerical classes of these curves in $N_1(X)$ form an *extremal ray* on the *closed cone of curves*, the closed convex cone generated by the numerical classes of all curves. This is the origin of the name Cone Contraction Theorem (cf. [M1]).

When $B = 0$ and X has only $\mathbf{Q}$-factorial terminal singularities, Theorem 2 gives so-called Minimal Model Program (MMP, cf. [KMM]). A normal projective variety X with only $\mathbf{Q}$-factorial terminal singularities is called *minimal* if K_X is nef. By Theorem 2, if X is not minimal, then there exists a contraction morphism from X such that K_X is relatively negative.

The Proof of Theorem 2 uses the vanishing theorem of Kodaira type ([K1], [V]), so we need the assumption that the base field is of characteristic zero, e.g., $\mathbf{C}$. It also needs the assumption on the projectivity of X; then the resulting variety Y is again projective. It is interesting to extend it to the case where X is a Kähler variety or a Moishezon space.

Theroem 3. *([K6]) Under the same assumption as in Theorem 2, the exceptional locus*

$$E = \{x \in X | \varphi \quad \text{is not an isomorphism}\}$$

is covered by a family of rational curves $\{C_\lambda\}$ which are mapped to points by φ such that

$$(-(K_X + B).C_\lambda) \leqq 2(\dim E - \dim \varphi(E)) .$$

For example, if X is a nonsingular projective variety of general type whose canonical divisor K_X is not ample, then it contains rational curves. In fact, we can write $mK_X = H + D$ for a positive integer m, an ample divisor H, and an effective Cartier divisor D by the Kodaira lemma (cf. [K1]), so that $K_X + \varepsilon D$ is not nef. Then we apply Theorem 2 and 3 to the pair $(X, \varepsilon D)$.

If $(X, 0)$ has only log terminal singularities, $K_X \sim_{\text{num}} 0$ and if D is an effective Cartier divisor which is not nef, then Theorem 2 and 3 applied to the pair $(X, \varepsilon D)$ give rational curves on X (cf. [W]).

We consider the 3-dimensional case in the following.

Theorem 4 (Classification Theorem of Algebraic 3-fold). *Let X_0 be a complex algebraic variety of dimension 3. Then there exists a normal projective $\mathbf{Q}$-factorial variety X with only terminal singularities which is birationally equivalent to X_0 and satisfies one of the following conditions:*

(1) $\rho(X) = 1$, and $-K_X$ is ample (X is called a $\mathbf{Q}$-Fano 3-fold),

(2) $\rho(X) = 2$, and there is a surjective morphism $f : X \to C$ to a nonsingular projective curve C such that $-K_X$ is f-ample, so that the generic fiber X_η of f is a nonsingular del Pezzo surface ($f : X \to C$ is called a del Pezzo fibration),

(3) X is nonsingular, there is a surjective morphism $f : X \to S$ to a nonsingular projective surface S such that $-K_X$ is f-ample, so that the generic fiber X_η of f is a nonsingular conic in $\mathbf{P}^2$, and $\rho(X) = \rho(S) + 1$ ($f : X \to S$ is called a conic bundle),

(4) $mK_X \sim_{\text{lin}} 0$ for a positive integer m,

(5) there is a surjective morphism $f : X \to C$ to a nonsingular projective curve C such that $mK_X \sim_{\text{lin}} f^*D$ for a positive integer m and an ample Cartier divisor D on C, so that the generic fiber X_η of f is either a K3 surface, an Enriques surface, an abelian surface (without the origin), or a hyperelliptic surface,

(6) There is a surjective morphism $f : X \to S$ to a normal projective surface S with only quotient singularities such that $mK_X \sim_{\text{lin}} f^*D$ for a positive integer m and an ample Cartier divisor D on S, so that the generic fiber X_η of f is an elliptic curve (without the origin),

(7) there is a birational morphism $f : X \to Y$ to a normal projective variety Y with only canonical singularities such that K_Y is ample and $mK_X \sim_{\text{lin}} f^*(mK_Y)$ for a positive integer m (Y is called a canonical model of X_0).

Theorem 4 is proved as in the following way. By Theorem 2, we have MMP, so we have the cases (1) through (3) or a minimal model X, provided that we have the existence of flips (Flip Theorem, finally proved in [M3] when $\dim X = 3$ using [K5]) and the termination of flips (proved in [S1] also in this case). For a minimal 3-fold X, the *m-canonical system* $|mK_X|$ is free for some positive integer m (A bundance Theorem, finally proved in [K7] in dimension 3 using [MY1], [MY2] and [S2]), and we have the fiber space structure as in the cases (4) through (7). For singularities of varieties in the cases (3) and (6), see [MN] and [N], respectively.

For a given X_0, any two minimal models X and X' are related by a sequence of *flops* ([K5], [KJ]), which are special types of *log flips* ([S2]); X' is obtained from X by deleting some $\mathbf{P}^1$'s and then inserting other $\mathbf{P}^1$'s in "different directions". In particular, there are only finitely many minimal models in the case (7) ([KM]).

Calabi–Yau 3-folds are in the calss (4). For this class, we have the following structure theorem.

Theorem 5. *In the case (4) of Theorem 4, let*

$$r = \min\{m | mK_X \sim_{\text{lin}} 0\} \ .$$

Then there exists a finite Galois covering $\pi : \widetilde{X} \to X$ with the Galois group $\mathbf{Z}/r\mathbf{Z}$ which is etale outside $\text{Sing}(X)$ and such that $K_{\widetilde{X}} \sim_{\text{lin}} 0$. Moreover,

(1) *if $q(X) =_{\mathrm{def}} \dim H^1(X, \mathcal{O}_X) > 0$, then X is nonsingular and $\chi(\mathcal{O}_X) = 0$,*

(2) *if $\chi(\mathcal{O}_X) = 0$, then $r|m_0$ with $m_0 = 2^5.3^3.5^2.7.11.13.17.19$,*

(3) *if $\chi(\mathcal{O}_X) = 1$, then $r|120$,*

(4) *if $\chi(\mathcal{O}_X) \geqq 2$, then $r|12$.*

By [K3], the Albanese map is an etale fiber bundle if $K_X \sim_{\mathrm{num}} 0$, so we have (1). The rest of the theorem is proved in [MD].

References

[K1] Y. Kawamata, *A generalization of Kodaira-Ramanujam's vanishing theorem*, Math. Ann. **261** (1982) 43-46.

[K2] Y. Kawamata, *The cone of curves of algebraic varieties*, Ann. of Math. **119** (1984) 606-633.

[K3] Y. Kawamata, *Minimal models and the Kodaira dimension of algebraic fiber spaces*, J. reine angew. Math. **363** (1985) 1-46.

[K4] Y. Kawamata, *On the plurigenera of minimal algebraic 3-folds with $K \sim_{\mathrm{num}} 0$*, Math. Ann. **275** (1986) 539-546.

[K5] Y. Kawamata, *Crepant blowing-up of 3-dimensional canonical singularities and its application to degenerations of surfaces*, Ann. of Math. **127** (1988) 93-163.

[K6] Y. Kawamata, *On the length of an extremal rational curve*, Invent. Math. **105** (1991) 609-611.

[K7] Y. Kawamata, *Abundance theorem for minimal algebraic threefolds*, Invent. Math. (to appear).

[KMM] Y. Kawamata, K. Matsuda and K. Matsuki, *Introduction to the minimal model problem*, Adv. St. Pure Math. **10** Algebraic Geometry Sendai 1985 (1987) 283-360.

[KM] Y. Kawamata and K. Matsuki, *The number of minimal models for a 3-fold of general type is finite*, Math. Ann. **276** (1987) 595-598.

[KJ] J. Kollár, *Flops*, Nagoya Math. J. **113** (1989) 15-36.

[MN] M. Miyanishi, *Algebraic methods in the theory of algebraic threefolds*, Adv. St. Pure Math. 1 Algebraic Varieties and Analytic Varieties (1983) 69-99.

[MY1] Y. Miyaoka, *The Chern classes and the Kodaira dimension of minimal variety*, Adv. St. Pure Math. 10 Algebraic Geometry Sendai 1985 (1987) 449-476.

[MY2] Y. Miyaoka, *On the Kodaira diension of minimal threefolds*, Math. Ann. **288** (1988) 325-332.

[MM] Y. Miyaoka and S. Mori, *A numerical criterion for uniruledness*, Ann. of Math. **124** (1986) 65-69.

[M1] S. Mori, *Threefolds whose canonical bundles are not numerically effective*, Ann. of Math. **116** (1982) 133-176.

[M2] S. Mori, *On 3-dimensional terminal singularities*, Nagoya Math. J. **98** (1985) 43-66.

[M3] S. Mori, *Flip theorem and the existence of minimal models for 3-folds*, J. Amer. Math. Soc. **1** (1988) 117-253.

[MD] D. Morrison, *A remark on [K4]*, Math. Ann. **275** (1986), 547-553.

[N] N. Nakayama, *The singularity of the canonical model of compact Kähler manifolds*, Math. Ann. **280** (1988) 509-512.

[R] M. Reid, *Canonical 3-folds*, Géométrie Algébriques Angers 1979 (1980) 273-310.

[S1] V.V. Shokurov, *The nonvanishing theorem*, Math. USSR. Izv. **26** (1986) 591-604.

[S2] V.V. Shokurov, *3-fold log flips*, to appear.

[V] E. Viehweg, *Vanishing theorems*, J. reine angew. Math. **335** (1982) 1-8.

[W] P.M.H. Wilson, *Calabi–Yau manifolds with large Picard number*, Invent. Math. **98** (1989) 139-155.

AMS/IP Studies in Advanced Mathematics
Volume **9**, 1998

Rational Curves on Calabi–Yau Threefolds

Sheldon Katz[1]

Department of Mathematics

Oklahoma State University

Stillwater, OK 74078

In the conformal field theory arising from the compactification of strings on a Calabi–Yau threefold X, there naturally arise fields corresponding to harmonic forms of types $(2, 1)$ and $(1, 1)$ on X [6]. The uncorrected Yukawa couplings on $H^{2,1}$ and $H^{1,1}$ are cubic forms that can be constructed by techniques of algebraic geometry – [31] contains a nice survey of this in a general context in a language written for mathematicians. The cubic form on the space $H^{p,1}$ of harmonic $(p, 1)$ forms is given by the intersection product $(\omega_i, \omega_j, \omega_k) \mapsto \int_X \omega_i \wedge \omega_j \wedge \omega_k$ for $p = 1$, while for $p = 2$ there is a natural formulation in terms of infinitesimal variation of Hodge structure [31, 9]. The Yukawa couplings on $H^{2,1}$ are exact, while those on $H^{1,1}$ receive instanton corrections. In this context, the *instantons* are non-constant holomorphic maps $f : \mathbb{CP}^1 \to X$. The image of such a map is a *rational curve* on X, which may or may not be smooth. If the rational curve C does not move inside X, then the contribution of the instantons which are generically $1 - 1$ (i.e. birational) maps with image C can be written down explicitly – this contribution only depends on the topological type of C, or more or less equivalently, on the

---

[1]Supported in part by NSF grant DMS-9311386.

© 1998 American Mathematical Society
and International Press

integrals $\int_C f^* J$ for $(1,1)$ forms J on X.

If the conformal field theory could also be expressed in terms of a "mirror manifold" X', then the uncorrected Yukawa couplings on $H^{2,1}(X')$ would be the same as the corrected Yukawa couplings on $H^{1,1}(X)$. So if identifications could be made properly, the infinitesimal variation of Hodge structure on X' would give information on the rational curves on X. In a spectacular paper [7], Candelas et. al. do this when X is a quintic threefold. Calculating the Yukawa coupling on $H^{2,1}(X')$ as the complex structure of X' varies gives the Yukawa coupling on $H^{1,1}(X)$ as J varies. The coefficients of the resulting Fourier series are then directly related to the instanton corrections.[2]

To the mathematician, there are some unanswered questions in deducing the number of rational curves of degree d on X from this [31]. I merely cite one problem here, which I will state a little differently. It is not yet known how to carry out the calculation of the instanton correction associated to a continuous family of rational curves on X. If there were continuous families of rational curves on a general quintic threefold X, this would complicate the instanton corrections to the Yukawa coupling on the space of $(1,1)$ forms. The question of whether or not such families exist has not yet been resolved. But even worse, this complication is present in any case – any degree m mapping of $\mathbb{CP}^1$ to itself may be composed with any birational mapping of $\mathbb{CP}^1$ to X to give a new map from $\mathbb{CP}^1$ to X; and this family has more moduli than mere reparametrizations of the sphere. In [7] this was explicitly recognized; the assertion was made that such a family counts $1/m^3$ times. This has since been verified by Aspinwall and Morrison [3].[3]

My motivation in writing this note is to give a general feel for the mathematical meaning of "the number of rational curves on a Calabi–Yau threefold", in particular, how to "count" a family of rational curves. I expect that these notions will directly correspond to the not as yet worked out procedure for calculating instanton corrections associated to general families of

---

[2] *Added in second edition*: There are now many more examples where this has been carried out, e.g. [8,16,20,27,30,32].

[3] *Added in second edition*: Y. Ruan has an interesting approach to giving a mathematically precise meaning to this statement [35].

rational curves. More satisfactory mathematical formulations are the subject of work still in progress.

I want to take this opportunity to thank David Morrison for his suggestions, to the Mathematics Department at Duke University for its hospitality while this manuscript was being prepared, and to the organizers of the Mirror Symmetry Workshop for providing a wonderful opportunity for mathematicians and physicists to learn from each other.

1. Rational Curves, Normal Bundles, Deformations

Consider a Calabi–Yau threefold X containing a smooth rational curve $C \simeq \mathbb{CP}^1$. The normal bundle $N_{C/X}$ of C in X is defined by the exact sequence

$$0 \to T_C \to T_X|_C \to N_{C/X} \to 0 . \tag{1}$$

$N = N_{C/X}$ is a rank 2 vector bundle on C, so $N = \mathcal{O}(a) \oplus \mathcal{O}(b)$ for some integers a, b. Now $c_1(T_C) = 2$, and $c_1(T_X) = 0$ by the Calabi–Yau condition. So the exact sequence (1) yields $c_1(N) = -2$, or $a + b = -2$. One "expects" $a = b = -1$ in the general case. This is because there is a moduli space of deformations of the vector bundle $\mathcal{O}(a) \oplus \mathcal{O}(b)$, and the general point of this moduli space is $\mathcal{O}(-1) \oplus \mathcal{O}(-1)$, no matter what a and b are, as long as $a + b = -2$.

Let $\mathcal{M}$ be the moduli space of rational curves in X. The tangent space to $\mathcal{M}$ at C is given by $H^0(N)$ [28, §12]. In other words, $\mathcal{M}$ may be locally defined by finitely many equations in $\dim H^0(N)$ variables.

Definition. *C is infinitesimally rigid if $H^0(N) = 0$.*

Infinitesimal rigidity means that C does not deform inside X, not even to first order. Note that $H^0(N) = 0$ if and only if $a = b = -1$. Thus

- C is infinitesimally rigid if and only if $a = b = -1$.
- C deforms, at least infinitesimally, if and only if $(a, b) \neq (-1, -1)$.

$\mathcal{M}$ can split up into countably many irreducible components. For instance, curves with distinct homology classes in X will lie in different components of $\mathcal{M}$. However, there will be at most finitely many components of $\mathcal{M}$ corresponding to rational curves in a fixed homology class.

There certainly exist Calabi–Yau threefolds X containing positive dimensional families of rational curves. For instance, the Fermat quintic threefold $x_0^5 + \cdots + x_4^5 = 0$ contains the family of lines given parametrically in the homogeneous coordinates (u, v) of $\mathbb{P}^1$ by $(u, -u, av, bv, cv)$, where (a, b, c) are the parameters of the plane curve $a^5 + b^5 + c^5 = 0$. However, suppose that all rational curves on X have normal bundle $\mathcal{O}(-1) \oplus \mathcal{O}(-1)$. Then since $\mathcal{M}$ consists entirely of discrete points, the remarks above show that there would be only finitely many rational curves in X in a fixed homology class. Since a quintic threefold has $H_2(X, \mathbb{Z}) \simeq \mathbb{Z}$, the degree of a curve is essentially the same as its homology class. This discussion leads to Clemens' conjecture [11]:

Conjecture (Clemens). *A general quintic threefold contains only finitely many rational curves of degree d, for any $d \in \mathbb{Z}$. These curves are all infinitesimally rigid.*

Clemens' original constant count went as follows [10]: a rational curve of degree d in $\mathbb{P}^4$ is given parametrically by 5 forms $\alpha_0(u, v), \ldots, \alpha_4(u, v)$, each homogeneous of degree d in the homogeneous coordinates $(u : v)$ of $\mathbb{P}^1$. These α_i depend on $5(d + 1)$ parameters. On the other hand, a quintic equation $F(x_0, \ldots, x_4) = 0$ imposes the condition $F\big(\alpha_0(u, v), \ldots, \alpha_4(u, v)\big) \equiv 0$ for the parametric curve to be contained in this quintic threefold. This is a polynomial equation of degree $5d$ in u and v. Since a general degree $5d$ polynomial $\Sigma a_i u^i v^{5d-i}$ has $5d + 1$ coefficients, setting these equal to zero results in $5d + 1$ equations among the $5(d + 1)$ parameters of the α_i. If F is *general*, it seems plausible that these equations should impose independent conditions, so that the solutions should depend on $5d + 5 - (5d + 1) = 4$ parameters. However, any curve has a 4-parameter family of reparametrizations $(u, v) \mapsto (au + bv, cu + dv)$, so there are actually a zero dimensional, or finite number, of curves on the general $F = 0$.

The conjecture is known to be true for $d \leq 7$ [23].[4] For any d, it can even be proven that there exists an infinitesimally rigid curve of degree d on a general X. Similar conjectures can

---

[4] *Added in second edition*: This result has recently been extended to $d = 8$ and $d = 9$ by T. Johnsen and S. Kleiman [21].

be stated for other Calabi–Yau threefolds.[5]

There are many kinds of non-rational curves which appear to occur in finite number on a general quintic threefold. For instance, elliptic cubic curves are all planar. The plane P that one spans meets the quintic in a quintic curve containing the cubic curve. The other component must be a conic curve. This sets up a $1 - 1$ correspondence between elliptic cubics and conics on any quintic threefold. Hence the number of elliptic cubics on a general quintic must be the same as the number of conics, 609250 [23]. Finiteness of elliptic quartic curves has been proven by Vainsencher [37]; the actual number has not yet been computed.[6]

On the other hand, there are infinitely many plane quartics on *any* quintic threefold: take any line in the quintic, and each of the infinitely many planes containing the line must meet the quintic in the original line union a quartic.

If a curve has $N \simeq \mathcal{O} \oplus \mathcal{O}(-2)$, then $C \subset X$ deforms to first order. In fact, since $H^0(N)$ is one-dimensional, there is a family of curves on X parametrized by a single variable t, subject to the constraint $t^2 = 0$. In other words, start with a rational curve given parametrically by forms $\alpha_0, \ldots, \alpha_4$, homogeneous of degree d in u and v. Take a perturbation $\alpha_i(u, v; t) = \alpha_i(u, v) + t\alpha_i'(u, v)$, still homogeneous in (u, v). Form the equation $F(\alpha_0, \ldots, \alpha_4) = 0$ and formally set $t^2 = 0$; the resulting equation has a 5 dimensional space of solutions for the α_i', which translates into a unique solution up to multiples and reparametrizations of $\mathbb{P}^1$. The curve C but may or may not deform to second order. C deforms to n^{th} order for all n if and only if C moves in a 1-parameter family. A pretty description of the general situation is given in [34].

If C deforms to n^{th} order, but not to $(n + 1)^{\text{th}}$ order, then one sees that while C is an isolated point in the moduli space of curves on X, it more naturally is viewed as the solution

---

[5]*Added in second edition*: The existence of an infinitesimally rigid curve of arbitrary degree d on a general $(2, 4)$ complete intersection Calabi-Yau threefold in $\mathbb{P}^5$ has been proven recently by K. Oguiso [33].

[6]*Added in second edition*: This number has recently been computed by G. Ellingsrud and S.A. Strømme to be 3718024750 (private communication). They use the description of the parameter space for elliptic curves given by Avritzer-Vainsencher and a formula of Bott, whose use they explained in [15]. Also, a method for predicting instantons of higher genus has been introduced by Bershadsky, Cecotti, Ooguri, and Vafa in a series of two papers [4,5]. The predictions are compatible with the calculation of Ellingsrud and Strømme as well as with previously known numbers. This method has been recently extended to multiparameter families as part of a paper by Candelas, de la Ossa, Font, Katz, and Morrison; the method has been applied again in a companion paper [8].

to the equation $t^{n+1} = 0$ in one variable t. So C should be viewed as a rational curve on X with multiplicity $n + 1$.

If a curve has $N \simeq \mathcal{O}(1) \oplus \mathcal{O}(-3)$, C has a 2 parameter space of infinitesimal deformations, and the structure of $\mathcal{M}$ at C is correspondingly more complicated. An example is given in the next section. The general situation has not yet been worked out.

2. Counting Rational Curves

In this section, a general procedure for calculating the number of smooth rational curves of a given type is described. Alternatively, a canonical definition of this number can be given using the Hilbert scheme (this is what was done by Ellingsrud and Strømme in their work on twisted cubics [14]); however, it is usually quite difficult to implement a calculation along these lines.

Embed the Calabi–Yau threefold X in a larger compact space $\mathbb{P}$ (which may be thought of as a projective space, a weighted projective space, or a product of such spaces). $\mathcal{M}_\lambda$ will denote the moduli space parametrizing smooth rational curves in $\mathbb{P}$ of given topological type or degree λ. $\mathrm{Def}(X)$ denotes the irreducible component of X in the moduli space of Calabi–Yau manifolds in $\mathbb{P}$ (here the Kähler structure is ignored). In other words, $\mathrm{Def}(X)$ parametrizes the deformations of X in $\mathbb{P}$.

(1) Find a compact moduli space $\overline{\mathcal{M}}_\lambda$ containing $\mathcal{M}_\lambda$ as a dense open subset, such that the points of $\overline{\mathcal{M}}_\lambda - \mathcal{M}_\lambda$ correspond to degenerate curves of type λ. $\overline{\mathcal{M}}_\lambda$ parametrizes degenerate deformations of the smooth curve (not the mapping from $\mathbb{CP}^1$ to X). It is better for $\overline{\mathcal{M}}_\lambda$ to be smooth.

(2) Find a rank $r = \dim(\mathcal{M}_\lambda)$ vector bundle $\mathcal{B}$ on $\overline{\mathcal{M}}_\lambda$ such that

    (a) To each $X' \in \mathrm{Def}(X)$ there is a section $s_{X'}$ of $\mathcal{B}$ which vanishes at $C \in \overline{\mathcal{M}}_\lambda$ if and only if $C \subset X'$.

    (b) There exists an $X' \in \mathrm{Def}(X)$ such that $s_{X'}(C) = 0$ if and only if $C \in \mathcal{M}_\lambda$ and $C \subset X'$.

    (c) C is an isolated zero of $s_{X'}$ with $\mathrm{mult}_C(s_{X'}) = 1$ if and only if $N_{C/X} \simeq \mathcal{O}(-1) \oplus \mathcal{O}(-1)$.

Working Definition. *The number of smooth rational curves n_λ of type λ is given by the rth Chern class $c_r(\mathcal{B})$.*

Why is this reasonable definition? Suppose that an $X' \in \text{Def}(X)$ can be found with the properties required above, with the additional property that there are only finitely many curves of type λ on X', and that they are all infinitesimally rigid. Then it can be checked that the number of (possibly degenerate) curves of type λ on X' is independent of the choice of X' satisfying the above properties, and is also equal to $c_r(\mathcal{B})$. This last follows since $c_{\text{rank}(E)}(E)$ always gives the homology class of the zero locus Z of any section of any bundle E on any variety Y, whenever $\dim(Z) = \dim(Y) - \text{rank}(E)$. In our case, $0 = \dim(\{lines\}) = \dim(\mathcal{M}_\lambda) - \text{rank}(\mathcal{B})$. In essentially all known cases, the number of curves has been worked out by the method of this working definition. Examples are given below.

There is a potential problem with this working definition. For families of Calabi–Yau threefolds such that no threefold in the family contains finitely many curves of given type, it may be that the "definition" depends on the choice of compactification and/or vector bundle, i.e. this is not well-defined. My reason for almost calling this method a definition is that it *does* give a finite number corresponding to an infinite family of curves, which *is* well-defined in the case that the Calabi–Yau threefold in question belongs to a family containing some other Calabi–Yau threefold with only finitely many rational curves of the type under consideration.

In the case where the general X contains irreducible singular curves which are the images of maps from $\mathbb{P}^1$, a separate but similar procedure must be implemented to calculate these, since they give rise to instanton corrections as well. For example, there is a 6-parameter family of two-planes in $\mathbb{P}^4$. A two plane P meets a quintic threefold X in a plane quintic curve. For general P and X, this curve is a smooth genus 6 curve. But if the curve acquired 6 nodes, the curve would be rational. This being 6 conditions on a 6 parameter family, one expects that a general X would contain finitely many 6-nodal rational plane quintic curves, and it can be verified that this is indeed the case.[7] The problem is that there is no way to deform a smooth

---

[7] *Added in second edition*: The number of such curves has recently been calculated to be 17,601,000 by I. Vainsencher [38].

rational curve to such a singular curve – the dimension of $H^1(\mathcal{O}_C)$ is zero for a smooth rational curve C, but positive for such singular curves, and this dimension is a deformation invariant [19, Theorem 9.9]. For example, if one tried to deform a smooth twisted cubic curve to a singular cubic plane curve by projecting onto a plane, there would result an "embedded point" at the singularity, creating a sort of discontinuity in the deformation process [19, Ex. 9.8.4].

Examples:

1. Let $X \subset \mathbb{P}^4$ be a quintic threefold. Take $\lambda = 1$, so that we are counting lines. Here $\mathcal{M}_1 = G(1,4)$ is the Grassmannian of lines in $\mathbb{P}^4$ and is already compact, so take $\overline{\mathcal{M}}_1 = \mathcal{M}_1$. Let $\mathcal{B} = \mathrm{Sym}^5(U^*)$, where U, the universal bundle, is the rank 2 bundle on $\mathcal{M}_1$ whose fiber over a line L is the 2-dimensional subspace $V \subset \mathbb{C}^5$ yielding $L \subset \mathbb{P}^4$ after projectivization. Note that $\mathrm{rank}(\mathcal{B}) = \dim(\mathcal{M}_1) = 6$. $\mathrm{Def}(X) \subset \mathbb{P}(H^0(\mathcal{O}_{\mathbb{P}^4}(5)))$ is the subset of smooth quintics. A quintic X induces a section s_X of $\mathcal{B}$, since an equation for X is a quintic form on $\mathbb{C}^5$, hence induces a quintic form on V for $V \subset \mathbb{C}^5$ corresponding to L. Clearly $s_X(L) = 0$ if and only if $L \subset X$. The above conditions are easily seen to hold. $c_6(\mathcal{B}) = 2875$ is the number of lines on X. This calculation is essentially the same as that done for cubics in [2, Thm. 1.3], where dual notation is used, so that the U^* used here becomes the universal quotient bundle Q in [2]. The number 2875 agrees with the result of Candelas et. al. [7].

(2) Continuing with the quintic, take $\lambda = 2$. Any conic C necessarily spans a unique 2-plane containing C. Let $G = G(2,4)$ be the Grassmannian of 2-planes in $\mathbb{P}^4$, and let U be the rank 3 universal bundle on G. Put $\overline{\mathcal{M}}_2 = \mathbb{P}(\mathrm{Sym}^2(U^*))$ be the projective bundle over G whose fiber over a plane P is the projective space of conics in P. Clearly $\mathcal{M}_2 \subset \overline{\mathcal{M}}_2$ (but they are not equal; $\overline{\mathcal{M}}_2$ also contains the union of any two lines or a double line in any plane). Let $\mathcal{B} = \mathrm{Sym}^5(U^*)/(\mathrm{Sym}^3(U^*) \otimes \mathcal{O}_{\mathbb{P}}(-1))$ be the bundle on $\overline{\mathcal{M}}_2$ of quintic forms on the 3 dimensional vector space $V \subset \mathbb{C}^5$, modulo those which factor as any cubic times the given conic. ($\mathcal{O}_{\mathbb{P}}(-1)$ is the line bundle whose fiber over a conic is the one dimensional vector space of equations for the conic within its supporting plane. The quotient is relative to the natural embedding $\mathrm{Sym}^3(U^*) \otimes \mathcal{O}_{\mathbb{P}}(-1) \to \mathrm{Sym}^5(U^*)$ induced by multiplication.) $\mathrm{rank}(\mathcal{B}) = \dim(\overline{\mathcal{M}}_2) = 11$. $c_{11}(\mathcal{B}) = 609250$. See [23] for more details, or [12] for an

analogous computation in the case of quartics. The number 609250 agrees with the result of Candelas et. al.

(3) Again consider the quintic, this time with $l = 3$. In [14], Ellingsrud and Strømme take $\overline{\mathcal{M}}_3$ to be the closure of the locus of smooth twisted cubics in $\mathbb{P}^4$ inside the Hilbert scheme. This space has dimension 16. $\mathcal{B}$ is essentially the rank 16 bundle of degree 15 forms on $\mathbb{P}^1$ induced from quintic polynomials in P^4 by the degree 3 parametrization of the cubic (it must be shown that this makes sense for degenerate twisted cubics as well). The equation of a general quintic gives a section of $\mathcal{B}$, vanishing precisely on the set of cubics contained in X. $c_{16}(\mathcal{B}) = 317206375$, again agreeing with the result of Candelas et. al.

The key to the first two calculations are the Schubert calculus for calculating in Grassmannians [18, Ch. 1.5] and standard formulas for projective bundles [19, Appendix A.3]. The third calculation is more intricate.

Regarding complete intersection Calabi–Yau manifolds, similar examples are found in [29,24] for lines and [36] for conics.[8]

Note that the number of curves $c_r(\mathcal{B})$ in no way depends on the choice of X, even if X is a degenerate Calabi–Yau threefold, or contains infinitely many rational curves of type λ. It turns out that a natural meaning can be assigned to this number.[9] In fact, the moduli space of curves of type λ on X splits up into "distinguished varieties" Z_i [17], and a number, the *equivalence* of Z_i, can be assigned to each distinguished variety (the number is 1 for each $\mathcal{O}(-1) \oplus \mathcal{O}(-1)$ curve). This number is precisely equal to the number of curves on X in Z_i which arise as limits of curves on X' and X' approaches X in a 1-parameter family [17, Ch. 11].

Examples:

(1) If a quintic threefold is a union of a hyperplane and a quartic, then the quintic contains infinitely many lines and conics. However, in the case of lines, given a general 1-parameter

---

[8] *Added in second edition*: These numbers agree with the numbers calculated by A. Libgober and J. Teitelbaum using proposed mirrors for these manifolds.

[9] *Added in second edition*: Y. Ruan has shown along these lines that this number can be given a precise meaning by deforming X to an almost complex manifold [35].

family of quintics approaching this reducible quintic, 1275 lines approach the hyperplane, and 1600 lines approach the quartic [22]. So the infinite set of lines in the hyperplane "count" as 1275, while those in the quartic count as 1600. For counting conics, the 609250 conics distribute themselves as 187850 corresponding to the component of conics in the hyperplane, 258200 corresponding to the component of conics in the quartic, and 163200 corresponding to the component of conics which degenerate into a line in the hyperplane union an intersecting line in the quartic [25]. Note that this reducible conic lies in $\overline{\mathcal{M}}_2 - \mathcal{M}_2$, and illustrates why $\mathcal{M}_2$ itself is insufficient for calculating numbers when there are infinitely many curves. Most of these numbers have been calculated recently by Xian Wu [39] using a different method.

(2) If a quintic threefold is a union of a quadric and a cubic, the lines on the quadric count as 1300, and the lines on the cubic count as 1575 [22, 39]. The conics on the quadric count as 215950, while the conics which degenerate into a line in the quadric union an intersecting line in the cubic count as $609250 - (215950 + 243900) = 149400$, but this has not been checked directly yet.

(3) There are infinitely many lines on the Fermat quintic threefold $x_0^5 + \cdots + x_4^5 = 0$. These divide up into 50 cones, a typical one being the family of lines given parametrically in the homogeneous coordinates (u, v) of $\mathbb{P}^1$ by $(u, -u, av, bv, cv)$, where (a, b, c) satisfy $a^5 + b^5 + c^5 = 0$. Each of these count as 20. There are also 375 special lines, a typical one being given by the equations $x_0 + x_1 = x_2 + x_3 = x_4 = 0$ (these lines were also noticed in [13]). These lines L count with multiplicity 5. Note that $50 \cdot 20 + 375 \cdot 5 = 2875$ [1]. This example illustrates the potential complexity in calculating the distinguished varieties Z_i – some components can be embedded inside others. This may be understood as well by looking at the moduli space of lines on the Fermat quintic locally at a line corresponding to a special line. Since $N_{L/X} \simeq \mathcal{O}(1) \oplus \mathcal{O}(-3)$, $H^0(N_{L/X})$ has dimension 2, and the moduli space of lines is a subset of a 2-dimensional space. A calculation shows that it is locally defined inside 2 dimensional (x, y) space by the equations $x^2 y^3 = x^3 y^2 = 0$. The x and y axes correspond to lines on each of the 2 cones, each occurring with multiplicity 2; there

would be just one equation $x^2 y^2 = 0$ if the special line corresponding to $(0,0)$ played no role; since this is not the case, it can be expected to have its own contribution; i.e. each special line is a distinguished variety.

(4) Examples for lines in complete intersection Calabi–Yau threefolds were worked out in [24].

These numbers can also be calculated by intersection-theoretic techniques. Let $s(Z_i, \overline{\mathcal{M}}_\lambda)$ denote the *Segre class* of Z_i in $\overline{\mathcal{M}}_\lambda$. If Z_i is smooth, this is simply the formal inverse of the total Chern class $1 + c_1(N) + c_2(N) + \ldots$ of the normal bundle N of Z_i in $\overline{\mathcal{M}}_\lambda$. Then if Z_i is a connected component of the zero locus of s_X, the equivalence of Z_i is the zero dimensional part of $c(\mathcal{B}) \cap s(Z_i, \overline{\mathcal{M}}_\lambda)$ [17, Prop. 9.1.1].

If Z_i is an irreducible component which is not a connected component, then this formula is no longer applicable. However, I have had recent success with a new method that supplies "correction terms" to this formula. The method is currently ad hoc (the most relevant success I have had is in calculating the "number" of lines on a cubic surface which is a union of three planes), but I expect that a more systematic procedure can be developed.

This of course is a reflection of the situation in calculating instanton corrections to the Yukawa couplings. If there is a continuous family of instantons, then calculating the corrections will be more difficult. If the parameter space for instantons is smooth, this should make the difficulties more manageable. If the space is singular, the calculation is more difficult. I expect that the calculation would be even more difficult if instantons occur in at least 2 families that intersect.

Of course, in the calculation of the Yukawa couplings via path integrals, there is no mention of vector bundles on the moduli space. This indicates to me that there should be a mathematical definition of the equivalence of a distinguished variety that does not refer to an auxiliary bundle.[10] Along these lines, one theorem will be stated without proof.

Let Z be a k-dimensional unobstructed family of rational curves on a Calabi–Yau threefold X. There is the total space $\mathcal{Z} \subset Z \times X$ of the family, with projection map $\pi : \mathcal{Z} \to Z$ such

---

[10] *Added in second edition:* The current situation is that a good definition exists as long as the dimension of $H^1(N)$ is independent of all curves in the connected component [26].

that $\pi^{-1}(z)$ is the curve in X corresponding to z, for each $z \in Z$. Let N be the normal bundle of $\mathcal{Z}$ in $Z \times X$. Define the equivalence $e(Z)$ of Z to be the number $c_k(R^1\pi_*N)$. For example, if $k = 0$, then the curve is infinitesimally rigid, and $e(Z) = 1$.

Theorem. [11] *Let Z be an unobstructed family of rational curves of type λ on a Calabi–Yau threefold X. Suppose that X deforms to a Calabi–Yau threefold containing only finitely many curves of type λ. Then precisely $e(Z)$ of these curves (including multiplicity) approach curves of Z as the Calabi–Yau deforms to X.*

Some of the examples given earlier in this section can be redone via this theorem. Also, as anticipated by [7], it can be calculated that a factor of $1/m^3$ is introduced by degree m covers by a calculation similar in spirit to that found in [3].

---

[11] *Added in second edition*: An explanation of this theorem appears in the second part of [8].

References

1. A. Albano and S. Katz, *Lines on the Fermat quintic threefold and the infinitesimal generalized Hodge conjecture*, Trans. AMS **324** (1991) 353–368.

2. A. Altman and S. Kleiman, *Foundations of the theory of Fano schemes*, Comp. Math. **34** (1977) 3–47.

3. P. Aspinwall and D. Morrison, *Topological field theory and rational curves*, Comm. Math. Phys. **151** (1993) 245–262.

4. M. Bershadsky, S. Cecotti, H. Ooguri and C. Vafa. *Holomorphic Anomalies in Topological Field Theories*, with an appendix by S. Katz. To appear in Nuc. Phys. B.

5. M. Bershadsky, S. Cecotti, H. Ooguri and C. Vafa. *Kodaira-Spencer theory of gravity*, preprint, HUTP–93/A025.

6. P. Candelas, G. Horowitz, A. Strominger and E. Witten, *Vacuum configurations for superstrings*, Nuc. Phys. **B258** (1985) 46–74.

7. P. Candelas, X.C. de la Ossa, P.S. Green and L. Parkes, *A pair of Calabi–Yau manifolds as an exactly soluble superconformal theory*, Nuc. Phys. **B359** (1991) 21–74.

8. P. Candelas, X.C. de la Ossa, A. Font, S. Katz and D.R. Morrison, *Mirror symmetry for two-parameter models — I*, to appear in Nuc. Phys. B.
   P. Candelas, A. Font, S. Katz and D.R. Morrison, *Mirror symmetry for two-parameter models — II*, in preparation.

9. J. Carlson, M. Green, P. Griffiths and J. Harris, *Infinitesimal variations of Hodge structure, I.*, Comp. Math. **50** (1983) 109–205.

10. H. Clemens, *Homological equivalence, modulo algebraic equivalence, is not finitely generated*, Publ. Math. IHES **58** (1983) 19–38.

11. H. Clemens, *Some results on Abel-Jacobi mappings*, In: Topics in Transcendental Algebraic Geometry, Princeton Univ. Press, Princeton, NJ 1984.

12. A. Collino, J. Murre, G. Welters, *On the family of conics lying on a quartic threefold*, Rend. Sem. Mat. Univ. Pol. Torino **38** (1980).

13. M. Dine, N. Seiberg, X.G. Wen, and E. Witten, *Non-perturbative effects on the string world*

sheet, Nucl. Phys. **B278** (1987) 769–789.

14. G. Ellingsrud and S.A. Strømme, *On the number of twisted cubics on the general quintic threefold*, University of Bergen preprint 63-7-2-1992.

15. G. Ellingsrud and S.A. Strømme, *The number of twisted cubic curves on Calabi-Yau complete intersections*, Dyrkolbotn lectures, August 1993.

16. A. Font, *Periods and duality symmetries in Calabi-Yau compactifications*, Nucl. Phys. **B391** (1993) 358.

17. W. Fulton, *Intersection Theory*, Springer-Verlag, Berlin-Heidelberg-New York 1984.

18. P. Griffiths and J. Harris, *Principles of Algebraic Geometry*, Wiley-Interscience 1978.

19. R. Hartshorne, *Algebraic Geometry*, Springer-Verlag, Berlin-Heidelberg-New York 1977.

20. S. Hosono, A. Klemm, S. Theisen and S.-T. Yau, *Mirror symmetry, mirror map, and an application to Calabi-Yau hypersurfaces*, preprint, HUTMP–93/0801.

21. T. Johnsen and S. Kleiman, *On Clemens' conjecture*, preprint.

22. S. Katz, *Degenerations of quintic threefolds and their lines*, Duke Math. Jour. **50** (1983) 1127–1135.

23. S. Katz, *On the finiteness of rational curves on quintic threefolds*, Comp. Math. **60** (1986) 151–162.

24. S. Katz, *Lines on complete intersection threefolds with $K = 0$*, Math. Zeit. **191** (1986) 293–296.

25. S. Katz, *Iteration of multiple point formulas and applications to conics*, Algebraic Geometry, Sundance 1986. Lecture Notes in Math. **1311** 147–155, Springer-Verlag, Berlin-Heidelberg-New York 1988.

26. S. Katz, *Curves on Calabi-Yau manifolds*, Dyrkolbotn lectures, August 1993.

27. A. Klemm and S. Theisen, *Considerations of one-modulus Calabi-Yau compactifications — Picard-Fuchs equations, Kähler potentials and mirror maps*, Nucl. Phys. **B389** (1993) 153.

28. K. Kodaira and D.C. Spencer, *On deformations of complex analytic structures, I.*, Annals of Mathematics **67** (1958) 328–401.

29. A. Libgober, *Numerical characteristics of systems of straight lines on complete intersec-*

tions, Math. Notes, **13** (1973) 51–56, Plenum Publishing, translated from Math. Zametki **13** (1973) 87–96.

30. A. Libgober and J. Teitelbaum, *Lines on Calabi-Yau complete intersections, mirror symmetry, and Picard-Fuchs equations*, Int. Math. Res. Not. **1** (1993) 29–39.

31. D.R. Morrison, *Mirror symmetry and rational curves on quintic threefold: a guide for mathematicians*, Jour. AMS **6** (1993) 223–247.

32. D.R. Morrison, *Picard-Fuchs equations and mirror maps for hypersurfaces*, these proceedings.

33. K. Oguiso, *Two remarks on Calabi-Yau Moishezon 3-folds*, preprint.

34. M. Reid, *Minimal models of canonical threefolds*, in: Algebraic Varieties and Analytic Varieties. Adv. Stud. Pure Math. **1**. North-Holland, Amsterdam-New York and Kinokuniya, Tokyo, 1983, 131–180.

35. Y. Ruan, *Topological sigma model and Donaldson type invariants in Gromov theory*, preprint.

36. S.A. Strømme and D. van Straten, private communication to the author by Strømme, 1990.

37. I. Vainsencher, *Elliptic quartic curves on a 5ic 3fold*, Proceedings of the 1989 Zeuthen Symposium, Cont. Math. **123** (1991) 247–257.

38. I. Vainsencher, *Enumeration of n-fold tangent hyperplanes to a surface*, preprint.

39. X. Wu, *Chern classes and degenerations of hypersurfaces and their lines*, Duke Math. J. **67** (1992) 633–652.

AMS/IP Studies in Advanced Mathematics
Volume **9**, 1998

PICARD-FUCHS EQUATIONS AND MIRROR MAPS
FOR HYPERSURFACES

DAVID R. MORRISON

Department of Mathematics
Duke University
Durham, NC 27708
drm@math.duke.edu

ABSTRACT. We describe a strategy for computing Yukawa couplings and the mirror map, based on
the Picard-Fuchs equation. (Our strategy is a variant of the method used by Candelas, de la Ossa,
Green, and Parkes [5] in the case of quintic hypersurfaces.) We then explain a technique of Griffiths
[14] which can be used to compute the Picard-Fuchs equations of hypersurfaces. Finally, we carry
out the computation for four specific examples (including quintic hypersurfaces, previously done by
Candelas et al. [5]). This yields predictions for the number of rational curves of various degrees on
certain hypersurfaces in weighted projective spaces. Some of these predictions have been confirmed by
classical techniques in algebraic geometry.

INTRODUCTION

The phenomenon of mirror symmetry dramatically caught the attention of mathematicians with the
recent work of P. Candelas, X. C. de la Ossa, P. S. Green, and L. Parkes [5]. Starting with a particular pair
of "mirror manifolds", calculating certain period integrals, interpreting the results as Yukawa couplings,
and then re-interpreting those results in light of the "mirror manifold" phenomenon, Candelas et al.
were able to give predictions for the numbers of rational curves of various degrees on the general quintic
threefold. In fact, algebraic geometers have had a difficult time verifying these predictions, but all
successful attempts to calculate the numbers of curves have eventually confirmed the predictions.

What is so striking about this work is that the calculation which predicts the numbers of rational
curves on quintic threefolds is in reality a calculation about the variation of Hodge structure on a
completely *different* family of Calabi-Yau threefolds. An asymptotic expansion is made of a function
which comes from that variation, and the coefficients in the expansion are then used to predict numbers
of rational curves.

In [22], we interpreted the calculation of Candelas et al. [5] in terms of variation of Hodge structure.
Here we take a more down to earth approach, and work directly with period integrals and their properties.
(This is perhaps closer in spirit to the original paper.) We have found a way to modify the computational
strategy employed in [5]. Our modified method computes a bit less (there are two unknown "constants
of integration"), but it is easier to actually carry out the computation. We in fact carry it out in three
new examples. This leads to new predictions about numbers of rational curves on certain Calabi-Yau
threefolds.

© 1998 American Mathematical Society
and International Press

185

Our strategy for computing Yukawa couplings is based on the Picard-Fuchs equation for the periods of a one-parameter family of algebraic varieties. We explain in sections 1 and 2 how this equation can be used to compute Yukawa couplings and the mirror map for a family of Calabi-Yau threefolds with $h^{2,1} = 1$. We then go on in section 3 to review a method of Griffiths [14] for calculating Picard-Fuchs equations of hypersurfaces. Related ideas have also been introduced into the physics literature in [2, 4, 12, 20].

In sections 4 and 5, we carry out the computation in four examples, including the quintic hypersurface. The resulting predictions about numbers of rational curves are discussed in section 6.

1. The Picard-Fuchs equation and monodromy

Let $\bar{\pi} : \overline{\mathcal{X}} \to \overline{C}$ be a family of n-dimensional projective algebraic varieties, parameterized by a compact Riemann surface $\overline{C}$. Let $C \subset \overline{C}$ be an open subset such that the induced family $\pi : \mathcal{X} \to C$ has smooth fibers. If we choose topological n-cycles $\gamma_0, \ldots, \gamma_{r-1}$ which give a basis for the nth homology of one particular fiber X_0, and choose a holomorphic n-form ω on X_0, then the *periods* of ω are the integrals

$$\int_{\gamma_0} \omega, \ldots, \int_{\gamma_{r-1}} \omega.$$

Since the fibration $\pi : \mathcal{X} \to C$ is differentiably locally trivial, a local trivialization can be used to extend the cycles γ_i from X_0 to cycles $\gamma_i(z)$ on X_z which depend on z, where z is a local coordinate on C. The holomorphic n-form ω can also be extended to a family of n-forms $\omega(z)$ which depend on the parameter z. If this is done in an algebraic way, then $\omega(z)$ extends to a meromorphic family of n-forms (i.e. poles are allowed) over the entire space $\overline{\mathcal{X}}$.

The cycles $\gamma_i(z)$ determine homology classes which are locally constant in z. However, an attempt to extend these cycles globally will typically lead to monodromy: for each closed path in C, there will be some linear map T represented by a matrix T_{ij} such that transporting γ_i along the path produces at the end a cycle homologous to $\sum T_{ij}\gamma_j$. The same phenomenon will hold for the periods: for a globally defined meromorphic family of n-forms $\omega(z)$, the local periods $\int_{\gamma_i(z)} \omega(z)$ extend by analytic continuation to multiple-valued functions of z, transforming according to the same monodromy transformations T as do the homology classes of the cycles.

The periods $\int_{\gamma(z)} \omega(z)$ satisfy an ordinary differential equation called the *Picard-Fuchs equation* of ω. The existence of this equation can be explained as follows. Choose a local coordinate z on some open set $U \subset C$, and consider the vector

$$v_j(z) := [\frac{d^j}{dz^j} \int_{\gamma_0(z)} \omega(z), \ldots, \frac{d^j}{dz^j} \int_{\gamma_{r-1}(z)} \omega(z)] \in \mathbb{C}^r.$$

For generic values of the parameter z, the dimensions

$$d_j(z) := \dim(\operatorname{span}\{v_0(z), \ldots, v_j(z)\})$$

must be constant. Since $d_j(z) \leq r$, these spaces cannot continue to grow indefinitely. There will thus be a smallest s such that

$$v_s(z) \in \operatorname{span}\{v_0(z), \ldots, v_{s-1}(z)\}$$

(for generic z). We can write

$$v_s(z) = -\sum_{j=0}^{s-1} C_j(z)v_j(z)$$

with the coefficients $C_j(z)$ depending on z. The *Picard-Fuchs equation*, satisfied by all the periods of $\omega(z)$, is then

$$\frac{d^s f}{dz^s} + \sum_{j=0}^{s-1} C_j(z) \frac{d^j f}{dz^j} = 0. \tag{1}$$

The precise form of the equation depends on both the local coordinate z on C, and the choice of holomorphic form $\omega(z)$. Note that the coefficients $C_j(z)$ may acquire singularities at special values of z.

When we approach a point P in $\overline{C} - C$, the Picard-Fuchs equation has (at worst) a *regular singular point* at P [15, 17, 8]. If we choose a parameter z which is centered at P (that is, $z = 0$ at P), then the coefficients $C_j(z)$ in the Picard-Fuchs equation typically will have poles at $z = 0$. However, if we multiply the Picard-Fuchs operator

$$\frac{d^s}{dz^s} + \sum_{j=0}^{s-1} C_j(z) \frac{d^j}{dz^j} \tag{2}$$

by z^s and rewrite the result in the form

$$(z\frac{d}{dz})^s + \sum_{j=0}^{s-1} B_j(z)(z\frac{d}{dz})^j \tag{3}$$

then the new coefficients $B_j(z)$ are holomorphic functions of z. (This is one of several equivalent definitions of "regular singular point".) We call eq. (3) the *logarithmic form* of the Picard-Fuchs operator.

The structure of ordinary differential equations with regular singular points is a classical topic in differential equations: a convenient reference is [7]. We can rewrite eq. (1) as a system of first-order equations, using the logarithmic form eq. (3), as follows: let

$$A(z) = \begin{bmatrix} 0 & 1 & & & & \\ & 0 & 1 & & & \\ & & \ddots & \ddots & & \\ & & & 0 & 1 \\ -B_0(z) & -B_1(z) & \cdots & \cdots & -B_{s-1}(z) \end{bmatrix}. \tag{4}$$

Then solutions $f(z)$ to the equation eq. (1) are equivalent to solution vectors

$$w(z) = \begin{bmatrix} f(z) \\ z\frac{d}{dz}f(z) \\ \vdots \\ (z\frac{d}{dz})^{s-1}f(z) \end{bmatrix}$$

of the matrix equation

$$z\frac{d}{dz}w(z) = A(z)w(z). \tag{5}$$

For a matrix equation such as eq. (5), the facts are these (see [7]). There is a constant $s \times s$ matrix R and a $s \times s$ matrix $S(z)$ of (single-valued) functions of z, regular near $z = 0$, such that

$$\Phi(z) = S(z) \cdot z^R$$

is a *fundamental matrix* for the system. This means that the columns of $\Phi(z)$ are a basis for the space of solutions at each nonsingular point $z \neq 0$. The multiple-valuedness of the solutions has all been put into R, since

$$z^R := e^{(\log z)R} = I + (\log z)R + \frac{(\log z)^2}{2!}R^2 + \cdots$$

is a multiple-valued matrix function of z. The local monodromy on the solutions given by analytic continuation along a path winding once around $z = 0$ in a counterclockwise direction is given by $e^{2\pi i R}$ (with respect to the basis given by the columns of Φ). The matrix R is by no means unique.

Theorem . *Suppose that $z\frac{d}{dz}w(z) = A(z)w(z)$ is a system of ordinary differential equations with a regular singular point at $z = 0$. Suppose that distinct eigenvalues of $A(0)$ do not differ by integers. Then there is a fundamental matrix of the form*

$$\Phi(z) = S(z) \cdot z^{A(0)}$$

and $S(z)$ can be obtained as a power series

$$S(z) = S_0 + S_1 z + S_2 z^2 + \cdots$$

by recursively solving the equation

$$z\frac{d}{dz}S(z) + S(z) \cdot A(0) = A(z) \cdot S(z)$$

for the coefficient matrices S_j. Moreover, any such series solution converges in a neighborhood of $z = 0$.

A proof can be found in [7], together with methods for treating the case in which eigenvalues of $A(0)$ *do* differ by integers.

We will be particularly interested in systems with *unipotent monodromy*: by definition, this means that $e^{2\pi i R}$ is a unipotent matrix, so that $(e^{2\pi i R} - I)^m \neq 0$, $(e^{2\pi i R} - I)^{m+1} = 0$ for some m called the *index*.

Corollary . *Suppose that $(z\frac{d}{dz})^s f(z) + \sum_{j=0}^{s-1} B_j(z)(z\frac{d}{dz})^j f(z)$ is an ordinary differential equation with a regular singular point at $z = 0$. If $B_j(0) = 0$ for all j, then the solutions of this equation have unipotent monodromy of index s.*

The corollary follows by calculating with eq. (4), setting $z = 0$ and $B_j(0) = 0$ to produce

$$e^{2\pi i A(0)} = \begin{bmatrix} 1 & 2\pi i & \frac{(2\pi i)^2}{2!} & \cdots & \frac{(2\pi i)^{s-1}}{(s-1)!} \\ & 1 & 2\pi i & \cdots & \frac{(2\pi i)^{s-2}}{(s-2)!} \\ & & \ddots & & \vdots \\ & & & 1 & 2\pi i \\ & & & & 1 \end{bmatrix} .$$

2. Computing the mirror map

Recall that a *Calabi-Yau manifold* is a compact Kähler manifold X of complex dimension n which has trivial canonical bundle, such that the Hodge numbers $h^{k,0}$ vanish for $0 < k < n$. Thanks to a celebrated theorem of Yau [27], every such manifold admits Ricci-flat Kähler metrics.

Suppose now that $\pi : \mathcal{X} \to C$ is a family of Calabi-Yau threefolds with $h^{2,1}(X) = 1$, which is not a locally constant family. The third cohomology group $H^3(X)$ has dimension $r = 4$. It follows that the Picard-Fuchs equation has order at most 4. (In fact, it is not difficult to show that it has order exactly 4.)

Let z be a coordinate on $\overline{C}$ centered at a point $P \in \overline{C} - C$. We say that P *is a point at which the monodromy is maximally unipotent* if the monodromy is unipotent of index 4. As we have seen in the corollary, if $B_j(0) = 0$ in the logarithmic form of the Picard-Fuchs equation, $z = 0$ will be such a point. We will assume for simplicity that our points of maximally unipotent monodromy have this form, leaving appropriate modifications for the general case to the reader.

We review the calculation of the Yukawa coupling, following [5]. Let $\omega(z)$ be a family of n-forms, and let

$$W_k := \int_{X_z} \omega(z) \wedge \frac{d^k}{dz^k}\omega(z).$$

A fundamental principle from the theory of variation of Hodge structure (cf. [16]) implies that W_0, W_1, and W_2 all vanish. The *Yukawa coupling* is the first non-vanishing term W_3. Candelas et al. show that the Yukawa coupling W_3 satisfies the differential equation

$$\frac{dW_3(z)}{dz} = -\frac{1}{2}C_3(z)W_3(z),$$

where $C_3(z)$ is a coefficient in the Picard-Fuchs equation (1).

The Yukawa coupling as defined clearly depends on the "gauge", that is, on the choice of holomorphic 3-form $\omega(z)$. If fact, if we alter the gauge by $\omega(z) \mapsto f(z)\omega(z)$, then W_k transforms as

$$W_k \mapsto f(z) \sum_{j=0}^{k} \binom{k}{j} \frac{d^j f(z)}{dz^j} W_{k-j}.$$

Since $W_0 = W_1 = W_2 = 0$, the change in the Yukawa coupling W_3 is simply $W_3 \mapsto f(z)^2 W_3$.

The Yukawa coupling also depends on the choice of coordinate z, and in fact is often denoted by κ_{zzz}. If we change coordinates from z to w, we must change the differentiation operator from d/dz to d/dw. The chain rule then imples that

$$\kappa_{www} = \left(\frac{dz}{dw}\right)^3 \kappa_{zzz}.$$

Candelas et al. [5] use physical arguments to set the gauge in this calculation, and to find an appropriate (multiple-valued) parameter t with which to compute. (The associated differentiation operator d/dt is single-valued.) What will be important for us are the following observations about their results.

The gauge used by Candelas et al. determines a family of meromorphic n-forms $\widetilde{\omega}(z)$ with the property that the period function

$$\int_\gamma \widetilde{\omega}(z) \equiv 1$$

for some cycle γ. Moreover, the parameter t determined by Candelas et al. is a parameter defined in an angular sector near $z = 0$ which has two crucial properties:

1. If we analytically continue along a simple loop around $z = 0$ in the counterclockwise direction, t becomes $t + 1$. (It will be convenient to also introduce $q = e^{2\pi i t}$, which remains single-valued near $z = 0$.)

2. There are cycles γ_0 and γ_1 such that $\int_{\gamma_0} \omega(z)$ is single valued near $z = 0$, and

$$t = \frac{\int_{\gamma_1} \omega(z)}{\int_{\gamma_0} \omega(z)}$$

   in an angular sector near $z = 0$.

Each period function $\int_\gamma \omega(z)$ is a solution to the Picard-Fuchs equation of the family. Translating the results of the previous section into the present context, we obtain the following:

Lemma . *Suppose that $z = 0$ is a point of maximally unipotent monodromy such that $B_j(0) = 0$, where $B_j(z)$ are the coefficients in the logarithmic form of the Picard-Fuchs equation. Then*

 1. *There is a period function for $\omega(z)$,*

$$f_0(z) := \int_{\gamma_0} \omega(z)$$

which is single-valued near $z = 0$. This period function is unique up to multiplication by a constant. (This implies that the cycle γ_0 is also unique up to a constant multiple.)

 In particular, the family of meromorphic n-forms

$$\widetilde{\omega}(z) := \frac{\omega(z)}{\int_{\gamma_0} \omega(z)}$$

will have the property that

$$\int_\gamma \widetilde{\omega}(z) \equiv 1$$

for some γ, and it is the unique such family up to constant multiple.

 2. *Fixing a choice of period function $f_0(z)$ as in part (1), there is a period function*

$$f_1(z) := \int_{\gamma_1} \omega(z)$$

such that $\varphi(z) := f_1(z)/f_0(z)$ transforms as

$$\varphi(z) \mapsto \varphi(z) + 1$$

upon transport around $z = 0$ in the counterclockwise direction. The ratio $\varphi(z)$ is unique up to the addition of a constant.

This, then, is our alternate strategy for computing the Yukawa coupling: we find solutions of the Picard-Fuchs equation which have the properties specified in the lemma, and we use those to fix the gauge and specify the natural parameter, up to two unknown constants of integration.

3. PICARD-FUCHS EQUATIONS FOR HYPERSURFACES

We now review a method of Griffiths [14] for describing the cohomology of a hypersurface, which can be used to determine the Picard-Fuchs equation of a one-parameter family of hypersurfaces. Calculations of this sort were earlier made by Dwork [10, Sec. 8]. Griffiths' method was extended to the weighted projective case by Steenbrink [26] and Dolgachev [9], who we follow.

We denote a weighted projective n-space by $\mathbb{P}^{(k_0,\ldots,k_n)}$, where $k_0, \ldots, k_n$ are the weights of the variables $x_0, \ldots, x_n$. Weighted homogeneous polynomials can be identified with the aid of the Euler vector field

$$\theta = \sum k_j x_j \frac{\partial}{\partial x_j}$$

which has the property that $\theta P = (\deg P) \cdot P$ for any weighted homogeneous polynomial P. Contracting the volume from on $\mathbb{C}^{n+1}$ with θ produces the fundamental weighted homogeneous differential form (of "weight" $k := \sum k_j$)

$$\Omega := \sum_{j=0}^{n} (-1)^j k_j x_j \, dx_0 \wedge \cdots \wedge \widehat{dx_j} \wedge \cdots \wedge dx_n.$$

Rational differentials of degree n on $\mathbb{P}^{(k_0,\ldots,k_n)}$ can be described as expressions $P\Omega/Q$, where P and Q are weighted homogeneous polynomials with $\deg P + k = \deg Q$.

Suppose that Q is a weighted homogeneous polynomial defining a quasismooth hypersurface $\mathcal{Q} \subset \mathbb{P}^{(k_0,\ldots,k_n)}$. (That is, $Q = 0$ defines a hypersurface in $\mathbb{C}^{n+1}$ which is smooth away from the origin.) The middle cohomology of $\mathcal{Q}$ is then described by means of differential forms with poles (of all orders) along $\mathcal{Q}$. Each such form $P\Omega/Q^\ell$ is made into a cohomology class by a "residue" construction: for an $(n-1)$-cycle γ on $\mathcal{Q}$, the tube over γ (an S^1-bundle inside the (complex) normal bundle of $\mathcal{Q}$) is an n-cycle Γ on $\mathbb{P}^{(k_0,\ldots,k_n)}$ disjoint from $\mathcal{Q}$. We can then define the residue of $P\Omega/Q^\ell$ by

$$\int_\gamma \mathrm{Res}_\mathcal{Q} \left(\frac{P\Omega}{Q^\ell} \right) = \frac{1}{2\pi i} \int_\Gamma \frac{P\Omega}{Q^\ell}.$$

Since altering $P\Omega/Q^\ell$ by an exact differential does not change the value of these integrals, we see that the cohomology of $\mathcal{Q}$ is represented by equivalence classes of rational differential forms $P\Omega/Q^\ell$ modulo exact forms.

Here is Griffiths' "reduction of pole order" calculation which shows how to reduce modulo exact forms in practice. Let Q and A_j be weighted homogeneous polynomials, with $\deg Q = d$, $\deg A_j = \ell d + k_j - k$. Define

$$\varphi = \frac{1}{Q^\ell} \sum_{i<j} (k_i x_i A_j - k_j x_j A_i) dx_0 \wedge \cdots \wedge \widehat{dx_i} \wedge \cdots \wedge \widehat{dx_j} \wedge \cdots \wedge dx_n$$

and then calculate

$$d\varphi = \frac{\left(\ell \sum A_j \frac{\partial Q}{\partial x_j} - Q \sum \frac{\partial A_j}{\partial x_j} \right) \Omega}{Q^{\ell+1}} = \frac{\ell \sum A_j \frac{\partial Q}{\partial x_j} \Omega}{Q^{\ell+1}} - \frac{\sum \frac{\partial A_j}{\partial x_j} \Omega}{Q^\ell}. \tag{6}$$

Thus, any form whose numerator lies in the Jacobian ideal $J = (\partial Q/\partial x_0, \ldots, \partial Q/\partial X_n)$ is equivalent (modulo exact forms) to a form with smaller pole order.

This idea can be used to calculate Picard-Fuchs equations as follows. The cycles Γ do not change (in homology) when z varies locally. So we can differentiate under the integral sign

$$\frac{d^k}{dz^k} \int_\gamma \mathrm{Res}_\mathcal{Q} \left(\frac{P\Omega}{Q^\ell} \right) = \frac{1}{2\pi i} \int_\Gamma \frac{d^k}{dz^k} \left(\frac{P\Omega}{Q^\ell} \right)$$

when Q depends on a parameter z. (Note that Ω is independent of z.) The Picard-Fuchs operator (2) will have the property that

$$\left(\frac{d^s}{dz^s} + \sum_{j=0}^{s-1} C_j(z) \frac{d^j}{dz^j} \right) \left(\frac{P\Omega}{Q} \right) = d\varphi$$

is an exact form. To find it, take successive z-derivatives of the integrand $P\Omega/Q$ and use the reduction of order of pole formula [14] to determine a linear relation among those derivatives, modulo exact forms.

4. EXAMPLES: PICARD-FUCHS EQUATIONS

We will calculate the Picard-Fuchs equations for certain one-parameter families of Calabi-Yau three-folds. Our choice of families is motivated by the mirror construction of Greene and Plesser [13].

We choose weights $k_0, \ldots, k_4$ with $k_0 \geq k_1 \geq \cdots \geq k_4$ for a weighted projective 4-space such that $d_j := k/k_j$ is an integer, where $k := \sum k_j$. We also assume that $\gcd\{k_j \mid j \neq j_0\} = 1$ for every j_0. These assumptions then imply that $k = \mathrm{lcm}\{d_j\}$.

Consider the pencil of hypersurfaces $\mathcal{Q}_\psi \subset \mathbb{P}^{(k_0,\ldots,k_4)}$ defined by $Q(x,\psi) = 0$, where

$$Q(x,\psi) := \sum_{j=0}^{4} x_j^{d_j} - k\psi \prod_{j=0}^{4} x_j.$$

This pencil has a natural group of diagonal automorphisms preserving the holomorphic 3-form. To define it, let μ_m denote the multiplicative group of mth roots of unity (considered as a subgroup of $\mathbb{C}^\times$), and let

$$G = (\mu_{d_0} \times \cdots \times \mu_{d_4})/\mu_k,$$

where we embed μ_k in $\mu_{d_0} \times \cdots \times \mu_{d_4}$ by

$$\alpha \mapsto (\alpha^{k_0}, \ldots, \alpha^{k_4}).$$

Note that since $\sum k_j = k$, the formula

$$f(\alpha_0, \ldots, \alpha_4) = \left(\prod \alpha_j\right)^{-1}$$

determines a well-defined homomorphism $f : G \to \mathbb{C}^\times$. Let $G_0 = \ker(f)$.

We can regard $Q(x,\psi) = 0$ as defining a hypersurface $\mathcal{Q} \subset \mathbb{P}^{(k_0,\ldots,k_4)} \times \mathbb{C}$. The group G acts on $\mathbb{P}^{(k_0,\ldots,k_4)} \times \mathbb{C}$ by

$$(x_0, \ldots, x_4; \psi) \mapsto (\alpha_0 x_0, \ldots, \alpha_4 x_4; f(\alpha)\psi)$$

for $\alpha = (\alpha_0, \ldots, \alpha_4) \in G$. The polynomial $Q(x,\psi)$ is invariant under this action. Thus, the action preserves $\mathcal{Q}$, and maps $\mathcal{Q}_\psi$ isomorphically to $\mathcal{Q}_{f(\alpha)\psi}$. It follows that the group G_0 acts on $\mathcal{Q}_\psi$ by automorphisms, and that the induced action of $G/G_0 \cong \mu_k$ establishes isomorphisms between $\mathcal{Q}_\psi/G_0$ and $\mathcal{Q}_{\lambda\psi}/G_0$ for $\lambda \in \mu_k$.

The quotient space $\mathcal{Q}_\psi/G$ has only canonical singularities. By a theorem of Markushevich [21, Prop. 4] and Roan [23, Prop. 2], these singularities can be resolved to give a Calabi-Yau manifold $\mathcal{W}_\psi$. There are choices to be made in this resolution process; we do not specify a choice. By a theorem of Kollár [19], any two resolutions differ by a sequence of flops.

Note that the differential form Ω from the previous section transforms as $\Omega \mapsto (\prod \alpha_j)\Omega$ under the action of $\alpha \in G$. Thus, the rational differential

$$\omega_1 = \frac{\psi\Omega}{Q(x,\psi)}$$

is invariant under the action of G; we define $\omega(\psi) = \mathrm{Res}_{\mathcal{Q}_\psi}(\omega_1)$.

Since the holomorphic 3-forms $\omega(\psi)$ on $\mathcal{Q}_\psi$ are invariant on G_0, they induce holomorphic 3-forms on $\mathcal{W}_\psi$. Moreover, the homology group $H_3(\mathcal{W}_\psi)$ contains the G_0-invariant part $H_3(\mathcal{Q}_\psi)^{G_0}$ of the homology of $\mathcal{Q}_\psi$. If we know that the dimensions of these spaces agree, then they will coincide (at least for homology with coefficients in a field). In this case, the periods of $\mathcal{W}_\psi$ can actually be computed as periods of the holomorphic form $\omega(\psi)$ on $\mathcal{Q}_\psi$, over G_0-invariant cycles. Thanks to the isomorphisms between $\mathcal{Q}_\psi$ and $\mathcal{Q}_{\lambda\psi}$ for $\lambda \in \mu_k$ and the invariance of the rational differential ω_1 under G, these periods will be invariant under $\psi \mapsto \lambda\psi$. In particular, they will be functions of $z = \psi^{-k}$ alone.

It is likely that the resolutions $\mathcal{W}_\psi$ of $\mathcal{Q}_\psi/G_0$ could be chosen so that the action of G/G_0 would lift to isomorphisms between $\mathcal{W}_\psi$ and $\mathcal{W}_{\lambda\psi}$. (We verified this in the case of quintic hypersurfaces in [22].) In this case, there would be an actual family of Calabi-Yau threefolds for which z served as a parameter. It may be that such resolutions could be constructed by finding an appropriate partial resolution of $\mathcal{Q}/G$. However, we do not need the existence of this family to describe the computation of the Yukawa coupling.

k	$(k_0, \ldots, k_4)$	$Q(x, \psi)$
5	$(1,1,1,1,1)$	$x_0^5 + x_1^5 + x_2^5 + x_3^5 + x_4^5 - 5\psi x_0 x_1 x_2 x_3 x_4$
6	$(2,1,1,1,1)$	$x_0^3 + x_1^6 + x_2^6 + x_3^6 + x_4^6 - 6\psi x_0 x_1 x_2 x_3 x_4$
8	$(4,1,1,1,1)$	$x_0^2 + x_1^8 + x_2^8 + x_3^8 + x_4^8 - 8\psi x_0 x_1 x_2 x_3 x_4$
10	$(5,2,1,1,1)$	$x_0^2 + x_1^5 + x_2^{10} + x_3^{10} + x_4^{10} - 10\psi x_0 x_1 x_2 x_3 x_4$

TABLE 1. The hypersurfaces.

We will carry out the computation in four specific examples. These come from the lists of Candelas, Lynker and Schimmrigk [6]; they found that there are exactly four types of hypersurface in weighted projective four-space which are Calabi-Yau threefolds with Picard number one. The weights of the space are given in the second column of table 1. For each of those cases, Greene and Plesser's mirror construction [13] yields the family $\mathcal{W}_\psi$ which we have described above. And Roan's formula [24] for the Betti numbers verifies that b_3 is indeed 4 (with $h^{2,1} = 1$). The remaining columns in table 1 show the value of k, and give the equation $Q(x, \psi)$ explicitly.

We describe the G_0-invariant cohomology by means of the rational differential forms

$$\omega_\ell := \frac{(-1)^{\ell-1}(\ell-1)!\,\psi^\ell(\prod x_i^{\ell-1})\Omega}{Q(x,\psi)^\ell}.$$

These are chosen because of the evident G-invariance in the numerator; the coefficients were adjusted so that the formula

$$-\frac{1}{k}\psi\frac{d}{d\psi}\omega_\ell = -\frac{\ell}{k}\omega_\ell + \omega_{\ell+1} \tag{7}$$

would not be overly burdened with constants. We compute with the differential operator $-\frac{1}{k}\psi\frac{d}{d\psi}$ because it coincides with $z\frac{d}{dz}$.

A basis for the G_0-invariant cohomology is then given by the residues of ω_1, ω_2, ω_3, ω_4. To compute the Picard-Fuchs equation, we must find an expression for ω_5 as a linear combination of $\omega_1, \ldots, \omega_4$ modulo exact forms. That expression, combined with (7), will then yield the desired differential equation.

We carried out this calculation using the Gröbner basis algorithm [3], modifying an implementation written in MAPLE by Yunliang Yu (cf. [28]). We first calculated a Gröbner basis for the Jacobian ideal $J = (\partial Q/\partial x_0, \ldots, \partial Q/\partial x_4)$, working in the ring $\mathbb{C}(\psi)[x_0, \ldots, x_4]$ of polynomials whose coefficients are rational functions of ψ. The reduction of pole order was then achieved step by step as follows: given a form η_ℓ, the residue of a global form with a pole of order ℓ, we used the Gröbner basis to reduce the numerators of both η_ℓ and ω_ℓ to standard form. We could thus determine a coefficient $\varepsilon_\ell \in \mathbb{C}(\psi)$ such that the numerator of $\eta_\ell - \varepsilon_\ell\omega_\ell$ lies in J. Another application of Gröbner basis reduction produced explicit coefficients

$$\eta_\ell - \varepsilon_\ell\omega_\ell = \sum A_{\ell j}\frac{\partial Q}{\partial x_j}.$$

Then the Griffiths formula (6) determines forms φ_ℓ and $\eta_{\ell-1}$ such that

$$\eta_\ell - \varepsilon_\ell\omega_\ell = d\varphi_\ell + \eta_{\ell-1},$$

and $\eta_{\ell-1}$ has a pole of order $\ell - 1$.

Beginning with $\eta_5 = \omega_5$ and applying this procedure several times, one finds

$$\omega_5 = \varepsilon_1\omega_1 + \cdots + \varepsilon_4\omega_4 + d\varphi.$$

k	ε_1	ε_2	ε_3	ε_4
5	$\dfrac{1}{625(z-1)}$	$\dfrac{-3}{25(z-1)}$	$\dfrac{1}{(z-1)}$	$\dfrac{-2}{(z-1)}$
6	$\dfrac{1}{324(z-4)}$	$\dfrac{-5}{18(z-4)}$	$\dfrac{-(z-50)}{18(z-4)}$	$\dfrac{-(z+20)}{3(z-4)}$
8	$\dfrac{1}{16(z-256)}$	$\dfrac{-15(z+256)}{512(z-256)}$	$\dfrac{-5(3z-1280)}{64(z-256)}$	$\dfrac{-(3z+1280)}{4(z-256)}$
10	$\dfrac{5}{4(z-12500)}$	$\dfrac{-(7z+37500)}{200(z-12500)}$	$\dfrac{-(7z-62500)}{20(z-12500)}$	$\dfrac{-(z+12500)}{(z-12500)}$

TABLE 2. The results of the Gröbner basis calculation.

The results of this computation for our four examples are summarized in table 2. The coefficients ε_ℓ are in fact functions of $z = \psi^{-k}$ (as expected from our earlier discussion), and have been displayed as such.

The differential equation for $[\omega_1, \ldots, \omega_4]$ determined by this procedure has the form

$$z\frac{d}{dz}\begin{bmatrix} \omega_1 \\ \omega_2 \\ \omega_3 \\ \omega_4 \end{bmatrix} = \begin{bmatrix} -\frac{1}{k} & 1 & 0 & 0 \\ 0 & -\frac{2}{k} & 1 & 0 \\ 0 & 0 & -\frac{3}{k} & 1 \\ \varepsilon_1 & \varepsilon_2 & \varepsilon_3 & \varepsilon_4 - \frac{4}{k} \end{bmatrix}\begin{bmatrix} \omega_1 \\ \omega_2 \\ \omega_3 \\ \omega_4 \end{bmatrix}.$$

To calculate the Picard-Fuchs equation, we must change basis via

$$\begin{bmatrix} \omega_1 \\ z\frac{d}{dz}\omega_1 \\ (z\frac{d}{dz})^2\omega_1 \\ (z\frac{d}{dz})^3\omega_1 \end{bmatrix} = \begin{bmatrix} 1 & 0 & 0 & 0 \\ -\frac{1}{k} & 1 & 0 & 0 \\ \frac{1}{k^2} & -\frac{3}{k} & 1 & 0 \\ -\frac{1}{k^3} & \frac{7}{k^2} & -\frac{6}{k} & 1 \end{bmatrix}\begin{bmatrix} \omega_1 \\ \omega_2 \\ \omega_3 \\ \omega_4 \end{bmatrix}.$$

This determines an equation in the form (4), with

$$\begin{aligned} B_0(z) &= -\varepsilon_1(z) - \tfrac{1}{k}\varepsilon_2(z) - \tfrac{2}{k^2}\varepsilon_3(z) - \tfrac{6}{k^3}\varepsilon_4(z) + \tfrac{24}{k^4} \\ B_1(z) &= -\varepsilon_2(z) - \tfrac{3}{k}\varepsilon_3(z) - \tfrac{11}{k^2}\varepsilon_4(z) + \tfrac{50}{k^3} \\ B_2(z) &= -\varepsilon_3(z) - \tfrac{6}{k}\varepsilon_4(z) + \tfrac{35}{k^2} \\ B_3(z) &= -\varepsilon_4(z) + \tfrac{10}{k}. \end{aligned} \tag{8}$$

As can be directly verified in each of our cases, $B_j(0) = 0$. It follows that the monodromy at $z = 0$ is maximally unipotent. (In the case of quintics ($k = 5$), this had been shown in [5]; cf. [22].)

5. Examples: Mirror maps

We next compute the mirror maps for our four examples, based on their Picard-Fuchs equations. Expanding eqs. (2) and (3), one finds that the coefficient $C_3(z)$ coincides with $(6 + B_3(z))/z$. Moreover, in our four examples, a straightforward computation based on eq. (8) and table 2 shows that $B_3(z) = 2z/(z - \lambda)$, where $\lambda = 1, 4, 256, 12500$ when $k = 5, 6, 8, 10$, respectively. Thus,

$$C_3(z) = \frac{6 + B_3(z)}{z} = \frac{6}{z} + \frac{2}{z - \lambda}.$$

The Yukawa coupling κ_{zzz} in the gauge $\omega(z)$ is therefore given by a function $W_3(z)$ which satisfies the differential equation

$$\frac{dW_3(z)}{dz} = \left(\frac{-3}{z} + \frac{-1}{z - \lambda} \right) W_3(z).$$

Thus, in the gauge $\omega(z)$ we have

$$\kappa_{zzz} = \frac{c_1}{(2\pi i)^3 z^3 (z - \lambda)}.$$

Here $c_1/(2\pi i)^3$ is the first "constant of integration": we have introduced a factor of $(2\pi i)^3$ in order to simplify a later formula.

In order to determine the natural gauge, we must find a solution $f_0(z)$ of the Picard-Fuchs equation which is regular near $z = 0$. Using the corresponding vector $w_0(z)$ of which $f_0(z)$ is the first component, we want a solution to the vector equation

$$z \frac{d}{dz} w_0(z) = A(z) w_0(z) \tag{9}$$

which is regular near $z = 0$. ($A(z)$ is given by eqs. (4), (8), and table 2.) This can be found using power-series techniques, and there is a solution with $f_0(0) \neq 0$ in each of our four cases. We normalize so that $f_0(0) = 1$; alternatively, we could have absorbed the leading term of $f_0(z)$ into the constant of integration c_1.

As a result, the gauge-fixed value of κ_{zzz} takes the form

$$\kappa_{zzz} = \frac{c_1}{(2\pi i)^3 z^3 (z - \lambda)(f_0(z))^2},$$

where the constant c_1 has yet to be determined.

We now search for the good parameter t. We should locate a second solution $f_1(z)$, or its corresponding vector $w_1(z)$, which is multiple-valued and has the correct monodromy properties. The monodromy will be such that if we introduce

$$v(z) := 2\pi i w_1(z) - (\log z) w_0(z)$$

and its first component

$$g(z) := 2\pi i f_1(z) - (\log z) f_0(z),$$

then $v(z)$ will be single-valued and regular near $z = 0$. It is easy to calculate that the matrix equation satisfied by $v(z)$ is

$$z \frac{d}{dz} v(z) = A(z) v(z) - w_0(z). \tag{10}$$

Solutions to this equation can be found by power-series techniques. We normalize the solution so that $g(0) = 0$. The parameter t is then given by

$$t = \frac{1}{2\pi i} \log c_2 + \frac{1}{2\pi i} \log z + \frac{g(z)}{f_0(z)}$$

($\frac{1}{2\pi i}\log c_2$ is the second "constant of integration") and the associated parameter q is

$$q = e^{2\pi i t} = c_2 z e^{g/f_0}.$$

Let us define

$$\delta(z) = 1 + z\frac{d}{dz}\left(\frac{g(z)}{f_0(z)}\right),$$

so that

$$\frac{dq}{dz} = c_2\delta(z)e^{g/f_0}.$$

Then by the chain rule,

$$\frac{dz}{dt} = \frac{dq/dt}{dq/dz} = \frac{2\pi i z}{\delta(z)}.$$

It follows that the gauge-fixed value of κ_{ttt} is

$$\kappa_{ttt} = \left(\frac{dz}{dt}\right)^3 \kappa_{zzz} = \frac{c_1}{(\delta(z))^3(z - \lambda)(f_0(z))^2}.$$

Finally we express this normalized κ_{ttt} as a power series in q. The constants c_1 and c_2 have yet to be determined; however, we can define

$$h_0(z) = \frac{1}{(\delta(z))^3(z - \lambda)(f_0(z))^2} \tag{11}$$

$$h_j(z) = \frac{1}{\delta(z)e^{g/f_0}} \cdot \frac{dh_{j-1}(z)}{dz} \tag{12}$$

and find that

$$h_j(z) = \frac{(c_2)^j}{c_1}\left(\frac{d}{dq}\right)^j \kappa_{ttt},$$

so that

$$\kappa_{ttt} = \sum_{j=0}^{\infty} \frac{c_1}{(c_2)^j}\frac{h_j(0)}{j!}\, q^j.$$

Proposition . *The numbers $h_j(0)$ are rational numbers.*

Proof. The coefficient matrix $A(z)$ in the vector equation (9) has entries in $\mathbb{Q}(z)$; if written out in power series, all the power series coefficients will be rational numbers. Finding a power series solution to (9) then involves solving linear equations with rational coefficients at each step: the solutions will be rational. Thus, $w_0(z)$ and $f_0(z)$ are power series in z with rational coefficients.

Similarly, $v(z)$ and $g(z)$ are power series with rational coefficients, since they come from equation (10). Furthermore, since exponentiating a power series with rational coefficients (whose constant term is zero) again gives a power series with rational coefficients, e^{g/f_0} and $\delta(z)$ are power series in z with rational coefficients.

But then by (11), $h_0(z)$ is clearly a power series in z with rational coefficients; similarly for $h_j(z)$ by (12). It follows that each $h_j(0)$ is a rational number. Q.E.D.

k	n_0	n_1	n_2	n_3	n_4
5	5	2875	609250	317206375	242467530000
6	3	7884	6028452	11900417220	34600752005688
8	2	29504	128834912	1423720546880	23193056024793312
10	2	462400	24431571200	3401788732948800	700309317702649312000

TABLE 3. The predicted numbers of curves.[1]

6. CHOOSING THE CONSTANTS AND PREDICTING THE NUMBERS OF RATIONAL CURVES

Calabi-Yau threefolds with $h^{2,1} = 1$ are conjectured to be the "mirrors" of other Calabi-Yau threefolds with $h^{1,1} = 1$. In the four examples we have considered, this mirror property can be realized by a construction of Greene and Plesser [13]. The threefolds $\mathcal{W}_\psi$ are mirrors of threefolds $\mathcal{M} \subset \mathbb{P}^{(k_0,\ldots,d_4)}$, which are hypersurfaces of weighted degree $k = \sum k_j$. The Picard group of $\mathcal{M}$ is cyclic, generated by some ample divisor H.

Mirror symmetry and topological field theory predict that the q-expansion of the gauge-fixed Yukawa coupling

$$\kappa_{ttt} = a_0 + a_1 q + a_2 q^2 + \cdots$$

will have integers as coefficients. Moreover, by a formula conjectured in [5] and established in [1], if this q-expansion is written in the form

$$\kappa_{ttt} = n_0 + \sum_{j=1}^{\infty} \frac{n_j j^3 q^j}{1 - q^j} = n_0 + n_1 q + (2^3 n_2 + n_1) q^2 + \cdots . \tag{13}$$

then the coefficients n_j should also be integers. The first term n_0 is predicted to coincide with H^3 (the absolute degree of $\mathcal{M}$), and n_j is predicted to be the number of rational curves C on $\mathcal{M}$ with $C \cdot H = j$, assuming that all rational curves on $\mathcal{M}$ are disjoint and have normal bundle $\mathcal{O}(-1) \oplus \mathcal{O}(-1)$.

These two predictions can be used to choose the constants of integration in our examples. First, the absolute degree d is the lowest order term which appears in the polynomial $Q(x, \psi)$; to ensure that $n_0 = d$ we must take $c_1 = -\lambda d$. Second, the formula (13) puts very strong divisibility constraints on the coefficients a_j, and it seems likely that there will be a unique choice of c_2 which satisfies all of these constraints.

We have calculated the first 20 coefficients (using MATHEMATICA) in each of our four examples. There does indeed appear to be a unique choice for c_2 which produces integers for $n_1, \ldots n_{20}$: that choice turns out to be $c_2 = k^{-k}$ in each of our examples. Making this choice leads to the values for n_j displayed in table 3.

Table 3 therefore contains predictions about numbers of rational curves on the weighted projective hypersurfaces. For a general hypersurface in $\mathcal{M} \subset \mathbb{P}^{(k_0,\ldots,d_4)}$ of degree $k = \sum k_j$, the prediction is that there should be n_j rational curves C with $C \cdot H = j$, where H generates $\text{Pic}(\mathcal{M})$.

The first line of the table reproduces the predictions made by Candelas et al. about quintic threefolds. Several of these have been verified: the number of lines was known classically, the number of conics was computed by Katz [18], and the number of twisted cubics n_3 has recently been computed by Ellingsrud and Strømme [11]—all of these results agree with the predictions.

---

[1]Note added in second edition: The normalization used in the final line of table 3 is incorrect: all of the entries n_j in that line should be divided by 2. I thank Anamaría Font who first drew this to my attention.

Of the remaining predictions in the table, we have only checked one. Each hypersurface from the third family (the case $k = 8$) can be regarded as a double cover of $\mathbb{P}^3$ branched on a surface of degree 8. The entry 29504 in the third line of the table can be interpreted as follows: for a general surface of degree 8 in $\mathbb{P}^3$, there should be 14752 lines which are 4-times tangent to the surface. (These lines will then split into pairs of rational curves on the double cover.) After we had obtained this number, Steve Kleiman was kind enough to locate a 19th-century formula of Schubert [25, Formula 21, p. 236], which states that the number of lines in $\mathbb{P}^3$ 4-times tangent to a general surface of degree n is

$$\frac{1}{12}n(n - 4)(n - 5)(n - 6)(n - 7)(n^3 + 6n^2 + 7n - 30).$$

Substituting $n = 8$, we find the predicted number 14752.

Acknowledgements

Several of the ideas explained in this paper arose in conversations with Sheldon Katz—it is a pleasure to acknowledge his contribution. I also benefited greatly from the chance to interact with physicists which was provided by the Mirror Symmetry Workshop. I wish to thank the M.S.R.I. as well as the organizers and participants in the Workshop.

This work was partially supported by NSF grant DMS-9103827.

References

1. P. S. Aspinwall and D. R. Morrison, *Topological field theory and rational curves*, Comm. Math. Phys. **151** (1993), 245–262.

2. B. Blok and A. Varchenko, *Topological conformal field theories and the flat coordinates*, Int. J. Mod. Phys. A **7** (1992), 1467–1490.

3. B. Buchberger, *Gröbner bases: An algorithmic method in polynomial ideal theory*, Multidimensional Systems Theory (N. K. Bose, ed.), D. Reidel, Dordrecht, Boston, Lancaster, 1985, pp. 184–232.

4. A. C. Cadavid and S. Ferrara, *Picard-Fuchs equations and the moduli space of superconformal field theories*, Phys. Lett. B **267** (1991), 193–199.

5. P. Candelas, X. C. de la Ossa, P. S. Green, and L. Parkes, *A pair of Calabi-Yau manifolds as an exactly soluble superconformal theory*, Phys. Lett. B **258** (1991), 118–126; Nuclear Phys. B **359** (1991), 21–74.

6. P. Candelas, M. Lynker, and R. Schimmrigk, *Calabi-Yau manifolds in weighted $\mathbb{P}_4$*, Nuclear Phys. B **341** (1990), 383–402.

7. E. A. Coddington and N. Levinson, *Theory of ordinary differential equations*, McGraw-Hill, New York, Toronto, London, 1955.

8. P. Deligne, *Equations différentielles à points singuliers réguliers*, Lecture Notes in Math., vol. 163, Springer-Verlag, Berlin, Heidelberg, New York, 1970.

9. I. Dolgachev, *Weighted projective varieties*, Group Actions and Vector Fields (J. B. Carrell, ed.), Lecture Notes in Math., vol. 956, Springer-Verlag, Berlin, Heidelberg, New York, 1982, pp. 34–71.

10. B. Dwork, *On the Zeta function of a hypersurface, II*, Ann. of Math. (2) **80** (1964), 227–299.

11. G. Ellingsrud and S. A. Strømme, *The number of twisted cubic curves on the general quintic threefold*, Math. Scand. **76** (1995), 5–34.

12. S. Ferrara, *Calabi-Yau moduli space, special geometry and mirror symmetry*, Modern Phys. Lett. A **6** (1991), 2175–2180.

13. B. R. Greene and M. R. Plesser, *Duality in Calabi-Yau moduli space*, Nuclear Phys. B **338** (1990), 15–37.

14. P. A. Griffiths, *On the periods of certain rational integrals, I*, Ann. of Math. (2) **90** (1969), 460–495.

15. ______, *Periods of integrals on algebraic manifolds: Summary of main results and discussion of open problems*, Bull. Amer. Math. Soc. **76** (1970), 228–296.

16. ______, ed., *Topics in transcendental algebraic geometry*, Ann. of Math. Stud., vol. 106, Princeton University Press, Princeton, 1984.

17. N. M. Katz, *Nilpotent connections and the monodromy theorem: Applications of a result of Turrittin*, Inst. Hautes Études Sci. Publ. Math. **39** (1970), 175–232.

18. S. Katz, *On the finiteness of rational curves on quintic threefolds*, Compositio Math. **60** (1986), 151–162.

19. J. Kollár, *Flops*, Nagoya Math. J. **113** (1989), 15–36.

20. W. Lerche, D.-J. Smit, and N. P. Warner, *Differential equations for periods and flat coordinates in two dimensional topological matter theories*, Nuclear Phys. B **372** (1992), 87–112.

21. D. G. Markushevich, *Resolution of singularities (toric method)*, appendix to: D. G. Markushevich, M. A. Olshanetsky, and A. M. Perelomov, *Description of a class of superstring compactifications related to semi-simple Lie algebras*, Comm. Math. Phys. **111** (1987), 247–274.

22. D. R. Morrison, *Mirror symmetry and rational curves on quintic threefolds: A guide for mathematicians*, J. Amer. Math. Soc. **6** (1993), 223–247.

23. S.-S. Roan, *On the generalization of Kummer surfaces*, J. Differential Geom. **30** (1989), 523–537.

24. ______ , *The mirror of Calabi-Yau orbifold*, Internat. J. Math. **2** (1991), 439–455.

25. H. C. H. Schubert, *Kalkül der abzählenden Geometrie*, 1879, reprinted with an introduction by S. Kleiman, Springer-Verlag, 1979.

26. J. Steenbrink, *Intersection form for quasi-homogeneous singularities*, Compositio Math. **34** (1977), 211–223.

27. S. T. Yau, *On Calabi's conjecture and some new results in algebraic geometry*, Proc. Nat. Acad. Sci. U.S.A. **74** (1977), 1798–1799.

28. Y. Yu, *An improvement on the Gröbner basis algorithm*, unpublished.

AMS/IP Studies in Advanced Mathematics
Volume 9, 1998

Kähler Classes on Calabi–Yau threefolds – an informal survey

P.M.H. Wilson

Department of Pure Mathematics, University of Cambridge,

16 Mill Lane, Cambridge CB2 1SB, UK

Introduction

In this article I shall survey some of the theory developed in [11,12] concerning the Kähler classes on Calabi–Yau threefolds.[1] By a Calabi–Yau threefold I shall mean a complex 3-dimensional projective manifold X with trivial line bundle K_X (i.e. there exists a nowhere vanishing holomorphic 3-form) and with $h^1(\mathcal{O}_X) = 0$ (there do not exist holomorphic 1-forms on X). By Serre duality, this implies that $h^2(\mathcal{O}_X) = 0$ (there are no holomorphic 2-forms). It is precisely this kind of threefold which slips through the net with standard birational classification techniques from algebraic geometry [10]. If X does not have finite fundamental group $\pi_1(X)$, we do know that X has a finite unramified cover which is either a complex torus or the product of an elliptic curve and a K3 surface [2], and so usually we assume that we are not in these cases (for which it might be noted that the euler number $e(X) = 0$). In terms of holonomy, this is just saying that the holonomy is the whole of $SU(3)$ rather than a subgroup.

From the point of view of superstrings, we are interested in Calabi–Yau threefolds together with a Calabi–Yau (i.e. Ricci flat Kähler) metric. By the famous theorem of Yau, for a given complex structure on X, such a metric corresponds to choosing the associated Kähler class in $\mathrm{H}^2(X, \mathbf{R})$. For a given complex structure we are therefore interested in the Kähler cone $\mathcal{K} \subset \mathrm{H}^2(X, \mathbf{R})$, the open cone of Kähler classes. This cone is also of interest to the algebraic geometer, and its closure $\overline{\mathcal{K}}$ consists of classes $D \in \mathrm{H}^2(X, \mathbf{R})$ with $D \cdot C \geq 0$ for all holomorphic curves C on X (where $D \cdot C$ represents the standard product induced from the integral pairing $\mathrm{H}^2(X, \mathbf{Z}) \times \mathrm{H}_2(X, \mathbf{Z}) \to \mathbf{Z}$, where C represents a class in $\mathrm{H}_2(X, \mathbf{Z})$; in terms of forms, this corresponds to integrating a 2-form over C). In the language of algebraic geometry such classes are known as *nef* classes, and $\overline{\mathcal{K}}$ is often

© 1998 American Mathematical Society
and International Press

referred to as the *nef cone*. It is the dual of the cone studied by Mori, the closed cone $\overline{NE}(X)$ generated by real convex combinations of curves on X. The theory of Mori tells us that the part of $\overline{NE}(X)$ on which the canonical class K_X is negative will be locally rational polyhedral. For the case we consider however, K_X is zero and so no information concerning $\overline{NE}(X)$ is obtained. Thus we see that neither Mori theory nor classification theory via holomorphic forms applies directly to Calabi–Yau threefolds.

The technique employed in the papers [11,12,19] is to exploit the cubic form $\mu : \mathrm{H}^2(X, \mathbf{Z}) \to \mathbf{Z}$ given by cup-product. On $\mathrm{H}^2(X, \mathbf{C}) = \mathrm{H}^{1,1}$, this form is just the topological (i.e. unquantized) Yukawa coupling. One indication of the mathematical significance of μ is the fact that for simply connected Calabi–Yau threefolds X, the diffeomorphism class is determined up to finitely many possibilities by the cubic form μ, the linear form $c_2 : \mathrm{H}^2(X, \mathbf{Z}) \to \mathbf{Z}$ defined by cup-product with the 2nd Chern class $c_2(X) \in \mathrm{H}^4(X, \mathbf{Z})$, and the middle cohomology $\mathrm{H}^3(X, \mathbf{Z})$. Indeed, if the middle cohomology has no torsion, the diffeomorphism class is determined uniquely by the above information. Usually, we shall just write $\mu(D)$ as D^3. As long as X does not have a complex torus as a finite unramified cover, we know that the linear form c_2 is non-trivial, and moreover induces a non-trivial linear form $\mathrm{H}^2(X, \mathbf{R}) \to \mathbf{R}$. For certain proofs this form will also play a vital role, although it seems to play a relatively minor role in most of the physics literature.[2]

In Section 1 we shall describe the boundary $\partial\overline{\mathcal{K}}$ of the cone $\overline{\mathcal{K}}$. This is of relevance to the question of Mirror Symmetry, since degenerating the complex structure of the mirror manifold $\check{X}$ should correspond to degenerating the Kähler structure on X (i.e. letting the Kähler class tend to the boundary of $\overline{\mathcal{K}}$). This is precisely what happens in the example by Aspinwall et al. in [1]. In Section 2, we shall apply the information obtained about $\partial\overline{\mathcal{K}}$ to the question of the existence of birational contraction morphisms on X. This is of relevance to the question of connecting moduli spaces of Calabi–Yau threefolds (by means of degenerations of complex and Kähler structures, see for instance [8,6,4,14,16]), and also to the existence of rational curves on X (worldsheet instantons in physics-speak). It is the rational curves which are involved when modifying the Yukawa coupling on $\mathrm{H}^{1,1}$ into the quantized Yukawa coupling, which under Mirror Symmetry it is hoped will correspond to the Yukawa coupling on $\mathrm{H}^{1,2}(\check{X})$ of the mirror $\check{X}$ (see [5,1]). We shall recover the fact that rational curves will always exist if $b_2(X)$ is large enough.[3] In Section 3, we study the question of whether the Kähler structure and the complex structure can (at least locally)

---

[2] This is no longer the case – see for instance [15]. For more discussion of c_2, the reader is referred to [21].

[3] $b_2 > 13$ will suffice [19].

be varied independently. This question translates into the question of whether the Kähler cone is invariant under small deformations, and this is shown to be the case if and only if none of the threefolds in question contains an elliptic *quasi-ruled surface* (by definition, this is either a $\mathbf{P}^1$-bundle over an elliptic curve, or a conic bundle over an elliptic curve, all of whose fibres are line pairs).

The aim of this survey is to present the results in such a way that they may be used by the non-specialist, and no detailed proofs are given. These may be found in the two papers [11,12]. Much of the research described was undertaken when the author was at Kyoto University, supported on a Fellowship from the Japan Society for the Promotion of Science. He wishes to thank the JSPS for this support. He also wishes to thank the organisers of the Mirror Symmetry Workshop at MSRI Berkeley for the invitation to talk on this work, this article being an expanded version of the talk given.

1. The boundary of the Kähler cone

Using the fact that $\mathrm{h}^1(\mathcal{O}_X) = \mathrm{h}^2(\mathcal{O}_X) = 0$ for a Calabi–Yau threefold X, it is a standard result that $\mathrm{H}^2(X, \mathbf{Z}) \cong \mathrm{Pic}(X)$, the group of isomorphism classes of line bundles on X. Given a line bundle $\mathcal{L}$ with $\mathrm{h}^0(X, \mathcal{L}) \neq 0$, we can choose a basis of sections $s_0, \ldots, s_N$ over $\mathbf{C}$ and define a rational map $\phi_{\mathcal{L}} : X - \to \mathbf{P}^N$ by

$$\phi_{\mathcal{L}}(x) = (s_0(x) : s_1(x) : \ldots : s_N(x)) \in \mathbf{P}^N.$$

The map is well-defined, except perhaps at points at which all the s_i vanish. More generally, if $\mathrm{h}^0(X, \mathcal{L}^{\otimes n}) \neq 0$, we have a corresponding map which, for ease of notation, I shall denote $\phi_{n\mathcal{L}}$ (this is consistent with the algebraic geometers' habit of thinking in terms of divisor classes rather than line bundles). We say that $\mathcal{L}$ is *very ample* if $\phi_{\mathcal{L}}$ is an embedding with $\mathcal{L}$ the pullback of the hyperplane line bundle on $\mathbf{P}^N$; we say that $\mathcal{L}$ is *ample* if $\mathcal{L}^{\otimes n}$ is very ample for some $n > 0$.

A cohomology class in $\mathrm{H}^2(X, \mathbf{Q})$ is called ample if a multiple in $\mathrm{H}^2(X, \mathbf{Z})$ corresponds to an ample line bundle. It follows from the Kodaira Embedding theorem that the ample classes are precisely the rational Kähler classes, and so $\mathcal{K}$ is the cone generated by all ample classes. It follows from this that its closure $\overline{\mathcal{K}}$ is the nef cone as described in the Introduction, and that $D^3 \geq 0$ for all $D \in \overline{\mathcal{K}}$. A rather more difficult fact is that the form c_2 is non-negative on $\overline{\mathcal{K}}$, i.e. $D \cdot c_2 \geq 0$ for all $D \in \overline{\mathcal{K}}$.

Inside the $\rho = b_2(X)$ dimensional real vector space $\mathrm{H}^2(X, \mathbf{R})$ we have another cone of importance, the *cubic cone*

$$W^* = \{D \in \mathrm{H}^2(X, \mathbf{R}) \; ; \; D^3 = 0\}.$$

Associated to W^* we have a corresponding real hypersurface $W \subset \mathbf{P}^{\rho-1}(\mathbf{R})$, and much of the geometry of X is reflected in the geometry and arithmetic of W. There are certain weak geometric restrictions on W (it cannot be a real cone and cannot have a real line of singularities for instance). We should note that the classes $D \in \mathrm{H}^2(X, \mathbf{R})$ for which the class $D^2 \in \mathrm{H}^4(X, \mathbf{R})$ is trivial correspond precisely to the singularities of W^* (and hence if $D \neq 0$ to a singularity of W). This statement is essentially trivial, since D being a singularity of W^* is just saying that for all $L \in \mathrm{H}^2(X, \mathbf{R})$, the line $\{D + xL \; ; \; x \in \mathbf{R}\}$ has a multiple intersection with W^* at $x = 0$, i.e. that $D^2 \cdot L = 0$.

In the light of these observations, we might for instance (when $\rho = 3$) have our two cones $\overline{\mathcal{K}}$ and W^* in $\mathrm{H}^2(X, \mathbf{R})$ as in Figure 1. In order not to complicate the picture, the hyperplane $c_2 = 0$ has been omitted, but in should be born in mind that $\overline{\mathcal{K}}$ is on the positive side of this hyperplane (although possibly touching it).

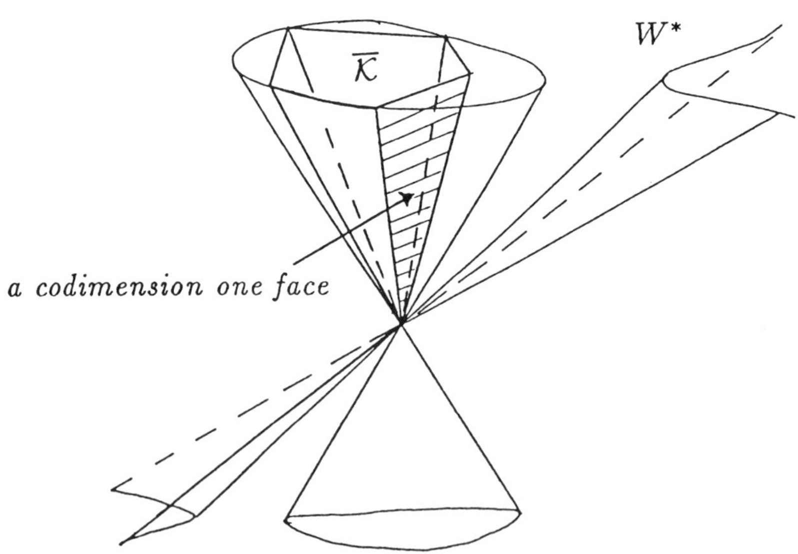

Figure 1.

In [11] and [12, Sect. 1], there is a proof of the following two facts concerning the boundary $\partial\overline{\mathcal{K}}$ of $\overline{\mathcal{K}}$.

Fact 1. *The cone $\overline{\mathcal{K}}$ is locally rational polyhedral away from W^*, the codimension one faces corresponding to primitive birational contractions on X.*

A birational morphism of normal varieties will be called *primitive* if it cannot be factored further into birational morphisms of normal varieties. Given a codimension one face of $\overline{\mathcal{K}}$ not contained in W^*, the corresponding contraction morphism is obtained as ϕ_{nD} for D any integral class in the interior of the face and suitable large integer n. The reader is referred to [3] for an example where $\overline{\mathcal{K}}$ has infinitely many codimension one faces (and incidentally for which the hyperplane $c_2 = 0$ does touch $\overline{\mathcal{K}}$). It is not known whether there do exist examples (with finite fundamental group) with $\partial\overline{\mathcal{K}}$ containing an open subset of rays in W^* (as shown in Figure 1). As will be seen in Section 2, it is an important question whether $\partial\overline{\mathcal{K}}$ can be the cone on a component of W.

Fact 2. *Non-singular rational points of W^* which are on $\overline{\mathcal{K}}$ but not on the linear space defined by $c_2 = 0$ give rise to elliptic fibre space structures on X, i.e. morphisms $\phi : X \to S$ with S a normal surface and generic fibre an elliptic curve. If the rational point in question is $D \in \mathrm{H}^2(X, \mathbf{Q})$, then the morphism ϕ is just ϕ_{nD} for some appropriate large n.*

The reader should observe that the statements of the two facts differ in an important respect. Fact 1 includes a statement about the existence of many rational points on $\partial\overline{\mathcal{K}} \setminus W^*$, whilst Fact 2 is assuming the existence of an appropriate rational point. The boundary $\partial\overline{\mathcal{K}}$ and hence $\overline{\mathcal{K}}$ is of course determined by W^* together with the codimension one faces described in Fact 1.

We now describe the primitive contraction morphisms (corresponding to the codimension one faces not contained in W^*) which can occur. These are divided into three types, all of which do occur in examples.

Type I.   This is the case when the contraction morphism $\phi : X \to \bar{X}$ contracts only a finite number of curves, these curves all being isomorphic to $\mathbf{P}^1$. This case has been studied by a number of authors – see [12] for references.

Type II.   The contraction morphism $\phi : X \to \bar{X}$ contracts an irreducible surface E down to a singular point and is otherwise an isomorphism. The reader is referred to [7,17,22] for a discussion of this case; the surface E will be an example of what are known as *generalized Del Pezzo surfaces*.

Type III.   The contraction $\phi : X \to \bar{X}$ contracts an irreducible surface E to a curve C of singularities. Here it can be shown [12, §2], [22, §3] that C is a smooth curve and that E is a conic bundle over C.[4] A fibre of the morphism $E \to C$ is therefore isomorphic to $\mathbf{P}^1$, or to two lines meeting at a point, or to a line taken with multiplicity two. If the generic fibre is isomorphic to $\mathbf{P}^1$, the only possible singularities of E are Du Val singularities (i.e. rational double points) of type A_n (locally analytically of the form $x^2 + y^2 + z^{n+1} = 0$) at the point where distinct components of a singular fibre meet, or a singularity of type D_n (locally analytically of the form $x^2 + y^2 z + z^{n-1}$, $n \geq 2$) appearing on a double fibre (including, when $n = 2$ the case of two A_1 singularities). If the generic fibre of $E \to C$ is isomorphic to a line pair, then E is non-normal with a curve of double points, locally analytically of the form $x^2 = y^2$, except for *pinch points* (locally analytically of the form $x^2 = y^2 z$) when the fibre degenerates into a double line [22, §3]. If all the fibres are line pairs, or $E \to C$ is a $\mathbf{P}^1$-bundle, then we say that E is *quasi-ruled* over C – in the former case, the normalization $\tilde{E}$ is then a a $\mathbf{P}^1$-bundle over an unramified cover $\tilde{C}$ of C.

As indicated in the Introduction, the mildly singular variety $\bar{X}$ which results from a primitive contraction may be viewed in terms of degenerating the Kähler class to the corresponding codimension one face of $\overline{\mathcal{K}}$. An obvious question now arises : When can $\bar{X}$ be deformed to a smooth projective threefold? The reader is referred to the paper [9] for a discussion of this in the case when for Type I contractions the threefold $\bar{X}$ has only ordinary double points as singularities (i.e. the curves Z contracted are disjoint and all have normal bundle of the form $\mathcal{O}_Z(-1) \oplus \mathcal{O}_Z(-1)$).[5]

If $\bar{X}$ has a smooth projective deformation X_1, then we have connected the moduli spaces of the Calabi–Yau threefolds X and X_1 by means of degenerating the Kähler structure on X and the complex structure on X_1. The speculation that the moduli spaces of any two Calabi–Yau threefolds might be connected by a sequence of such operations appeared first in [8], and has been investigated in more detail in the papers [6,4,14,16]. In the next Section, we shall consider the question of whether $\partial\overline{\mathcal{K}}$ contains points not on W^*, and hence whether there exist codimension one faces of $\overline{\mathcal{K}}$ giving rise to birational contraction morphisms of the types described above.

----

[4]   The proof of this statement was in fact only completed in [22].

[5]   Since the 1st edition, the question has been answered completely for all primitive contractions [20,17,18].

2. Existence of contraction maps and rational curves

It is an immediate consequence of Fact 1 from Section 1 that if $\partial\overline{\mathcal{K}}$ is not contained in W^*, then there exists a primitive contraction morphism $\phi : X \to \bar{X}$ of one of the three types described. For each type, it is easy to see that X contains rational curves (in the exceptional locus). Thus X contains rational curves unless $\partial\overline{\mathcal{K}} \subset W^*$.

Let us therefore suppose that $\partial\overline{\mathcal{K}}$ is contained in W^*. Suppose also that X has finite fundamental group, and so in particular $c_2(X)$ is non-trivial. We consider the case when the hypersurface $W \subset \mathbf{P}^{\rho-1}(\mathbf{R})$ satisfies the arithmetical property that its rational points are dense (essentially what the number theorists call weak approximation). Under these assumptions, it is not difficult to show (see [11]) that there are rational points satisfying the conditions of Fact 2 from Section 1. There therefore exists an elliptic fibre space structure $\phi : X \to S$ on X. Not all the fibres of ϕ can be smooth elliptic curves, since otherwise the j-invariant would provide a holomorphic map from the projective surface S to $\mathbf{C}$, which would therefore have to be constant. In this circumstance, X has a finite unramified cover which is $\tilde{S} \times E$, where E is the elliptic curve obtained as any fibre of ϕ, and $\tilde{S}$ is a K3 or abelian surface. This is ruled out by the assumption on the fundamental group. Contained in any singular fibre of ϕ however we have a rational curve. Putting these facts together, we deduce:

Proposition. *If X has finite fundamental group and the hypersurface $W \subset \mathbf{P}^{\rho-1}(\mathbf{R})$ satisfies the condition that its rational points are dense, then there exists a rational curve on X.*

The arithmetical condition is however satisfied for $\rho(X) = b_2(X)$ large enough by results from analytic number theory (the Hardy-Littlewood circle method). In [11], the author slightly adapts a proof of Davenport to show that when W arises as above from the cubic form on $\mathrm{H}^2(X, \mathbf{Z})$, the arithmetical condition is satisfied provided $\rho(X) > 19$. From this it follows that any Calabi–Yau threefold X with $b_2(X) > 19$ must contain a rational curve.[6] There is no suggestion that this is a sensible bound on $\rho(X)$, and the author has at least a heuristic argument why the general method should work with a very much smaller bound. It should however be noted that the argument given will never work for say $\rho = 1$ (where as in the case of the quintic hypersurface, we still expect rational curves), and that

---

[6] Using an improved argument [19], the number 19 may be replaced by 13.

a completely new idea will be needed if the conjecture that all Calabi–Yau threefolds with non-trivial c_2 contain rational curves is to be proved. What the methods described here are essentially doing is detecting those rational curves which can be contracted projectively (as opposed to being contractible in the context of analytic spaces).

With a little more work, we can in fact prove a stronger result than the one stated above [11,19].[7]

Theorem. *If X is a Calabi–Yau threefold with $\rho(X) > 13$, then there exists a birational contraction morphism $\phi : X \to \bar{X}$ with $\bar{X}$ a normal projective threefold with Picard number $\rho(\bar{X}) \leq 13$.*

In the statement of the theorem, the Picard number $\rho(\bar{X})$ is defined to be the rank of $\mathrm{Pic}(\bar{X})$; for singular $\bar{X}$, this will not in general be the same as $b_2(\bar{X})$. It suffices to add however that the exceptional locus of ϕ will contain rational curves, and so the existence of rational curves follows as a corollary. The observation that all known examples of Calabi–Yau threefolds with largish $b_2(X)$ arose as desingularizations of mildly singular threefolds was the starting point of the investigation contained in [11]. In order to apply induction on the Picard number, one works with what are called *Calabi–Yau models* in [11], which by definition are normal projective varieties with a (smooth) Calabi–Yau threefold as a desingularization. Such singular Calabi–Yau threefolds seem to be playing an increasingly important role in the physics literature.

3. Deformation properties of the Kähler cone

The information obtained in Section 1 on the boundary of the Kähler cone of a Calabi–Yau threefold X also enables us to answer the question (raised in [4]) as to whether we can (at least locally) independently vary the complex and Kähler structures on X. This question translates into the question of whether the Kähler cone is locally invariant under small deformations (and so whether the parameter space we are interested in is locally a product).

For a Calabi–Yau threefold X, it follows from work of Tian, Todorov and others, that there is a family of deformations (the *versal* or *Kuranishi* family) $\pi : \mathcal{X} \to B$, with B a

---

[7] Here, we have incorporated the improvement from [19] into the text.

polydisc around the origin in $\mathrm{H}^1(T_X)$ (the fibres being denoted X_b, with $X_0 = X$), from which any small deformation of X may be obtained as a pullback. We need therefore ask about the invariance of the Kähler cone just for this family.

In passing I should point out that we are implicitly identifying the cohomology groups $\mathrm{H}^2(X_b, \mathbf{Z}) \cong \mathrm{H}^2(\mathcal{X}, \mathbf{Z}) \cong \mathrm{H}^2(X_0, \mathbf{Z})$ for all $b \in B$, and so the Kähler cones $\mathcal{K}(b)$ may be considered in a fixed vector space $\mathrm{H}^2(X, \mathbf{Z})$. The Kähler cone will be locally invariant in the family if we can show :

(∗)   For any rational D in the interior of a codimension face (not contained in W^*) of $\overline{\mathcal{K}}$, D is in $\partial\overline{\mathcal{K}}(b)$ for all b in some neighbourhood of zero.

It is a standard piece of mathematics that if D is nef on X_0 then it will be nef on X_b for all b in some neighbourhood of zero, and so the Kähler cone can only become larger under a small deformation. It is clear that any $D \in \mathrm{H}^2(X, \mathbf{R})$ on $W^* \cap \partial\overline{\mathcal{K}}$ must remain on $W^* \cap \partial\overline{\mathcal{K}}(b)$ under a small deformation (since $D^3 = 0$), and so we need only check whether the same remains true for the D on $\partial\overline{\mathcal{K}} \setminus W^*$. Since the codimension one faces of $\overline{\mathcal{K}}$ are rational, we need only answer the question for rational D in the interior of codimension one faces.

We saw in Section 1 that for an appropriate large n, we had a primitive contraction morphism $\phi_{nD} : X \to \bar{X}$, which was one of the three types described. It turns out that, except in the case when ϕ_{nD} is contracting a quasi-ruled surface over an elliptic curve along its fibres, part of the exceptional locus of the contraction always deforms sideways under small deformations. Therefore, on a neighbouring fibre X_b, we can always find a curve Z_b with $D \cdot Z_b = 0$, and in particular D is not an ample class on X_b. Therefore, except when the codimension one face corresponds to the contraction of such an *elliptic* quasi-ruled surface, we see that (∗) always holds.

When X does contain an elliptic quasi-ruled surface E, it is easily verified that there is a primitive contraction of E and a corresponding face of the Kähler cone. A slightly technical argument is needed to show that in this case the condition (∗) cannot hold for a general small deformation.

Details of the above arguments may be found in [12][8] – the resulting theorem may be

---

[8]   One does however need the result proved in [22] that the exceptional locus of a Type III contraction is a conic bundle over a smooth curve.

stated as follows.

Theorem. *The Kähler cone is invariant in the Kuranishi family $\pi : \mathcal{X} \to B$ of $X = X_0$ if and only if none of the threefolds X_b contains an elliptic quasi-ruled surface. More generally, it is invariant over the dense subset of $b \in B$ for which X_b contains no elliptic quasi-ruled surfaces, this subset being the complement of at most countably many codimension one submanifolds.*

In many ways it seems to the author that perhaps the physicists should just ignore elliptic quasi-ruled surfaces on Calabi–Yau threefolds – the important cone to consider would seem to be the Kähler cone at the *general point of the moduli space,* and this we see from the Theorem is a well-defined concept.

The elliptic quasi-ruled surfaces have the property that they disappear without trace under a general small deformation. Putting it another way, if $\phi : X \to \bar{X}$ denotes the contraction of an elliptic quasi-ruled surface E, then the smoothings of X and $\bar{X}$ coincide. Thus degenerating the Kähler class to the face corresponding to E does not enable us to connect up with a different family of threefolds.

In passing, I also mention that if X contains a ruled surface E over a smooth curve C of genus $g > 1$, a first order calculation performed in Section 4 of [12] suggests that E will correspond to $2g - 2$ rational curves on a general small deformation of X; what can be proved here is that this is true for a general almost complex deformation of X, since the Gromov–Witten invariants can be calculated using a cobordism argument of Ruan.[9]

Mathematically, the case of X containing elliptic quasi-ruled surfaces is of interest. Suppose X is a Calabi–Yau threefold containing such surfaces, say $E_1, E_2, \ldots$ (with the corresponding fibres having classes $l_1, l_2, \ldots$ in $\mathrm{H}^4(X, \mathbf{R})$). In the vector space $\mathrm{H}^2(X, \mathbf{R})$ we have what might be termed the *general Kähler Cone* $\mathcal{K}_1$, viz the Kähler cone of a general small deformation (well-defined by the Theorem). The cone $\mathcal{K}_1$ is by definition invariant under deformation, and so is a candidate for distinguishing diffeomorphic Calabi–Yau threefolds which are not deformation equivalent.

In the above situation, the actual Kähler cone $\mathcal{K}_0$ of X is the subcone of $\mathcal{K}_1$ defined by the extra inequalities $D \cdot l_i > 0$ for $i = 1, 2, \ldots$ (see [13]). It turns out that the theory of

---

[9]   The author can now calculate the Gromov–Witten invariants associated to an arbitrary Type III contraction [24], but things are less clear in the case of a Type II contraction.

elliptic ruled surfaces on Calabi–Yau threefolds is very analogous to the theory of rational curves of self-intersection -2 on a K3 surface, and that one has a theory of flops on families of Calabi–Yau threefolds inducing reflections on $\mathrm{H}^2(X, \mathbf{R})$ (see [13] for a detailed account of this theory).[10] There is one respect however in which there is a difference from the K3 case. With K3 surfaces, there is always a small deformation containing a rational curve of self-intersection -2. For Calabi–Yau threefolds X, it will be quite rare for there to be any deformation which contains an elliptic ruled surface. This is because an elliptic ruled surface on some deformation will determine a rational point P on the hypersurface $W \subset \mathbf{P}^{\rho-1}(\mathbf{R})$ which lies on the hyperplane $c_2 = 0$ and such that the tangent hyperplane to W at P intersects W in a *cubic cone* (such points are usually known as *Eckardt* points). This property is deformation invariant, and it will be unusual for such points of W to occur. I conclude however with a simple example to show that the situation described is not impossible.

Example. We let $\bar{X}$ be the normal projective threefold

$$\begin{pmatrix} \mathbf{P}^2 & \| & 2 & 0 & 1 \\ \mathbf{P}^2 & \| & 0 & 2 & 1 \\ \mathbf{P}(1,1,2) & \| & 0 & 0 & 4 \end{pmatrix}.$$

The notation here indicates a complete intersection of three hypersurfaces in the product $\mathbf{P}^2 \times \mathbf{P}^2 \times \mathbf{P}(1,1,2)$, the hypersurfaces having degrees (in each set of coordinates) $(2,0,0)$, respectively $(0,2,0)$, respectively $(1,1,4)$. It is easily verified that $\bar{X}$ has an elliptic curve C of A_1 singularities; when we blow C up, we obtain a Calabi–Yau threefold X with $b_2(X) = 4$ and $\mathrm{h}^{1,2}(X) = 68$, and containing a smooth ruled surface over C.

Both $\bar{X}$ and X have a small deformation X_1 which is a non-singular hypersurface in $\mathbf{P}^1 \times \mathbf{P}^1 \times \mathbf{P}^1 \times \mathbf{P}^1$ defined by a polynomial which is homogeneous of degree two in each set of variables. This follows since both the quadratic cone $\mathbf{P}(1,1,2)$ and its minimal desingularization F_2 have $\mathbf{P}^1 \times \mathbf{P}^1$ as a small deformation. Thus $\bar{X}$ and X both have as a small deformation

$$\begin{pmatrix} \mathbf{P}^2 & \| & 2 & 0 & 1 \\ \mathbf{P}^2 & \| & 0 & 2 & 1 \\ \mathbf{P}^1 & \| & 0 & 0 & 2 \\ \mathbf{P}^1 & \| & 0 & 0 & 2 \end{pmatrix}$$

---

[10] Similar statements are true for elliptic quasi-ruled surfaces – see final Remark in [23].

which is isomorphic to the one claimed.

The closure of the Kähler cone $\overline{\mathcal{K}}$ on X_1 is generated by the divisors D_i $(i = 1, \ldots, 4)$ obtained by pulling back a point from one of the $\mathbf{P}^1$ factors and restricting to X_1. The cubic form on X_1 (and hence on X) is therefore just a multiple of

$$x_1 x_2 x_3 + x_2 x_3 x_4 + x_3 x_4 x_1 + x_4 x_1 x_2 \ .$$

The corresponding hypersurface $W \subset \mathbf{P}^{\rho-1}(\mathbf{R})$ is a cubic surface with four nodes at the points corresponding to the D_i.

If one considers the deformation of F_2 to $\mathbf{P}^1 \times \mathbf{P}^1$, one observes that one of the rulings of $\mathbf{P}^1 \times \mathbf{P}^1$ corresponds to the ruling on F_2, and therefore represents a nef class, whilst the other ruling corresponds to a non-nef class on F_2. Thus three of our classes, say D_1, D_2, D_3, remain nef on $X = X_0$, whilst the remaining one D_4 is not nef on X. The closure of the Kähler cone on X is therefore the cone on the tetrahedron generated by D_1, D_2, D_3 and say $D_3 + D_4$, and under a general small deformation this jumps to the cone on the tetrahedron generated by all four D_i.

References

[1]   P.S. Aspinwall, C.A. Lütken, G.G. Ross, *Construction and couplings on mirror manifolds*, Phys. Letters B **241** (1990) 373-380.

[2]   A. Beauville, *Variétés Kähleriennes dont la première classe de Chern est nulle*, J. Differ. Geom. **18** (1983) 755-782.

[3]   C. Borcea, *On desingularized Horrocks–Mumford quintics*, J. Reine Angew. Math. **421** (1991) 23-41.

[4]   P. Candelas, P.S. Green, T. Hübsch, *Rolling among Calabi–Yau Vacua*, Nuclear Phys. B **330** (1990) 49-102.

[5]   P. Candelas, X.C. de la Ossa, P.S. Green, L. Parkes, *A pair of Calabi–Yau manifolds as an exactly soluble superconformal theory*, Nuclear Phys. B **359** (1991) 21-74.

[6]   P.S. Green, T. Hübsch, *Connecting moduli spaces of Calabi–Yau threefolds*, Comm. Math. Phys. **119** (1988) 431-441.

[7]   M. Reid, *Canonical 3-folds.* In : A. Beauville, (ed.) Géométrie algébrique, Angers 1979, pp. 273-310. Sijthoff and Noordhoff, The Netherlands 1990.

[8]   M. Reid, *The Moduli Space of 3-folds with K=0 may nevertheless be irreducible,* Math. Ann. **278** (1987) 329-334.

[9]   G. Tian, *Smoothing 3-folds with trivial canonical bundle and ordinary double points.* In : S.-T. Yau (ed.) Essays on Mirror Manifolds, pp 458-479. International Press. Hong Kong 1992.

[10]   P.M.H. Wilson, *Towards birational classification of algebraic varieties,* Bull. Lond. Math. Soc. **19** (1987) 1-48.

[11]   P.M.H. Wilson, *Calabi–Yau manifolds with large Picard number,* Invent. math. **98** (1989) 139-155.

[12]   P.M.H. Wilson, *The Kähler Cone on Calabi–Yau threefolds,* Invent. math. **107** (1992) 561-583. *Erratum,* Invent. math. **114** (1993) 231-233.

[13]   P.M.H. Wilson, *Elliptic ruled surfaces on Calabi–Yau threefolds,* Math. Proc. Cam. Phil. Soc. **112** (1992) 45-52.

References added in 2nd edition

[14]   A.C. Avram, P. Candelas, D. Jancic, M. Mandelberg, *On the connectedness of the moduli space of Calabi–Yau manifolds,* Nuclear Phys. B **465** (1996) 458-472.

[15]   M. Bershadsky, C. Cecotti, H. Ooguri, C. Vafa, *Holomorphic anomalies in topological field theories,* Nuclear Phys. B **405** (1993) 279-304.

[16]   T.-M. Chiang, B.R. Greene, M. Gross, Y. Kantor, *Black hole condensation and the web of Calabi–Yau manifolds,* Nuclear Phys. B Proc. Suppl. **46** (1996) 82-95.

[17]   M. Gross, *Deforming Calabi–Yau threefolds,* Math. Ann. (1997) to appear.

[18]   M. Gross, *Primitive Calabi–Yau threefolds,* J. Diff. Geom. (1997) to appear.

[19]   D.R. Heath-Brown, P.M.H. Wilson, *Calabi–Yau threefolds with $\rho > 13$,* Math. Ann. **294** (1992) 49-57.

[20]   Y. Namikawa, J.H.M Steenbrink, *Global smoothing of Calabi–Yau threefolds,* Invent. math. **122** (1995) 403-419.

[21]   P.M.H. Wilson, *The role of c_2 in Calabi–Yau classification - a preliminary survey.* In: B.R. Greene, S.-T. Yau (eds.) Essays on Mirror Symmetry II, pp. 370-381. International Press, Hong Kong and AMS, Providence 1996.

[22]   P.M.H. Wilson, *Symplectic Deformations of Calabi–Yau threefolds*, J. Diff. Geom. **45** (1997) to appear.

[23]   P.M.H. Wilson, *The existence of elliptic fibre space structures on Calabi–Yau threefolds II*, Math. Proc. Cam. Phil. Soc. **122** (1997) to appear.

[24]   P.M.H. Wilson, *Flops, Type III contractions and Gromov–Witten invariants on Calabi–Yau threefolds*, in preparation (1997).

AMS/IP Studies in Advanced Mathematics
Volume **9**, 1998

Automorphic Functions
and Special Kähler Geometry[†]

S. Ferrara[1)], C. Kounnas[2)], D. Lüst[1)] and F. Zwirner[1),3)]

1) CERN, CH-1211 Geneva 23, Switzerland

2) Ecole Normale Supérieure, 75231 Paris Cedex 05, France

3) On leave from INFN, Sezione di Padova, Italy

Abstract

In the context of the geometry of the moduli space of Calabi–Yau manifolds, we define automorphic functions in terms of norms of holomorphic sections of a line bundle (whose first Chern class $c_1(L) = [J]$). This procedure naturally generalizes the Dedekind funtion and other modular forms of multitoroidal compactifications. The relevance of these functions for the computation of physical quantities in heterotic string compactifications will be addressed.

[†]Work supported in part by the Department of Energy of USA under the contract DOE-AT03-88ER40 384, TASK E, and by the Grant-in-Aid for Scientific Research (International Scientific Research Program) of the Japanese Ministry of Education, Science and Culture (No. 02045009).

© 1998 American Mathematical Society
and International Press

We will report on some recent results which have been obtained on the geometrical structure of the moduli space of Calabi–Yau threefolds [1] and their application to compute couplings for effective Lagrangians of superstrings compactified [2] on Calabi–Yau spaces corresponding to $c = 9$ $(2,2)$ superconformal field theories [3]. From the lesson learned from toroidal and orbifold compactifications [4] we know that the moduli space of superconformal field theories is not a smooth manifold but rather a target space of superconformal field theories is not a smooth manifold but rather a target space orbifold, with fixed points. Specifically, the moduli space is obtained by modding out a manifold by some discrete (possibly infinite-dimensional) group Γ, called the duality group [5-8] which has a (stringy) quantum origin since it is related to the invariance of the underlying two-dimensional field theory under discrete transformations acting on the moduli fields.

The effective action for the moduli fields which is obtained after integrating out all massive string excitations has to obey the stringy duality symmetries. Therefore one expects that the effective action is determined by the automorphic functions of the duality group Γ. As an example let us consider the compactification of the ten-dimensional heterotic string on a six-dimensional orbifold which is based on the direct product of three two-dimensional tori. Then the compactification is generically described by three moduli T_i $(i = 1, 2, 3)$ corresponding to the sizes of the three subtori (plus their axionic partners), and the moduli space is locally given by $\mathcal{M} = [SU(1,1)/U(1)]^3$. The corresponding duality group is given (up to permutations) by the product of three modular groups: $\Gamma = [PSL(2, \mathbf{Z})]^3$. As shown in [6] the modular invariance of the $d = 4$, $N = 1$ effective supergravity action establishes a connection between the holomorphic superpotential and the theory of modular functions. Specifically, identifying the three moduli, i.e. $T_1 = T_2 = T_3 = T$, the Kähler function for T describes a $SU(1,1)/U(1) \simeq SL(2, \mathbf{R})/U(1)$ sigma-model. Any non-trivial superpotential would break the continuous $SL(2, \mathbf{R})$ symmetry, but the quantum string symmetry is only the discrete duality group $SL(2, \mathbf{Z})$. Since under $SL(2, \mathbf{Z})$ the Kähler potential transforms as $K \to K + \Lambda(T) + \overline{\Lambda}(\overline{T})$, any possible non-perturbative superpotential has to transform as $W \to We^{-\Lambda(T)}$ in order to have a modular invariant G-function, $G = K + \log |W|^2$. The constraint on W is satisfied by the

Dedekind eta function. i.e. the choice $W(T) \propto \eta(T)^{-6}$ respects the $SL(2, \mathbf{Z})$ duality invariance. In [7] it was discussed that exactly this type of non-perturbative superpotential emerges in the context of gaugino condensation.

The holomorphic automorphic functions are also relevant for the topological target space free energy F of string compactifications [9]. This quantity corresponds to the generating functional of the effective action after integrating out all the massive modes. F always appears when computing amplitudes with light states as external legs, and all massive states entering in loops. Therefore the free energy F must be a target-space duality invariant function of the moduli fields. We call the free energy a topological quantity because the path integral should be performed only over those massive states which are due to the compactification, like Kaluza–Klein and winding modes, but not over states which are also present in the uncompactified theory, like all massive oscillator states. Integration over the massive fermions leads to

$$F = \det \frac{|W_{ij}|^2}{Y} \ .\tag{1}$$

The determinant of the chiral masses, $\det W_{ij}$, is the second derivative of the superpotential of the massive states and must be an automorphic function of the duality group Γ. The factor Y in (1) is related to the non-canonical kinetic energy of the massive fields.

For toroidal/orbifold compactifications F can be explicitly computed from the effective field theory of the massive momentum and winding states using the results of [10]. For the mentioned case of orbifold compactifications with three moduli T_i the free energy can be written as a sum over six completely unconstrained integers m_A, n_A (two for each subsetor):

$$F\bigg|_{\mathrm{reg.}} = \left[\sum_{i=1}^{3}\sum_{m_i,n_i} \log \frac{|(m_i + in_iT_i)|^2}{(T_i + \overline{T}_i)}\right]_{\mathrm{reg.}} = \sum_{i=1}^{3} \log |\eta(T_i)|^4 (T_i + \overline{T}_i) \ .\tag{2}$$

This expression has already appeared in different places in the literature. For instance, the authors of ref. [11] obtained exactly this expression when calculating one-loop corrections to gauge coupling constants. Also in the recent work of ref. [12], F is the free energy of compactified $N = 2$ critical strings. Thus, the form of the orbifold free energy is completely analogous to the form of the G-function itself: $e^G = \dfrac{|W|^2}{Y}$.

Now let us discuss general (2,2) Calabi–Yau compactifications and consider the following function of the moduli fields T_i $(i = 1, \ldots, n)$:

$$\frac{|\Delta(T_i)|^2}{Y} \, , \tag{3}$$

Y is given by the Kähler potential of the moduli, $Y(T_i, \overline{T}_i) = e^{-K(T_i, \overline{T}_i)}$, where $\Delta(T_i)$ is a holomorphic function of the moduli fields T_i. Thus, $\Delta(T_i)$ provides an expression for the non-perturbative, purely moduli dependent superpotential, and we will also give arguments for the conjecture that $\Delta(T_i)$ is again related to the determinant of all chiral masses in Calabi–Yau compactification spectrum. $\Delta(T_i)$ is an automorphic function of the discrete target-space duality group Γ. More precisely, it is inversely proportional to some power of the cusp form (the generalization of $\eta(T)^{24}$) of the duality group, since only for infinite (zero) radius of the internal space does the contribution of the Kaluza-Klein (winding) modes create a pole for Δ.

More generally, $\Delta(T_i)$ can be regarded as a holomorphic section of a line bundle over the moduli space $\mathcal{M}$. In particular, since the target space duality transformations $\Gamma : T_i \to \widetilde{T}_i(T_j)$ act as Kähler transformations on the Kähler potential $K = -\log Y$,

$$Y \to Y |e^{-\Lambda(T_i)}|^2 \quad \text{for} \quad T_i \to \widetilde{T}_i \, , \tag{4}$$

the holomorphic section Δ has to transform as

$$\Delta(T_i) \to \Delta(T_i) e^{-\Lambda(T_i)} \, . \tag{5}$$

The duality invariant norm of the holomorphic section is defined by the Kähler potential:

$$\|\Delta(T_i)\|^2 = \Delta(T_i) e^K \overline{\Delta}(\overline{T}_i) \, . \tag{6}$$

Note that it is impossible to write the modular invariant function $\|\Delta\|^2$ as a perfect holomorphic square. The non-holomorphic part, namely e^K, describes a holomorphic anomaly and is related to a Kähler anomaly of the moduli space $\mathcal{M}$. More precisely, the non-holomorphicity of $e^{-F}|_{\text{reg.}}$ is due to a $U(1)$ anomaly (see also the discussion at the end of this section). This universal $U(1)$ is just related to the $U(1)$ symmetry of the internal (2,2) superconformal algebra. For the case of $\mathcal{M} = \dfrac{SU(1,1)}{U(1)}$ with modular group $SL(2,\mathbf{Z})$, this anomaly is given by Quillen's holomorphic anomaly, which means that the Eisenstein function $G_2 = \sum_{m,n} (m + inT)^{-2}$ can either be regularized in a holomorphic way, which however destroys the modular covariance (this regularization leads to $G_2(T)|_{\text{reg.}} = -4\pi \dfrac{\eta'(T)}{\eta(T)}$), or in a non-holomorphic way, which leads to an automorphic function with modular weight two (namely to $\hat{G}_2(\overline{T},T)|_{\text{reg.}} = -4\pi \dfrac{\eta'(T)}{\eta(T)} - \dfrac{2\pi}{T + \overline{T}}$). Thus, in general, the regularization procedure which is necessary to define $\Delta(M_i)$ will either destroy holomorphicity or modular invariance. However we have to insist on the latter.

We now construct the holomorphic section $\Delta(T_i)$ for all (2,2) symmetric Calabi–Yau compactifications. For this purpose we will use the important observation [13] that the metric of the moduli space of Calabi–Yau manifolds should not only be Kähler, as implied by $N = 1$ space-time supersymmetry, but also "special Kähler", as implied by $N = 2$ space-time supersymmetry when one regards the (2,2) theories as possible vacua for type II superstrings. Specifically, in type IIA theories the (1,1) moduli belong to vector multiplets, whereas the (2,1) moduli belong to hypermultiplets. In type IIB theories the situation is reversed. Since the particular form of the moduli space does not depend on whether we deal with a heterotic, type IIA or type IIB thoery, the couplings of the moduli fields are strongly constrained by the possible couplings of $N = 2$ vector multiplets.

In the $N = 2$ superconformal tensor calculus [14], one describes the couplings of n physical vector multiplets by first introducing $n + 1$ vector multiplets, whose scalar components we denote by L^I, $I = 0,1,\ldots,n$. The vector component of the extra (compensating) multiplet is the spin-one field of the $N = 2$ supergravity multiplet. The L^I are complex, covariantly holomorphic functions of the complex physical moduli fields T^i, $i = 1,\ldots,n : L^I = L^I(\overline{T^i},T^i)$ (see ref. [15] for details). The special Kähler potential can then be written as

$$K = -\log Y = -\log \frac{L^I \overline{L^J} N_{IJ}}{L^0 \overline{L^0}} \qquad (7)$$

where

$$N_{IJ} = \mathcal{F}_{IJ}(L) + \overline{\mathcal{F}}_{IJ}(\overline{L}) \tag{8}$$

$$\mathcal{F}_{IJ} = \frac{\partial}{\partial L^I}\frac{\partial}{\partial L^J}\mathcal{F}(L) \tag{9}$$

and $\mathcal{F}(L)$ is a homogeneous holomorphic function of degree 2 in the $n+1$ variables L. One can also introduce completely holomorphic fields (sections) $X^I = X^I(T^i)$,

$$X^I = e^{-K/2}L^I \ , \tag{10}$$

and the Kähler potential looks like (using the homogeneity properties of $\mathcal{F}$)

$$K = -\log(X^I\overline{\mathcal{F}}_I + \overline{X}^I\mathcal{F}_I) \ . \tag{11}$$

These formulas are independent of a particular coordinates system. However, to compare with the previous formulas it is very convenient to choose a "special" coordinate system, the "special" gauge, in the following way:

$$T^i = -iX^i/X^0 \ . \tag{12}$$

In this gauge the Kähler potential takes the form

$$K = -\log\left[\sum_i (T^i + \overline{T}^i)(f_i + \overline{f}_i) - 2(f + \overline{f})\right] \tag{13}$$

where $f(T) = (X^0)^{-2}\mathcal{F}(X)$.

After this brief introduction to the special Kähler geometry, we are now ready to construct the automorphic function $\Delta(M^i)$ of the Calabi–Yau modular group Γ [9]. Since the function $\frac{|\Delta|^2}{Y}$ must be duality-invariant, $\Delta(T^i)$ must be a holomorphic section of holomorphic degree one (Y has degree (1,1)). Holomorphic sections of exactly this degree are X^I and $\mathcal{F}_I$. Therefore the most natural ansatz for Δ is to take a product of all possible linear combinations of these two holomorphic functions:

$$\log\frac{|\Delta|^2}{Y} = \left[-\sum_{M_I, N^I} \log\frac{|M_I X^I + iN^I \mathcal{F}_I|^2}{X^I\overline{\mathcal{F}}_I + \overline{X}\mathcal{F}_I}\right]_{\text{reg.}} \ . \tag{14}$$

Using the covariantly holomorphic functions L^I and $P_I = e^{K/2} \mathcal{F}_I$, this expression can be written in a slightly simpler form:

$$\log \frac{|\Delta|^2}{Y} = \left[- \sum_{M_I, N^I} \log |M_I L^I + i N^I P_I|^2 \right]_{\text{reg.}} . \tag{15}$$

Finally, in the special gauge eq. (12) we obtain the following expression:

$$\log \frac{|\Delta|^2}{Y} = \left[- \sum_{M_I, N^I} \log \frac{|M_0 + M_i T^i + i N^0 f + i N^i f_i|^2}{(T^i + \overline{T^i})(f_i + \overline{f}_i) - 2(f + \overline{f})} \right]_{\text{reg.}} . \tag{14}$$

As already discussed, the holomorphic section Δ is an automorphic function, inversely proportional to the cusp form, of the Calabi–Yau duality group Γ. Δ has to transform under target space duality transformations in such a way that $\dfrac{|\Delta|^2}{Y}$ is a completely invariant function. This requirement implies that the product over M_I and N^I in eqs. (14), (15), (16) does not go over all but only over a restricted set of integers. This restriction is equivalent to the topological level matching condition $\vec{p}_L^2 - \vec{p}_R^2 = 0$ for orbifold compactifications. To understand this, recall (see for example [15]) that the target space duality transformations act (up to a holomorphic function) as particular $Sp(2n + 2)$ transformations on the vector $(X^I, i\mathcal{F}_I)$:

$$\begin{pmatrix} X^I \\ i\mathcal{F}_I \end{pmatrix} \rightarrow \begin{pmatrix} B & D \\ C & A \end{pmatrix} \begin{pmatrix} X^I \\ i\mathcal{F}_I \end{pmatrix} . \tag{17}$$

A, B, C, D are real constant matrices and provide a linear realization of the target space modular group, i.e. $\Gamma \subset Sp(2n + 2)$. The symplectic condition implies that A, B, C, D have to satisfy

$$\begin{aligned}
B^T C - C^T B &= 0 , \\
D^T A - A^T D &= 0 , \\
B^T A - C^T D &= \alpha 1 ,
\end{aligned} \tag{18}$$

where α is a real constant. These symplectic transformations on X^I, $i\mathcal{F}_I$ leave the Kähler potential invariant up to a Kähler transformation. Note that not any symplectic transformation corresponds to a duality transformation; the specific form of the matrices A, B, C, D depends on the particular holomorphic function $\mathcal{F}$ and on the relevant duality group. It is now evident that eqs. (14), (15), (16) provide a duality invariant expression only if the symplectic transformations on X^I and $i\mathcal{F}_I$ are accommpanied by the following transformations on the integers M_I and N^I:

$$\begin{pmatrix} M^I \\ -N^I \end{pmatrix} \rightarrow \begin{pmatrix} C & A \\ B & D \end{pmatrix} \begin{pmatrix} M_I \\ -N^I \end{pmatrix} . \tag{19}$$

This means that (M_I, N^I) must build proper representations of the symplectic duality transformations, i.e. they form a restricted set of integers. In other words, take a given pair of (M_I, N^I); then the action of the symplectic matrices A, B, C, D, which is in general of infinite order, generates all possible integers N_I and M^I which may contribute to Δ. This means that the product in (14), (15), (16) has to go over the specific orbits $\mathcal{O}$ $(M^I, N_I \in \mathcal{O})$ which are generated by the particular symplectic transformations.

As already said, it is very natural to regard $\Delta(T_i)^{-1} = [\prod_{M_I, N^I}(M_I X^I + iN^I \mathcal{F}_I)]$ as the determinant of all chiral masses of the compactified string which provides at the same time an expression for the non-perturbative, moduli-dependent superpotential. We shall now see that this identification becomes very plausible by discussing the form of the possible mass terms as they arise in the corresponding type II compactifications with enlarged $N = 2$ space-time supersymmetry. (Of course, the chiral masses and therefore also the free energies of the corresponding heterotic and type II theories are identical.)

Consider first the four-dimensional type IIA theories with the (1,1) moduli fields belonging to $N = 2$ vector multiplets. Then the chiral masses are described by the couplings between one $N = 2$ vector multiplet and two $N = 2$ hypermultiplets. This means that the relevant massive string states are hypermultiplets with masses given by the vev's of the moduli inside the (1,1) vector multiplets. The general field theoretical couplings among the vector- and hypermultiplets in "gauged" $N = 2$ supergravity theories were recently discussed in ref. [16] and also already some time ago in ref. [14]. These are of the form $X^I H H'$ where H, H' denote the massive hypermultiplet fields. Thus we see that the chiral $N = 2$ masses are just given by the holomorphic degree-one functions $X^I(T_i)$. However if one only considers these "field theoretical" masses X^I it is clear that the spectrum is not invariant under the target space duality transformations which mix X^I with the other degree-one functions X^J and $i\mathcal{F}_I$ in a non-trivial way. In fact, the duality transformations act on an infinite number of massive $N = 2$ hypermultiplets. This means only an infinite number of massive hypermultiplets provide a duality invariant $N = 2$ spectrum with masses given by the X^I and also by the "dual" fields $\mathcal{F}_I$. Therefore the most general ansatz for the hypermultiplet masses M is to take $M = M_I X^I + i N^I \mathcal{F}_I$, where the M_I, N^I are a restricted set of integers satisfying the constraints inferred by the symplectic modular transformations. It is important to note that this expression for the chiral masses exactly provides the duality invariant generalization of the $N = 4$ gauging procedure of ref. [10] for the case of the special Kähler geometry of $N = 2$ supergravity.

To obtain a clearer understanding about the generalized momentum (Kaluza-Klein) and winding spectrum of a Calabi–Yau compactification let us consider the large radius, i.e. field theory, limit (the weak coupling limit of the corresponding σ-model) of the Calabi–Yau compactification. In this limit $Y_{(1,1)}$ just becomes the volume of the six-dimensional Calabi–Yau space,

$$Y_{(1,1)} = \int J \wedge J \wedge J = d_{ijk}(T_i + \overline{T}_i)(T_j + \overline{T}_j)(T_k + \overline{T}_k) \,, \qquad (20)$$

where $J = \sum_{i=1}^{n}(T^i + \overline{T^i})V_i$ and V_i being the harmonic $(1,1)$ forms in $H^{(1,1)}$. The corresponding holomorphic function $f(T_i)$ which leads to (20) has the form $f(T_i) = d_{ijk}T_iT_jT_k$. In the field theory limit d_{ijk} coincide with the intersection matrices of $V_i : d_{ijk} = \int V_i \wedge V_j \wedge V_k$. For the case of a single modulus the Kähler potential becomes $K = -3\log(T + \overline{T})$ in the large radius limit. Now choose the special gauge eq. (12) and consider those "masses" in eq. (16) which have only non-vanishing M_0:

$$M_{M_0}^2 = \frac{M_0^2}{Y} = \frac{M_0^2}{\int J \wedge J \wedge J} \,. \qquad (21)$$

We recognize that in the field theory limit the masses $M_{M_0}^2$ are inversely proportional to the volume of the Calabi–Yau space. This fact strongly suggests that $M_{M_0}^2$ are just the ordinary field-theoretical Kaluza-Klein masses, i.e. that the momentum spectrum of the Calabi–Yau compactification is given by the states with only non-vanishing M_0.

On the other hand, for small radii, non-perturbative effects, i.e. instanton configurations [17] which give a non-trivial map of the world-sheet onto the Calabi–Yau space modify eqs. (20), (21). (They also break the continuous Peccei-Quinn symmetry $T^i \rightarrow T^i + iC^i$, $C^i \in \mathbf{R}$, to a discrete subgroup.) At the same time the generalized winding states become light. The symplectic duality invariance enforces to also include the winding states with non-vanishing M_i and N^I into the target space free energy. Thus, the masses of the winding states seem to be determined by $M_i T^i$ and in particular by $N^I \mathcal{F}_I$. In summary, just like for torodial/orbifold compactifications we can regard the states with non-vanishing M_0 as momentum states, and the states with non-vanishing M_i and N^I as generalized winding states (vortex configurations). (In the context of the following orbifold example M_0 is a pure momentum number whereas N^0 is a pure winding number. The M_i, N^i ($i = 1, \ldots, n$) are products of winding and momentum numbers.) Thus, the M_I and N^I define, in some sense, a generalized Calabi–Yau lattice.

A very similar interpretation is valid when regarding $\Delta(T_i)$ as the non-perturbative superpotential. As discussed in [18], the most general, "field theoretical" superpotential compatible with the special Kähler geometry has the form $\log W \sim - \log M_I X^I$. (In the special gauge, the constant superpotential is proportional to X^0.) Thus, Δ again provides the duality invariant completion of this already known expression. This also shows that the orbifold superpotential $\log W = - \sum_{m,n} \log(m + inT)^3 = -6 \log \eta(T)$ is compatible with the special Kähler structure of $N = 2$ supersymmetry (see also the following example).

So far we have been discussing the couplings of the (1,1) moduli fields. However, analogous arguments are also true for the (2,1) moduli fields. This comes from the fact that we can use the same Calabi–Yau background (the same (2,2) superconformal field theory) for type IIB compactifications, where the (2,1) moduli fields belong to $N = 2$ vector multiplets. Remember that the (2,1) moduli describe the deformations of the complex structure of the Calabi–Yau space. The moduli space for these fields reflects the homology of the underlying Calabi–Yau manifold. More precisely, the complex structure can be described by the periods of the holomorphic three-form Ω over a canonical homology basis. Choosing a symplectic basis (A^I, B_I) such that $A^I \cap B_J = \delta^A_B$, $A^I \cap A^J = 0$ and $B_I \cap B_J = 0$, then the $2n + 2$ functions $(X^I, \mathcal{F}_J)$ are nothing other than the periods of Ω:

$$X^I = \int_{A^I} \Omega \, , \quad i\mathcal{F}_J = \int_{B_J} \Omega \, . \tag{22}$$

The symplectic modular transformations on the X^I, $i\mathcal{F}_I$ therefore just describe the action of the target space duality group on the homology cycles. (This is in complete analogy to the action of the so-called Dehn twists around the non-trivial cycles of a higher genus Riemann surface.) Thus the holomorphic section Δ is just the product of all possible linear combinations of the periods of Ω, and the periods are just the chiral masses of the massive hypermultiplets in the corresponding $N = 2$, type IIB theory.

From eqs. (11) and (22) it follows that the Kähler potential of the (2,1) moduli is given as

$$K_{(2,1)} = -\log(-i \int \Omega \wedge \overline{\Omega}) \, . \tag{23}$$

Then the "masses" $M^2_{M_I, N^I}$ look like

$$M^2_{M_I, N^I} = \frac{|M_I \int_{A^I} \Omega + N^I \int_{B_I} \Omega|^2}{-i \int \Omega \wedge \overline{\Omega}} \, . \tag{24}$$

In constrast to the (1,1) moduli, the Kähler potential eq. (23) as well as the periods X^I, $i\mathcal{F}_I$ do not receive any quantum corrections from world-sheet instantons. They are entirely determined by their field-theoretical expressions. This is also a reflection of the fact that the duality transformation involving the (2,1) fields have a geometrical origin and do not involve the exchange of momentum and topological winding modes. Thus we expect that mass spectrum and the free energy in the (2,1) sector is also of pure field theoretical nature and can be computed by algebraic methods as demonstrated for a specific example in [8]. Then, for the compactifications which appear in mirror pairs the knowledge about the (2,1) mass spectrum provides important information about the winding spectrum in the (1,1) sector of the corresponding mirror image Calabi–Yau space.

Now we show that eq. (14) reproduces the results for the free energy as computed in the previous section for the case of the orbifold compactifications thus giving support for its general validity. First, consider the orbifolds with moduli space $\mathcal{M} = \left(\dfrac{SU(1,1)}{U(1)}\right)^3$. The corresponding holomorphic function $\mathcal{F}(X)$ takes the form [18]:

$$\mathcal{F}(X) = i\frac{X^1 X^2 X^3}{X^0} \ .$$

(25)

In the special gauge the standard moduli fields are given by

$$T_i = -i\frac{X^i}{X^0} \ , \quad i = 1,2,3 \ .$$

(26)

Then, using eq. (13), it is not difficult to show that $Y = \prod\limits_{i=1}^{3} (T_i + \overline{T}_i)$. Eq.(14) now leads to the following expression for $\dfrac{|\Delta|^2}{Y}$:

$$\log\frac{|\Delta|^2}{Y} = \left[-\sum_{M_I,N^I} \log\frac{1}{Y}|M_0 + iM_1T_1 + iM_2T_2 + iM_3T_3 \right.$$
$$\left. - iN^0T_1T_2T_3 + N^1T_2T_3 + N^2T_1T_3 + N^3T_1T_2|^2 \right]_{\text{reg.}} \ .$$

(27)

On the other hand, the corresponding orbifold calculation of the previous section led to the expression (2) for the free energy. The two formulae (2) and (27) coincide, provided the following relations among the M_I, N^I and the unrestricted integers m_i, n_i hold:

$$\begin{aligned}
M_0 &= m_1 m_2 m_3 \ , & N^0 &= n_1 n_2 n_3 \ , \\
M_1 &= n_1 m_2 m_3 \ , & N^1 &= -m_1 n_2 n_3 \ , \\
M_2 &= n_2 m_1 m_3 \ , & N^2 &= -m_2 n_1 n_3 \ , \\
M_3 &= n_3 m_1 m_2 \ , & N^3 &= -m_3 n_1 n_2 \ .
\end{aligned}$$

(28)

We recognize the M_0 is a product of three momentum numbers and N^0 is a product of three winding numbers, whereas the others contain both momentum and winding numbers. Thus the pure Kaluza-Klein spectrum is entirely described by M_0.

Now it is crucial to observe that this constraint on M_I and N^I is just the orbit $\mathcal{O}$ which is inferred by the symplectic modular transformations on M_I and N^I. Specifically, consider the action of the target space duality modular group $\Gamma = [SL(2,\mathbf{Z})]^3 \subset Sp(8,\mathbf{Z})$. First, the transformation $T_1 \to T_1 - i$ acts as the following $Sp(8,\mathbf{Z})$ transformation on M_I and N^I (see also [19]):

$$
\begin{aligned}
M_0 &\to M_0 - M_1 , & N^0 &\to N^0 , \\
M_1 &\to M_1 , & N^1 &\to N^0 + N^1 , \\
M_2 &\to M_2 + N^3 , & N^2 &\to N^2 , \\
M_3 &\to M_3 + N^2 , & N^3 &\to N^3
\end{aligned}
\tag{29}
$$

and similarly for the shifts of T_2 and T_3. Second, the inversion $T_1 \to \dfrac{1}{T_1}$ acts as the symplectic transformation

$$
\begin{aligned}
M_0 &\to -M_1 , & N^0 &\to -N^1 , \\
M_1 &\to M_0 , & N^1 &\to N^0 , \\
M_2 &\to N^3 , & N^2 &\to -M_3 , \\
M_3 &\to N^2 , & N^3 &\to -M_2
\end{aligned}
\tag{30}
$$

plus analogous transformations for the inversions of T_2 ans T_3. On the other hand, we know already that eh modular transformation $T_1 \to T_1 - i$ acts on the unconstrained integers as $m_1 \to m_1 - n_1$ where the remaining m_i, n_i are not transformed. Analogously the transformation $T_1 \to \dfrac{1}{T_1}$ induces $m_1 \to -n_1$, $n_1 \to m_1$. However this transformation on the unconstrained integers exactly induces the symplectic transformations (29) and (30) on M_I, N^I when using the constraint eq. (28). This shows that this identification exactly generates the set of integers M_I and N^I which correctly transform under the symplectic modular transformations. Therefore we conclude that the formula (27) with restricted M^I, N_I and the orbifold free energy (2) agree.

It is also illustrative to consider briefly the simplext case with moduli space $\mathcal{M} = \dfrac{SU(1,1)}{U(1)}$ of the overall modulus which is obtained by identifying all three moduli fields: $T = T_1 = T_2 = T_3$. The corresponding holomorphic function $\mathcal{F}(X)$ looks like $\mathcal{F} = i\dfrac{(X^1)^3}{X^0}$. Then the holomorphic section Δ takes the form:

$$\Delta(T) = \left[\prod_{M_I, N^I} (M_0 - iN^0 T^3 + iM_1 T + 3N^1 T^2)^{-1} \right]_{\text{reg.}} . \tag{31}$$

In order to reproduce the free energy with $\Delta(T) = [\prod_{m,n} (m + inT)^{-3}]_{\text{reg.}} = \dfrac{1}{\eta(T)^6}$ the product has to go over the following set of restricted integers:

$$
\begin{aligned}
M_0 &= m^3 \,, & N^0 &= n^3 \,, \\
M_1 &= 3m^2 n \,, & N^1 &= -mn^2 \,.
\end{aligned}
\tag{32}
$$

Again this restriction is exactly the one which defines the $SL(2, \mathbf{Z})$ orbit when demanding the proper transformation properties of M_I, N^I under the symplectic modular transformations. These are for $T \to T - i$

$$
\begin{aligned}
M_0 &\to M_0 - M_1 - N^0 - 3N^1 \,, \\
M_1 &\to M_1 + 3N^0 + 6N^1 \,, \\
N^0 &\to N^0 \,, \\
N^1 &\to N^0 + N^1
\end{aligned}
\tag{33}
$$

and for $T \to \dfrac{1}{T}$

$$
\begin{aligned}
M_0 &\to -N^0 \,, \\
M_1 &\to -3N^1 \,, \\
N^0 &\to M_0 \,, \\
N^1 &\to \frac{1}{3} M_1 \,.
\end{aligned}
\tag{34}
$$

These are exactly the symplectic transformations derived in ref. [20].

Let us emphasize that $\Delta(T_i)$ is in general not the unique automorphic function of the Calabi–Yau target space duality group Γ. Remember that $\Delta(T_i)$ is the generalization of the (inverse) Dedekind function such that $|\Delta|^2 / Y$ is duality invariant. However, apart from $Y = e^{-K}$, which determines the Kähler metric $G_{i\bar{j}} = -\partial_i \partial_{\bar{j}} \log Y$ for the moduli and also the massless matter fields, there exists another globally defined real function, now of degree zero, namely

$\det N_{IJ} = \det[\mathcal{F}_{IJ}(X) + \overline{\mathcal{F}}_{IJ}(\overline{X})]$. This function provides important information about the moduli-dependent tree-level Yukawa couplings among the charged matter fields:

$$\|C\|^2_{i\bar{j}} = C_{imp}\overline{C}_{\bar{j}\bar{n}\bar{q}}G^{m\bar{n}}G^{p\bar{q}} = -\partial_i\partial_{\bar{j}}\log\det -N_{IJ} \ . \tag{35}$$

Here C_{ijk} are covariantly holomorphic 3-index tensors which are related to the holomorphic Yukawa couplings W_{ijk} of the massless matter fields in the following way (see [15] for details): $W_{ijk}(T_i) = 2e^{-K}C_{ijk}(T_i,\overline{T}_i)$. Eq.(35) follows from the fact that the Riemann tensor of a special Kähler space has to satisfy the constraint

$$R_{i\bar{j}k\bar{l}} = G_{i\bar{j}}G_{k\bar{l}} + G_{i\bar{l}}G_{k\bar{j}} - C_{ikm}\overline{C}_{\bar{j}\bar{l}\bar{n}}G^{m\bar{n}} \ . \tag{36}$$

Now, using eq. (17), the target space duality transformations act on $\mathcal{F}_{IJ}$ in the following way (in matrix notation):

$$\mathcal{F} \rightarrow (A\mathcal{F} - iC)(iD\mathcal{F} + B)^{-1} \ . \tag{37}$$

Remembering that for the case of the (2,1) moduli the X^I, $i\mathcal{F}_I$ are nothing other than the periods of the holomorphic three form Ω it follows that $\mathcal{F}_{IJ}$ is just the symmetric period matrix of the Calabi–Yau space. (Again, all these features are in complete analogy to the symplectic modular transformations in the theory of higher genus Riemann surfaces.)

Using eq. (18) it is not difficult to show that $\det N_{IJ}$ transforms as

$$\det N_{IJ} \rightarrow |\det(iD\mathcal{F} + B)|^{-2}\det N_{IJ} \ . \tag{38}$$

Now, in analogy to our previous discussion, we are looking for automorphic functions $\Pi(T_i)$ of degree zero such that

$$\frac{|\Pi(T_i)|^2}{(\det N_{IJ}(T_i,\overline{T}_i))^r} \tag{39}$$

is duality invariant. Thus $\Pi(T_i)$ has to transform (up to a phase) as

$$\Pi(T_i) \rightarrow \det(iD\mathcal{F} + B)^{-r}\Pi(T_i) \ . \tag{40}$$

Thus r defines the weight of Π, its exact value depends on the particular model. ($\Pi(T_i)$ transforms analogously to the Riemann theta functions of more than one variable.) Since Π

is of holomorphic degree zero it must be a function of $\mathcal{F}_{IJ}$. To explicitly construct Π one can make the following formal ansatz:

$$\Pi(T_i) = [\det \prod_{M_{IJ}, N_{IK}} (M_{IJ} + iN_{IK}\mathcal{F}_{KJ})]^r_{\text{reg.}} \cdot \tag{41}$$

It follows that the target space duality transformation (37) induces the following transformation on M_{IJ} and N_{IJ}:

$$M \to MB + NC \ , \quad N \to MD + NA \ . \tag{42}$$

Thus, in order to obtain a duality invariant expression, the product again has to go over an infinite, but restricted set of integers M_{IJ}, N_{IJ}.

In general, the automorphic functions $\Pi(T_i)$ are not identical to the previously discussed automorphic functions $\Delta(T_i)$. However, considering maximally symmetric moduli spaces as they arise for example for all orbifold compactifications one can show that $(\det N_{IJ})^r$ is exactly proportional to Y. Thus in these cases $\Delta(T_i)$ and $\Pi(T_i)$ are expected to agree after a proper restriction on the integers M_{IJ} and N_{IJ}.

The non-holomorphic structures K and $\det N_{IJ}$ previously defined are relevant to discuss, on genral grounds, moduli-dependent corrections to the gauge couplings or to the dilaton Kähler protential due to one-loop (genus one) string effects. These corrections can be discussed in full generality for smooth Calabi–Yau manifolds, whilst they are somewhat model-dependent for orbifold compactifications. According to refs.[21], [22], [23], these moduli-dependent corrections can be interpreted as coming from supersymmetrization of anomalies for target-space duality transformations. They are due to a mixed $U(1)$ gauge anomaly for the gauge groups E_8 and E_6 (for general Calabi–Yau compactifications), where the $U(1)$ connection is a composite Kähler-holonomy connection. The Kähler connection is the vector component of the superfield, whose first component is the Kähler potential, whilst the $U(1)$ holonomy connection is the vector component of a superfield, whose first component is $\log \det G_{i\bar{j}}$, where $G_{i\bar{j}}$ is the Kähler metric. For general Calabi–Yau manifolds, the $U(1)$ composite connection has two contributions, since the moduli space is a product space for Kähler class ((1, 1) moduli) and complex structure deformations ((2,1) moduli), corresponding to E_6 27 and $\overline{27}$ families, respectively.

Let us discuss first the Kähler anomaly coming from the $U(1)$ Kähler connection, coupled, via a triangular graph, to the (E_8 and E_6) gauge fields. For the E_8 gauge group, since only gauginos circulate in the loop, one gets for the $F^2_{\mu\nu}$ term (the superpartner of the anomaly term), up to a normalization factor [21], [23]

$$\mathrm{Tr}\ F^2_{\mu\nu}\big|_{E_8} \times C_{(G=E_8)}(K_{(1,1)} + K_{(2,1)})\ , \tag{43}$$

where $K_{(1,1)} + K_{(2,1)}$ is the Kähler potential for the full moduli space and $C_{(G=E_8)}$ is the quadratic Casimir of the adjoint representation of E_8.

For the E_6 gauge group, three types of fermions can circulate in the loop: gauginos, 27 and $\overline{27}$. Let us first consider the contribution to the $U(1)$ anomaly coming form the 27 families. This can be separated into a curvature contribution and a Kähler contribution. The Kähler contribution is easily computed from the fact that, under a Kähler transformation of the moduli space, the 27 and $\overline{27}$ family multiplets transform as [24]

$$\phi_{27} \to \phi_{27} e^{-\frac{-\Lambda_{(1,1)}+\Lambda_{(2,1)}}{3}}\ , \quad \phi_{\overline{27}} \to \phi_{\overline{27}} e^{\frac{\Lambda_{(1,1)}-\Lambda_{(2,1)}}{3}}\ . \tag{44}$$

One then gets a contribution to the E_6 $R^2_{\mu\nu}$ term of the form [21], [25]

$$-T(R)\left(\frac{5}{3}K_{(1,1)} + \frac{1}{3}K_{(2,1)}\right) - T(\overline{R})\left(\frac{5}{3}K_{(2,1)} + \frac{1}{3}K_{(1,1)}\right)\ . \tag{45}$$

($T(R)$ is the quadratic Casimir of the representation R.) The gaugino contribution gives a term

$$C_{(G=E_6)}(K_{(1,1)} + K_{(2,1)})\ , \tag{46}$$

while the $U(1)$ curvature contribution gives a term [23]

$$2T(R)\log \det G^{(1,1)}_{i\bar{j}} + 2T(\overline{R})\log \det G^{(2,1)}_{i\bar{j}}\ , \tag{47}$$

where $G^{(1,1)}_{i\bar{j}}$, $G^{(2,1)}_{i\bar{j}}$ are the moduli metrices and we used the fact that, for smooth Calabi-Yau, the 27, $\overline{27}$ family metrices are parallel to the moduli metrices [24]

$$g^{27}_{i\bar{j}} = e^{\frac{-K_{(1,1)}+K_{(2,1)}}{3}}G^{(1,1)}_{i\bar{j}}\ , \quad g^{\overline{27}}_{i\bar{j}} = e^{\frac{K_{(1,1)}-K_{(2,1)}}{3}}G^{(2,1)}_{i\bar{j}}\ . \tag{48}$$

The above terms are not target space duality invariant. Duality invariance is restored by adding to the superfields K and $\log \det G_{i\bar{j}}$ appropriate holomorphic terms $\log \Delta$ and $\log \Pi$ as previously discussed, in such a way that the coefficients of the $F^2_{\mu\nu}$ terms become modular invariant. In the $F_{\mu\nu}\widetilde{F}^{\mu\nu}$ sector this fact can be interpreted as a cancellation of target-space duality anomalies [21], [22], [23], [26].

For example, in the case of the E_8 anomalies $K_{(1,1)}$ and $K_{(2,1)}$ have to be replaced by

$$K + \log \Delta + \log \overline{\Delta} \, , \tag{49}$$

where $\Delta = W$ is the Calabi–Yau automorphic form defined in eq. (14). For the E_6 gauge group, one must also consider the second automorphic function Π defined by eqs. (39), (40), (41), since for special geometry one has [18]

$$\log \det G^{(1,1)}_{i\bar{j}} = (n_{(1,1)} + 1)K_{(1,1)} + \log \det(-N^{(1,1)}_{IJ}) \, , \tag{50}$$

and a similar formula for $G^{(2,1)}_{i\bar{j}}$.

In summary, the free energy F we have computed is a topological index, in the sense that it takes into account only those massive states which are due to the compact target space, and not those states which are also present in a flat background, like all oscillator excitations. Moreover, the general form of the automorphic functions $\Delta(T_i)$, $\Pi(T_i)$ indicates a very close relationship between Calabi–Yau compactifications and topological superconformal field theories [27], [28]. This observation relies on the fact that $\Delta(T_i)$, $\Pi(T_i)$ depends only on the moduli fields and on the holomorphic function $\mathcal{F}[X^I(T_i)]$. One knows [28] that $\mathcal{F}(X^I)$ is entirely determined by the couplings of the chiral primary fields [29] of the underlying $(2,2)$ $\hat{c} = 6$ superconformal field theory. However, in the topological counterparts of the superconformal models one is essentially left with the chiral ring, and the $\mathcal{F}$-function plays the role of the generating functional [28] of the topological field theory. Then all S-matrix elements in the topological theory, like the three-point functions of the chiral primary fields (Yukawa couplings), are determined by $\mathcal{F}$. So we recognize that all the chiral masses $M_I X^I + iN^I \mathcal{F}_I$, and thus also the free energy of Calabi–Yau compactifications, are topological, since they are entirely determined by the $\mathcal{F}$-function, i.e. by the topological sector of the superconformal field theory. It is a very interesting problem to further clarify this relation between Calabi–Yau compactifications and topological field theories.

References

[1] E. Calabi, in *Algebraic geometry and topology* (Univ. Press, Princeton, 1957), p. 78; S.T. Yau, Proc. Nat. Acad. Sci. **74** (1977) 1798.

[2] P. Candelas, G. Horowitz, A. Strominger and E. Witten, Nucl. Phys. **B258** (1985) 46.

[3] N. Boucher, D. Friedan and A. Kent, Phys. Lett. **172B** (1986) 316; T. Banks and L. Dixon, Nucl. Phys. **B307** (1988) 33.

[4] K. Narain, Phys. Lett. **169B** (1986) 41; K. Narain, M. Sarmadi and E. Witten, Nucl. Phys. **B279** (1987) 369.

[5] K. Kikkawa and M. Yamasaki, Phys. Lett. **149B** (1984) 357;
N. Sakai and I. Senda, Progr. Theor. phys. **75** (1986) 692;
V.P. Nair, A. Shapere, A. Strominger and F. Wilczek, R. Dijkgraaf, E. Verlinde and H. Verlinde, Comm. Math. Phys. **115** (1988) 649; preprint THU-87/30;
A. Giveon, E. Rabinovici and G. Veneziano, Nucl. Phys. **B322** (1989) 167;
A. Shapere and F. Wilczek, Nucl. Phys. **B320** (1989) 669;
M. Dine, P. Huet and N. Seiberg, Nucl. Phys. **B322** (1989) 301;
J. Molera and B. Ovrut, Phys. Rev. **D40** (1989) 1146;
M. Duff, Nucl. Phys. **B335** (1990) 610;
E. Alvarez and M.A.R. Osorio, Phys. Rev. **D40** (1989) 1150;
J. Lauer, J. Mas and H.P. Nilles, Phys. Lett. **B226** (1989) 251; Nucl. Phys. **B351** (1991) 353;
W. Lerche, D. Lüst and N.P. Warner, Phys. Lett. **B231** (1989) 417;
S. Ferrara, D. Lüst and S. Theisen, Phys. Lett. **B233** (1989) 362;
L. Ibañez, W. Lerche, D. Lüst and S. Theisen, Nucl. Phys. **B352** (1991) 435;
M. Cvetic, J. Molera and B. Ovrut, Phys. Lett. **B248** (1990) 83.

[6] S. Ferrara, D. Lüst, A. Shapere and S. Theisen, Phys. Lett. **B225** (1989) 363.

[7] A. Font, L.E. Ibáñez, D. Lüst and F. Quevedo, Phys. Lett. **B245** (1990) 401;
S. Ferrara, N. Magnoli, T.R. Taylor and G. Veneaiano, Phys. Lett.**B245** (1990) 409;
H.P. Nilles and M. Olechowski, Phys. Lett. **B248** (1990) 268;

P. Binetruy and M.K. Gaillard, Phys. Lett. **B253** (1991) 119;

M. Cvetic, A. Font, L.E. Ibáñez, D. Lüst and F. Quevedo, *Target space duality, supersymmetry breaking and the stability of classical string vacua*, preprint CERN-TH.5999/91, UPR-0444-T, to appear in Nucl. Phys.

[8] P. Candelas, X. De La Ossa, P. Green and L. Parkes, *A pair of Calabi–Yau manifolds as an exactly soluble superconformal theory*, preprint UTTG-25-90; Phys. Lett. **B258** (1991) 118.

[9] S. Ferrara, C. Kounnas, D. Lüst and F. Zwirner, *Duality-invariant Partition Functions and Automorphic Superpotentials for (2,2) String Compactifications*, preprint CERN-TH.6090/91.

[10] A. Giveon and M. Porrati, Phys. Lett. **B246** (1990) 54; Nucl. Phys. **B355** (1991) 422.

[11] L. Dixon, v. Kaplunovsky and J. Louis, Nucl. Phys.**B355** (1991) 649.

[12] H. Ooguri and C. Vafa, Mod. Phys. Lett. **A5** (1990) 1389; *Geometry of $N = 2$ strings*, preprint EFI91/05, HUTP-91/A003.

[13] S. Cecotti, S. Ferrara and L. Giradello, Int. J. Mod. Phys. **4** (1989) 2475, Phys. Lett. **B213** (1988) 443;

S. Ferrara and A. Strominger, in Strings '89 (R. Arnowitt, R. Bryan, M.J. Duff, D. Nanopoulos and C.N. Pope eds., World Scientific, 1990), p. 245;

L. Dixon, V. Kaplunovsky and J. Louis, Nucl. Phys. **B329** (1990) 27;

A. Strominger, Comm. Math. Phys. **133** (1990) 163;

L. Castellani, R. D'Auria and S. Ferrara, Class. Quant. Grav. **7** (1990) 1767.

[14] B. De Wit, P.G. Lauwers and A. Van Proeyen, Nucl. Phys. **B255** (1985) 569.

[15] S. Ferrara, *Superstring effective field theories on (2,2) vacua*, preprint CERN-TH.5873/90.

[16] R. D'Auria, S. Ferrara and P. Fré, *Special and quaternionic isometries: general couplings in $N = 2$ supergravity and the scalar potential*, preprint CERN-TH.5954/90, DFPD-90-TH-35, SISSA 192/90/EP.

[17] M. Dine, N. Seiberg, X. Wen and E. Witten, Nucl. Phys. **B278** (1986) 769, Nucl. Phys. **B289** (1987) 357.

[18] E. Cremmer, C. Kounnas, A. Van Proeyen, J.P. Derendinger, S. Ferrara, B. de Wit and L. Girardello, Nucl. Phys. **B250** (1985) 385.

[19] D. Shevitz, Nucl. Phys. **B338** (1990) 283.

[20] S. Ferrara, D. Lüst and S. Theisen, Phys. Lett. **B242** (1990) 39.

[21] J.P. Derendinger, S. Ferrara, C. Kounnas and F. Zwirner, *On loop corrections to string effective field theories: field-dependent gauge couplings and sigma-model anomalies*, preprint CERN-TH.6004/91, LPTENS 91-4.

[22] G. Cardoso and B. Ovrut, *A Green-Schwarz Mechanism for $D = 4$, $N = 1$ super-gravity Anomalies*, preprint UPR-0464T (1991).

[23] J. Louis, *Non-harmonic gauge coupling constants in supersymmetry and superstring theory*, preprint SLAC-PUB-5527;

V. Kaplunovsky and J. Louis, SLAC preprint to appear (1991).

[24] L. Dixon, V. Kaplunovsky and J. Louis, Nucl. Phys. **B329** (1990) 27.

[25] L. Dixon, talk given at the Berkeley conference, May 1991.

[26] L. Ibanez and D. Lüst, *The Strong CP Problem and Target-space Modular Invariance in 4-D Strings*, preprint CERN-TH.6120/91 (1991).

[27] E. Witten, Comm. Math. Phys. **118** (1988) 411; Nucl. Phys. **B340** (1990) 281;

R. Dijkgraaf and E. Witten, Nucl. Phys. **B342** (1990) 486;

C. Vafa, Mod. Phys. Lett. **A6** (1991) 337.

[28] R. Dijkgraaf, E. Verlinde and H. Verlinde, *Topological strings in $d < 1$*, preprint PUPT-1204 (1990).

[29] W. Lerche, C. Vafa and N.P. Warner, Nucl. Phys. **B324** (1989) 427.

AMS/IP Studies in Advanced Mathematics
Volume **9**, 1998

Picard-Fuchs Equations and Flat Holomorphic Connectons From $N = 2$ Supergravity

S. Ferrara* and J. Louis

CERN, 1211 Geneva 23, Switzerland

Abstract

The geometry of the moduli space of Calabi-Yau manifolds is described in terms of special Kähler geometry of $N = 2$ space-time supergravity. We show that for this geometry the gauge variant part of the connection is holomorphic and flat (as a Riemannian connection), as is expected from its relation to $N = 2$ (world-sheet) topological field theories. A set of differential identities (Picard-Fuchs identities) are satisfied on a holomorphic bundle. The relationship with the linear differental equations obeyed by the periods of the holomorphic three form of Calabi-Yau manifolds is outlined.

---

*Work supported in part by the Department of Energy of the USA under the contract DOE-AT03-88ER40 384, Task E.

© 1998 American Mathematical Society
and International Press

1. Introduction

For the study of $N = 1$ superstring vacua in four space-time dimensions one is particularly interested in super conformal field theories (SCFT) with central charge $c = 9$ and worldsheet supersymmetry $(0,2)$ [1]. For the subset of string vacua where this worldsheet supersymmetry is enlarged to $(2,2)$ the low energy effective Lagrangian satisfies an additional constraint. The manifold spanned by certain massless scalar fields in the string spectrum (so called moduli) is promoted from being a Kähler manifold to a 'special Kähler manifold' [2]. Such a geometric structure first appeared in the context of $N = 2$ supergravity theories in four space-time dimensions as the scalar manifold spanned by the scalars of vectormultipletes [3,4], The moduli space of $(2,2)$ SCFT with central charge $c = 9$ obeys a similar constraint as can be understood from the relation between heterotic and type II string theories [5, 6] or directly from SCFT Ward identities [7]. Calabi-Yau threefolds are an example of $(2,2)$ SCFT and thus their moduli spaces are also special Kähler manifolds. This can also be derived by purely geometrical methods without ever referring to a SCFT [8, 9, 10].

Recently, for a special Calabi-Yau manifold (a quintic in CP_4) the exact Kähler and superpotential of the low energy theory was given without using the underlying SCFT [11]. This was made possible by explicitly constructing the mirror map [12] of this Calabi-Yau modulu space. In the course of this investigation it was shown that the periods of the holomorphic threeform Ω satisfy a linear differential equation. It was then realized that this equation is a particular case of Picard-Fuchs systems of differential equations, knwon to be obeyed by the periods of globally defined p-forms ($p = 3$ for a threefold) from algebraic geometry [13–15]. Furthermore, substantial information about $(2,2)$ SCFT can be obtained from a purely topological sector of the theory [16–19]. For $c = 3$ SCFT analogous differential equations were derived using methods developed in topological field theories (TFT) [20–22].

Since the moduli space of Calabi-Yau-threefolds or more generally the moduli space of $(2,2)$ SCFT of central charge $c = 9$ is constrained to display special geometry it should be possible to understand the Picard-Fuchs equations entirely within the framework of special geometry. This is the aim of the present paper. We first briefly remind the reader of the properties of special Kähler geometry. Then we show that the Christoffel as well as the Kähler connection consist of two distinctively difference pieces. A holomorphic part is responsible for the transformation properties of the connection whereas the non-holomorphic piece transforms like a tensor. The holomorphic part of the connection turns out to be flat (in a Riemannian sense). We believe that this holomorphic flat connection is the analogue of the flat connection encountered in TFT when it is restricted to marginal deformations [17, 19, 14].

Equipped with this understanding we then show how special geometry implies a set of covariant differential identities (which we call Picard-Fuchs identities). These identities are entirely equivalent to the identity that is satisfied by the Riemann tensor in special geometry but are instead of a purely holomorphic nature. Once the coefficient functions of these identities are specified a non-trivial linear differential equation arises which is equivalent to the Picard-Fuchs equation satisfied by the periods of the Calabi-Yau-manifold.

2. Summary of special geometry

The metric $g_{\alpha\overline{\beta}}$, $\alpha = 1, \ldots, n$ of a n-dimensional Kähler manifold is given by the second derivative of a Kähler potential K : $g_{\alpha\overline{\beta}} = \partial_\alpha\overline{\partial}_{\overline{\beta}}K$ ($\equiv K_{\alpha\overline{\beta}}$). K is a real function of the complex coordinates z^α and $\overline{z}^{\overline{\alpha}}$. $g_{\alpha\overline{\beta}}$ is invariant under the Kähler transformations $K(z,\overline{z}) \to K(z,\overline{z}) + f(z) + \overline{f}(\overline{z})$. A field Ψ of Kähler weight $(k,\overline{k})$ transforms according to $\Psi \to \Psi e^{kf} e^{\overline{k}\overline{f}}$. Its Kähler covariant derivative is defined by

$$D_\alpha\Psi = (\nabla_\alpha - kK_\alpha)\Psi$$
$$\overline{D}_{\overline{\alpha}}\Psi = (\overline{\nabla}_{\overline{\alpha}} - \overline{k}K_{\overline{\alpha}})\Psi \tag{1}$$

where ∇_α denotes the covariant derivative with repsect to reparametrizations of the Kähler manifold. Thus K_α, $K_{\overline{\alpha}}$ act as connections for Kähler transformations. This is a consequence of the fact that special manifolds are manifolds of the restricted type. This means that K_α is an Abelian connection of a holomorphic line bundle L whose first Chern class is the Kähler class ($c_1(L) = [J]$) [2].

240 S. FERRARA AND J. LOUIS

Special geometry is defined by an additional constraint on the Kähler metric. This constraint can be expressed as a covariant equation that the Riemann tensor of the Kähler manifold satisfies. It reads [4]

$$R^{\delta}_{\alpha\overline{\beta}\gamma} = g_{\alpha\overline{\beta}}\delta^{\delta}_{\gamma} + g_{\gamma\overline{\beta}}\delta^{\delta}_{\alpha} - C_{\alpha\gamma\epsilon}g^{\epsilon\overline{\epsilon}}\overline{C}_{\overline{\beta}\,\overline{\delta}\,\overline{\epsilon}}g^{\delta\overline{\delta}} \tag{2}$$

where $C_{\alpha\beta\gamma}$ is of Kähler weight $(-1,1)$, symmetric in its indices and satisfies

$$\overline{D}_{\overline{\epsilon}}C_{\alpha\beta\gamma} = 0 \ ,$$

$$D_{\epsilon}C_{\alpha\beta\gamma} - D_{\alpha}C_{\epsilon\beta\gamma} = 0 \ . \tag{3}$$

Using $D_{\alpha}e^{K} = 0$ eq. (3) can be solved by

$$C_{\alpha\beta\gamma} = e^{K}W_{\alpha\beta\gamma}(z) \ ,$$

$$W_{\alpha\beta\gamma} = D_{\alpha}D_{\beta}D_{\gamma}S(z,\overline{z}) \ , \tag{4}$$

$$\overline{\partial}_{\overline{\epsilon}}W_{\alpha\beta\gamma} = 0 \ .$$

Also, one can find K and S such that eq. (2) is satisfied in arbitrary coordinates [23]. This is achieved by introducing $n+1$ holomorphic sections $X^{A}(z)$, $A = 0, 1, \ldots, n$ which obey $\partial_{\overline{\alpha}}X^{A} = 0$. In terms of X^{A} the Kähler potential reads

$$K = -\ln Y$$

$$Y = X^{A}N_{AB}\overline{X}^{B} = X^{A}\overline{F}_{A} + \overline{X}^{A}F_{A} \tag{5}$$

where

$$N_{AB}(X,\overline{X}) = F_{AB}(X) + \overline{F}_{AB}(\overline{X})$$

$$F_{AB}(X) = \partial_{A}\partial_{B}F(X) \tag{6}$$

and $F(X)$ is a homogeneous function of degree two. $C_{\alpha\beta\gamma}$ and S are then given by

$$W_{\alpha\beta\gamma} = \partial_{\alpha}X^{A}\partial_{\beta}X^{B}\partial_{\gamma}X^{C}F_{ABC}$$

$$S = -\frac{1}{2}X^{A}N_{AB}X^{B} \ . \tag{7}$$

X^{A} transforms under Kähler transformations with weight $(-1,0)$, $\overline{X}^{A}$ with weight $(0,-1)$. N_{AB} is invariant as a consequence of F being homogeneous of degree two.

Let us define the $(2(n + 1)$ dimensional sympletic) holomorphic vector $V(z) = (X^A(z), iF_A(z))$. It is straightforward to show that as a consequence of eqs. (5)–(7) $V(z)$ satisfies the following set of identities (and their complex conjugate) [23]

$$
\begin{aligned}
D_\alpha V &= U_\alpha \\
D_\alpha U_\beta &= C_{\alpha\beta\gamma} g^{\gamma\overline{\gamma}} \overline{U}_{\overline{\gamma}} \\
D_\alpha \overline{U}_{\overline{\beta}} &= g_{\alpha\overline{\beta}} \overline{V} \\
D_\alpha \overline{V} &= 0
\end{aligned}
\tag{8}
$$

where the first equation defines U_α.* Eq. (2) now follows from eqs. (8), which is just an alternative way to characterize special geometry.

Finally, there is a simple set of coordinates given by $z^\alpha = X^\alpha/X^0$, $a = 1, \ldots, n$ [3, 23]. In these so called 'special coordinates' K and $C_{\alpha\beta\gamma}$ reduce to[†]

$$
\begin{aligned}
K &= -ln[2(F + \overline{F}) - (F_\alpha - \overline{F}_{\overline{\alpha}})(z^\alpha - \overline{z}^{\overline{\alpha}}] \\
C_{\alpha\beta\gamma} &= Y^{-1} F_{\alpha\beta\gamma} \ .
\end{aligned}
\tag{9}
$$

We will see in the next section that these coordinates are the flat coordinates for a holomorphic connection.

As we remarked in the introduction special Kähler geometry arises in $N = 2$ supergravity theories in four space-time dimensions as the manifold spanned by the scalars in the vector-multiplets [3, 4]. The moduli space of (2.2) SCFT (and thus the moduli space of Calabi-Yau-threefolds) also obeys eq. (2) where $C_{\alpha\beta\gamma}$ are the Yukawa couplings (the structure constants of the chiral ring [24]) of the assoicated matter multiples [7].

3. Holomorphic connection

Before we turn to the study of eqs. (8) let us first make a few observations about the connections in special geometry. We already remarked that in addition to the usual Christoffel

---

*The exact same set of identities were derived in a Calabi-Yau context in ref. 9,10.

[†]With a further (Kähler) gauge choice one can set $X^0 = 1$. However, special coordinates do not require this gauge.

connection $\Gamma^\gamma_{\alpha\beta}$ there also is a Kähler connection given by K_α. However, because of the additional constraint on the Kähler potential (eqs. (2), (8)) we find that both connections enjoy further properties in special geometry. These are most easily displayed by introducing Kähler invariant functions t^α as follows

$$t^a(z) = \frac{X^\alpha(z)}{X^0)(z)} \ , \qquad a = 1,\dots,n \ . \tag{10}$$

In terms of t^a and X^0 the Kähler connection K_α decomposes into a sum of a purely holomorphic term and a non-holomorphic piece

$$K_\alpha(z,\overline{z}) = \hat{K}_\alpha(z) = \mathcal{K}_\alpha(z,\overline{z}) \tag{11}$$

where

$$\mathcal{K}_\alpha(z,\overline{z}) = e^a_\alpha(z)K_a(z,\overline{z}) \equiv e^a_\alpha(z)\frac{\partial}{\partial t^a}K(t(z),\overline{t}(\overline{z}))$$
$$\hat{K}_\alpha(z) = -\partial_\alpha ln X^0(z) \tag{12}$$
$$e^a_\alpha(z) = \partial_a t^a(z)$$

The significance of the holomorphic piece $\hat{K}_\alpha$ is that it transforms as a connection whereas $\mathcal{K}_\alpha$ is invariant under Kähler transformation:

$$K_\alpha \to K_\alpha + f_\alpha \ , \qquad \hat{K}_\alpha \to \hat{K}_\alpha + f_\alpha \ , \qquad \mathcal{K}_\alpha \to \mathcal{K}_\alpha \ . \tag{13}$$

A very similar structure also occurs for $\Gamma^\gamma_{\alpha\beta}$. Let us first compute the metric in terms of t^a and X^0:

$$g_{\alpha\overline{\beta}} = e^a_\alpha \overline{e}^{\overline{b}}_{\overline{\beta}} g_{a\overline{b}} \equiv e^a_\alpha \overline{e}^{\overline{b}}_{\overline{\beta}} \frac{\partial}{\partial t^a}\frac{\partial}{\partial \overline{t}^{\overline{b}}}K(z,\overline{z}) \ . \tag{14}$$

(We see that $e^a_\alpha(z)$ can be viewed as a holomorphic vielbein.) As a consequence the Christoffel connecton ($\Gamma^\gamma_{\alpha\beta} \equiv g_{\alpha\overline{\gamma}\beta}g^{-1\overline{\gamma}\gamma}$) also splits in the following way

$$\Gamma^\gamma_{\alpha\beta}(z,\overline{z}) = \hat{\Gamma}^\gamma_{\alpha\beta}(z) + T^\gamma_{\alpha\beta}(z,\overline{z}) \tag{15}$$

where

$$T^\gamma_{\alpha\beta}(z,\overline{z}) = e^a_\alpha e^b_\beta g_{a\overline{d}b}g^{-1\overline{d}c}e^{-1\gamma}_c$$
$$\hat{\Gamma}^\gamma_{\alpha\beta}(z) = (\partial_\beta e^a_\alpha)e^{-1\gamma}_a \tag{16}$$

Again the holomorphic $\hat{\Gamma}^{\gamma}_{\alpha\beta}(z)$ transforms as a connection under reparametrizations whereas $T^{\gamma}_{\alpha\beta}$ enjoys tensorial transformation properties. Thus in both cases the transformation properties of the connections are entirely carried by the holomorphic objects $\hat{K}$ and $\hat{\Gamma}$. It is thus possible to introduce a purely holomorphic covariant derivative $\hat{D}$ where Γ and K are replaced by $\hat{\Gamma}$ and $\hat{K}$. As we will see in the next section the holomorphic (Picard-Fuchs) identities precisely use $\hat{D}$.

The (holomorphic) metric for which $\hat{\Gamma}$ is the connecton reads

$$\hat{g}_{\alpha\beta} = e^a_\alpha e^b_\beta \eta_{ab} \tag{17}$$

where η_{ab} is a constant (invertible) symmetric matrix. Note that $\hat{g}_{\alpha\beta}$ has two holomorphic indices in contrast to the Kähler metric $g_{\alpha\bar{\beta}}$. From eqs. (16) and (17) it comes as no surprise that $\hat{\Gamma}$ also satisfies

$$\hat{R}^{\gamma}_{\delta\alpha\beta} \equiv \partial_\delta \hat{\Gamma}^{\gamma}_{\alpha\beta} - \partial_\alpha \hat{\Gamma}^{\gamma}_{\delta\beta} + \hat{\Gamma}^{\mu}_{\alpha\beta}\hat{\Gamma}^{\gamma}_{\mu\delta} - \hat{\Gamma}^{\mu}_{\delta\beta}\hat{\Gamma}^{\gamma}_{\mu\alpha} = 0 \tag{18}$$

which means that it is a flat Riemannian connection. The flat coordinates are exactly the special coordinates $(t^a = z^a)$ introduced in the last section. In these coordinates we find

$$e^a_\alpha = \delta^a_\alpha \ , \qquad \hat{\Gamma}^{\gamma}_{\alpha\beta} = 0 \ , \qquad \hat{g}_{\alpha\beta} = \eta_{\alpha\beta} \ . \tag{19}$$

(The gauge choice $X^0 = 1$ then impliex $\hat{K}_\alpha = 0$.)

Before we turn to the derivative of the differential equation let us collect a few more formulas which we use in the next section. We observe that eq. (2) can also be expressed as

$$\partial_{\bar{\beta}}\left(\Gamma^{\delta}_{\alpha\gamma} - K_\alpha \delta^{\delta}_\gamma - K_\gamma \delta^{\delta}_\alpha + T^{\delta}_{\alpha\gamma}\right) = 0 \tag{20}$$

where

$$T^{\delta}_{\alpha\gamma} = C_{\alpha\gamma\epsilon}g^{\epsilon\bar{\epsilon}}(\overline{D}_{\bar{\delta}}\overline{D}_{\bar{\epsilon}}\overline{S})g^{\delta\bar{\delta}} = Y^{-1}D_\alpha X^A D_\gamma X^C \overline{D}_{\bar{\beta}}\overline{X}^B F_{ABC}g^{\bar{\beta}\delta} \ . \tag{21}$$

Inserting eqs. (15) and (11) into eq. (20) we learn that $T^{\delta}_{\alpha\gamma}$ can be further decomposed into

$$T^{\delta}_{\alpha\gamma} = \mathcal{K}_\alpha \delta^{\delta}_\gamma + \mathcal{K}_\gamma \delta^{\delta}_\alpha - \mathcal{T}^{\delta}_{\alpha\gamma} \tag{22}$$

Of course, this can also be verified directly from the definitions of $T^{\delta}_{\alpha\gamma}$ and $\mathcal{K}_\alpha$.

Another flat (non-holomorphic) connection exists on the flat symplectic bundle of ref. 9. What we find here is that a flat holomorphic connection exists on an n-dimensional space rather than a $2n + 2$ dimensional space as in ref. 9. In the deformation theory of the Hodge structure of Calabi-Yau-threefolds the components of the sympletic vector V are precisely the periods of the holomorphic three form Ω. We believe that $\hat{\Gamma}$ is the analogue of the flat connection of $N = 2$ TFT when it is restricted to the marginal deforomations [17, 19, 14]. (In general the connection of the TFT acts in the space of all topological deformations not only of the marginal ones.)

4. Holomorphic covariant Picard-Fuchs identities

Let us come back to the set of identities given in eq. (8). As we remarked above they are equivalent to the identity (2) satisfied by the Riemann tensor of special geometry. The Riemann tensor identity turns into a non-trivial differential equation for the Kähler potential K if the Yukawa couplings C_{abg} are specified. In the same spirit the identities (8) turn into a non-trivial differential equation once the coefficient functions are specified. As we will see in a moment the difference is that the latter leads to holomorphic differential equations which are exactly the Picard-Fuchs equations of the Calabi-Yau-manifold [25, 11, 13–15] and the analogous equations in TFT [20–22].

First we consider the case where $C_{\alpha\beta\gamma} = 0$ holds. Then eqs. (8) redude to

$$D_\alpha D_\beta V(z) = (\partial_\alpha \partial_\beta + \Lambda^\gamma_{\alpha\beta}\partial_\gamma + \Sigma_{\alpha\beta})V(z) = 0 \qquad (23)$$

where

$$\begin{aligned}
\Lambda^\gamma_{\alpha\beta} &= K_\alpha \delta^\gamma_\beta + K_\beta \delta^\gamma_\alpha - \Gamma^\gamma_{\alpha\beta} \,, \\
\Sigma_{\alpha\beta} &= K_{\alpha\beta} + K_\alpha K_\beta - \Gamma^\gamma_{\alpha\beta}K_\gamma \,.
\end{aligned} \qquad (24)$$

As we just discussed eq. (23) is an identity if we use the explicit form of K given by eqs. (5)–(7). However, the coefficients $\Lambda^\gamma_{\alpha\beta}$ and $\Sigma_{\alpha\beta}$ satisfy two noteworthy features. First, they transform under coordinate and Kähler transformations as to render eq. (23) covariant. This is obvious from eq. (23) but it can also be checked explicitly in eq. (24). Secondly, they satisfy

$\overline{\partial}_{\bar{\delta}} \Lambda^{\gamma}_{\alpha\beta} = \overline{\partial}_{\bar{\delta}} \Sigma_{\alpha\beta} = 0$ which holds as a consequence of eq. (2). Again, this has to be the case since $V(z)$ is holomorphic by definition. The considerations of the last section explain how it is possible to have manifestly holomorphic Λ, Σ which at the same time transform appropriately: they can only depend on the holomorphic connection $\hat{K}$ and $\hat{\Gamma}$. Indeed, inserting eqs. (11) and (15) into (24) one verifies

$$\Lambda^{\gamma}_{\alpha\beta} = \hat{\Lambda}^{\gamma}_{\alpha\beta} \equiv \hat{K}_{\alpha} \delta^{\gamma}_{\beta} + \hat{K}_{\beta} \delta^{\gamma}_{\alpha} - \hat{\Gamma}^{\gamma}_{\alpha\beta} \ ,$$
$$\Sigma_{\alpha\beta} = \hat{\Sigma}_{\alpha\beta} \equiv \hat{K}_{\alpha\beta} + \hat{K}_{\alpha} \hat{K}_{\beta} - \hat{\Gamma}^{\gamma}_{\alpha\beta} \hat{K}_{\gamma} \ . \tag{25}$$

Thus eq. (23) is covariant and holomorphic and the coefficients Λ and Σ depend on the holomorphic connections defined in the last section. Thus, if these holomorphic coefficients are given eq. (23) turns into a non-trivial differential equation for V. Of course in this simple case the solution is already known. In special coordinates (and in the gauge $X^0 = 1$) eq. (23) simplifies further and reads $\partial_{\alpha} \partial_{\beta} X^A = \partial_{\alpha} \partial_{\beta} F_A = 0$ which is consistent with our input $C_{\alpha\beta\gamma} = F_{\alpha\beta\gamma} = 0$. The solutions are (at least locally) the homogeneous spaces $SU(1, n)/SU(n) \times U(1)^*$ with $F = c_{AB} X^A X^B$ where c_{AB} is some constant symmetric matrix [3, 26].

Examples of this situation occur frequently in string compactifications. For the quintic in CP_4 there are 101 deformations of the complex structure. Certain subspaces of this 101 dimensional moduli space do have the property $C_{\alpha\beta\gamma} = 0$, where x denotes all 101 deformation [7]. As a concrete example consider the defining polynomial

$$p = \sum_{k=1}^{5} x_k^5 - \rho x_4^3 x_5^2 - \sigma x_4^2 x_5^3 \tag{26}$$

where ρ and σ denote the parameters of the deformation. By using the methods developed in ref. 13–15 one can derive the following system of coupled second order differential equations

$$\left((1 - 4\sigma\rho) \frac{\partial^2}{\partial\rho^2} - 3\rho \frac{\partial^2}{\partial\sigma^2} - 6\sigma^2 \frac{\partial^2}{\partial\rho\partial\sigma} - \left(6\sigma + \frac{6\sigma}{1 - 4\rho\sigma} \right) \frac{\partial}{\partial\rho} - \frac{12\rho^2}{1 - 4\rho\sigma} \frac{\partial}{\partial\sigma} \right) V = 0 \tag{27}$$

where the second equation is obtained from (27) by interchanging ρ and σ. This is in agreement with eq. (23).

---

*Globally, there can be discrete identification which leads to the quotient of the homogeneous space with the duality group. The latter is related to the monodromy group of eq. (23) [13, 14].

Let us turn to the case of non vanishing $C_{\alpha\beta\gamma}$. Now $D_\alpha D_\beta V$ is non-zero and Λ, Σ are no longer holomorphic. Instead they satisfy

$$\Lambda_{\alpha\beta}^\gamma = \hat{\Lambda}_{\alpha\beta}^\gamma + T_{\alpha\beta}^\gamma$$
$$\Sigma_{\alpha\beta} = \hat{\Sigma}_{\alpha\beta} + T_{\alpha\beta}^\gamma K_\gamma + \Sigma_{\alpha\beta}^{(3)} \tag{28}$$

where

$$\overline{\partial}_{\overline{\gamma}} \Sigma_{\alpha\beta}^{(3)} = -T_{\alpha\beta}^\gamma g_{\gamma\overline{\gamma}} \ .$$

Now a holomorphic identity appears if one use the full set (8). This is particularly simple if one considers a one dimensional manifold. In this case eqs. (8) are equivalent to

$$DDW^{-1}DDV = \sum_{n=0}^{4} a_n(z)\partial^n V = 0 \ . \tag{30}$$

The coefficients a_n are given by[†]

$$a_4 = W^{-1}$$
$$a_3 = 2\partial W^{-1}$$
$$a_2 = W^{-1}(\partial\Lambda - \Lambda^2 + 2\Sigma) + \partial W^{-1}\Lambda + \partial^2 W^{-1} \tag{31}$$
$$a_1 = W^{-1}(\partial^2\Lambda + 2\partial\Sigma - 2\Lambda\partial\Lambda) + \partial W^{-1}(2\Sigma + 2\partial\Lambda - \Lambda^2) + \partial^2 W^{-1}\Lambda$$
$$a_0 = W^{-1}(\Sigma^2 - \Sigma\partial\Lambda - \Lambda\partial\Sigma + \partial^2\Sigma) + \partial W^{-1}(2\partial\Sigma - \Lambda\Sigma) + \partial^2 W^{-1}\Sigma$$

where

$$\Lambda = 2\partial K - \Gamma$$
$$\Sigma = \partial^2 K + (\partial K)^2 - \Gamma\partial K \ . \tag{32}$$

As before the a_n's have the appropriate transformation properties and satisfy $\overline{\partial} a_n = 0$ as a consequence of eq. (2). Inserting eqs. (15) and (11) into (31) we find that in all coefficients a_n K and Γ are replaced by their holomorphic pieces $\hat{K}$ and $\hat{\Gamma}$. Thus in eq. (31) instead of Γ and Σ appear

$$\hat{\Lambda} = 2\partial\hat{K} - \hat{\Gamma} \ ,$$
$$\hat{\Sigma} = \partial^2\hat{K} + (\partial\hat{K})^2 - \hat{\Gamma}\partial\hat{K} \ . \tag{33}$$

[†]The dependence of a_3 on W was also found in ref. 11.

Now the meaning of eq. (30) becomes transparent. As it stands it is a covariant identity in special geometry. The coefficient functions a_n are holomorphic functions of the Yukawa couplings W and the holomorphic connections $\hat{K}$, $\hat{\Gamma}$ (or equivalently the einbein e and X^0). If these three quantities are given as 'data' eq. (30) turns into a linear differential equation for V. (Given a_n eqs. (31) also imply a non-linear differential equation for e and X^0.)

From eq. (31) we learn that the a_n are related in the following form

$$a_3 = 2\partial a_4 \ , \qquad a_1 = \partial a_2 - \frac{1}{2}\partial^2 a_3 \ . \tag{34}$$

Indeed, for the specific example of the quintic in CP_4 the a_n's (in the Landau-Ginzburg coordinates) are given by [11, 13, 14]

$$a_4 = 1 - \psi^5 \qquad a_3 = -10\psi^4 \qquad a_2 = -25\psi^3$$
$$a_1 = -16\psi^2 \qquad a_0 = -\psi \tag{35}$$

and they obey (34).

In special coordinates (with the additional gauge choice $X^0 = 1$) eq. (30) takes a particularly simple form, it coefficients are

$$a_4 = W^{-1} \ , \qquad a_3 = 2\partial W^{-1} \ ,$$
$$a_2 = \partial^2 W^{-1} \ , \qquad a_1 = a_0 = 0 \ . \tag{36}$$

The reason for the vanishing of a_1 and a_0 is that in these coordinates $\partial X^0 = 0$ and $\partial X^1 = 1$ holds. Thus for consistency $a_0 = a_1 = 0$ is required. (Also, we can use (36) and $W = \partial^3 F$, $V = (1, X^1, i(2F - X^1\partial F), i\partial F)$ to verify that (30) is identically satisfied.)

Finally, it is possible to remove the Kähler connection $\hat{K}$ from eq. (30) altogether. If we rescale V by $1/X^0$ and define a Kähler invariant $\mathcal{V} = V/X^0$ the differential equation reads

$$\sum_{n=0}^{4} a'_n(z)\partial^n \mathcal{V} = 0 \ . \tag{37}$$

In this basis, a'_n is independent of $\hat{K}$ and all X^0 dependence appears through the ratio $W/(X^0)^2$. In this basis $a_0 = 0$ holds which again is required for consistency. Inverting this procedure produces a relation between a_0 and a_1, a_2, a_3.

5. Conclusion

In this letter we found that the connections in special geometry can be split into a holomorphic piece which transforms a connection and a non-holomorphic piece with tensorial transformation properties. This fact allows for the existence of holomorphic covariant identities in special geometry. When certain 'data' (the Yukawa coupling and the holomorphic connection) are given these identities turn into linear differential equations which are the exact analogue of the Picard-Fuchs equations for the periods of the holomorphic threeform Ω in Calabi-Yau manifolds.

We treated in some detail the case of a one dimensional Kähler manifold but clearly our consierations can be extended a arbitary dimension. From eq. (8) it is obvious that again holomorphic identities occur whose coefficient functions depend on the holomorphic connections.

Our method is particularly powerful when applied to the deformation theory of Calabi-Yau threefolds and the $N = 2$ TFT. Indeed, it allows Picard-Fuchs equations to be obtained in arbitrary coordinate system and to give a precise relation between the Kähler geometry of the moduli space and the flat geometry of the deformation of the chiral ring restricted to marginal operators.

Various issues discussed recently in the literature can be further investigated. In particular it is known that nth-order Picard-Fuchs equations are associated with classical W_n-algebras. For $n = 4$ (the case discussed at the end of the previous section) the W_n generators ($n = 2$, 3, 4) can be explicitly constructed in terms of the coefficients a_n given in eq. (31). One finds generically $W_3 = 0$ as a consequence of eq. (34) [27], as is the case for the K_3 surface [14]. W_2 can always be set to zero by a combined choice of Kähler gauge and coordinate transformations. Therefore the coordinate free content of the dynamic inherent to a given Calabi-Yau space seems to reside in the W_4 generator. Constant Yukawa couplings give $W_4 = 0$ while instanton contributions to Yukawa couplings (for the $(1,1)$ deformations), generally expressed as nonperturbative corrections due to rational curves of degree k on the Calabi-Yau manifold give $W_4 \sim k^4$. In word-sheet terminology W_4 vanishes to any finite order in perturbation theory of the Calabi-Yau σ-model. Thus W_4 seems to characterize the instanton effects of the quantum Calabi-Yau manifold, i.e. the quantum distortion from a cubic prepotential F for the moduli space of Kähler class deformations of a classical Calabi-Yau manifold. We also suspect that the W_4 generator plays an important role of elucidating the target space duality group of generic Calabi-Yau manifolds.

For the case of a higher dimensional moduli space and Picard-Fuchs equations become a coupled set of fourth order partial differential equations. Relation between its coefficients, generalizing eq. (34), are conjectured to lead to a generalization of covariant holomorphic tensors and therefore to a generalized notion of W_n algebras. These topics are discussed in detail in ref. 27.

Acknowledgements:

It is a pleasure to thank R. D'Auria, A. Ceresole, M. Cvetič, W. Lerche, M. Peskin, R. Schimmrigk, D. Smit and N. Warner for useful discussions and in particular P. Candelas and X. de la Ossa for communicating some unpublished work. J.L. thanks the Aspen Center of Physics for providing a stimulating atmosphere where this investigation started.

Some of the computations were performed using MAPLE.

References

[1] For a review see L. Dixon, in Proc. of the 1987 ICTP Summer Workshop in High Energy Physics and Cosmology, Trieste, ed. G. Furlan, J.C. Pati. D.W. Sciama, E. Sezgin and Q. Shafi and references therein.

[2] For a review see S. Ferrara, Mod. Phys. Lett. **A6** (1991), 2175 and references therein.

[3] B. de Wit and A. van Proeyen, Nucl. Phys. **B245** (1984), 89;
B. de Wit, P. Lauwers and A. van Proeyen, Nucl. Phys. **B255** (1985), 569.

[4] E. Cremmer, C. Kounnas, A. van Proeyen, J.P. Derendinger, S. Ferrara, B. de Wit and L. Girardello, Nucl. Phys. **B250** (1985), 385.

[5] N. Seiberg, Nucl. Phys. **B303** (1988), 206.

[6] S. Cecotti, S. Ferrara and L. Girardello, Int. J. Mod. Phys. 4 (1989), 2475, Phys. Lett. **B213** (1988), 443.

[7] L. Dixon, V. Kaplunovsky and J. Louis, Nucl. Phys. **B329** (1990), 27.

[8] S. Ferrara and A. Strominger, in *Strings 89*, eds. R. Arnowitt, R. Bryan, M. Duff, D. Nanopoulos and C. Pope, World Scientific, Singapore, 1989.

[9] A. Strominger, Comm. Math. Phys. **133** (1990), 261.

[10] P. Candelas and X. de la Ossa, Nucl. Phys. **B342** (1990), 246.

[11] P. Candelas, X. de la Ossa, P. Green and L. Parkes, Phys. Lett. **B258** (1991), 118, Nucl. Phys. **B359** (1991), 21.

[12] B. Greene and M. Plesser, Nucl. Phys. **B338** (1990), 15.

[13] A. Cadavid and S. Ferrara, Phys. Lett. **B267** (1991), 193.

[14] W. Lerche, D.J. Smit and N. Warner, Caltech preprint CALT-68-1738.

[15] D. Morrison, Duke preprint DUK-M-91-14.

[16] R. Dijkgraaf, E. Verlinde and H. Verlinde, Nucl. Phys. **B352** (1991), 59.

[17] B. Blok and A. Varchenko, Princeton Preprint IASSNS-HEP-91/5.

[18] S. Cecotti, S. Girardello and A. Pasquinucci, Nucl. Phys. **B328** (1989), 701, Int. J. Mod. Phys. **A6** (1991) 2427.

[19] S. Cecotti and C. Vafa, Harvard preprint HUTP-91/A031.

[20] E. Verlinde and N. Warner, Phys. Lett. **B269** (1991), 96.

[21] Z. Maassarani, USC preprint USC-91/023.

[22] A. Klemm, M.G. Schmidt and S. Theisen, Karlsruhe preprint KA-THEP-91-09.

[23] L. Castellani, R. D'Auria and S. Ferrara, Phys. Lett. **B241** (1990), 57 Class. Quant. Grav. **1** (1990), 317;

R. D'Auria, S. Ferrara and P. Fré, Nucl. Phys. **B359** (1991), 705.

[24] W. Lerche, C. Vafa and N. Warner, Nucl. Phys. **B324** (1989), 427.

[25] P. Candelas and X. de la Ossa, private communication.

[26] E. Cremmer and A. van Proeyen, Class. Quant. Grav. **2** (1985), 445.

[27] A. Ceresole, R. D'Auria, S. Ferrara, W. Lerche and J. Louis, paper in preparation.

AMS/IP Studies in Advanced Mathematics
Volume **9**, 1998

A New Geometry From Superstring Theory

P.S. Aspinwall And C.A. Lütken

Department of Theoretical Physics

University of Oxford

1 Keble Road, Oxford OX1 3NP

U.K.

June 1991

Abstract

We review the main ideas associated with problems concerning "geometrical" interpretations of the information contained in 2–dimensional $N = 2$ superconformal field theories. At stake is a new way of looking at space-time which is uniquely stringy and which goes a long way towards removing the classical dichotomy between space and matter, thus opening up the possibility of a truly unified picture of natural phenomena. We try to bridge the gap between mathematicians and physicists by mixing languages from both cultures. For technical reasons and for the purpose of illustration most of the discussion is centered on a specific calculational example, related to quintics in $\mathbb{P}^4$, which remains the only tractable case found to date. Of central interest is the "classical" or "large radius limit" in which the conformal field theory reduces to a non-linear σ-model with target space of the Calabi-Yau type. The possibility of "flopping" between manifolds of different topology implies that the large radius limit of a given conformal model is ambiguous, and that the instantons in string theory could smooth out some of the singularities present in the classical moduli space.

© 1998 American Mathematical Society
and International Press

1. Introduction

The ultimate goal of high energy physics is to give a unified description of gravity, which in Einstein's approach is modeled using Riemannian geometry, and the elementary particles, which in the standard model are described by quantum fields living in a flat 3+1 dimensional Minkowski space-time. In order to achieve this unification space-time should be promoted to a more active dynamical role, while the description of particles should become more geometrical. This is exactly what happens in string theory, and is the main justification for the enormous effort currently being invested in strings.

In order for 4-dimensional space-time to be a concept derived from, rather than a prerequisite for, dynamics, it must be purged from the fundamental formulation of the theory. This is currently done by regarding strings not as extended objects propagating through an ambient space-time, but as a 2-dimensional quantum field theory. It must satisfy very restrictive consistency conditions which amount to requiring conformal symmetry and complex supersymmetry. Physicists think of the 2-manifold as the "world-sheet" of the string. Mathematicians are perhaps better served by simply taking "string" to be a mnemonic for the kind of data which go into the precise definition of an object we call an "$N = 2$ superconformal quantum field theory". The underlying structure of this object is an algebra known as the $N = 2$ superconformal algebra. This construct has a surprisingly rich mathematical structure which we are only beginning to understand. The solutions of the field equations span a large variety of geometrical objects. In fact the space which parameterizes the solutions (the moduli space of the field theory which is itself a Kähler manifold) interpolates in a continuous manner between Kähler manifolds with radically different topologies.

All this fits in nicely with the ideas behind algebraic geometry. That is, we try not to think of geometric ideas directly but use algebraic structures to model geometry. String theory appears to take this a step further by demanding that the geometric concepts are dictated by the underling algebra and that conventional geometry arises only as an approximation in some limit. This structure has been dubbed "quantum algebraic geometry"[1] [1]. The simplest way to see that conventional geometry arises as some limit is to naively interpret string theory as saying that all point-like objects should be replaced by small loops of string. These loops of string have a very large tension which confines them to being very small at normal energies. Now, conventional geometry is all based on the notion of space as being a collection of points. If string theory is going to agree with geometry we should make sure that all our scales are sufficiently large as to approximate the string loop to a point. This is generally referred to as the "large-radius limit" and in this limit we expect to regain conventional geometry, Einsteinian gravity and normal particle physics.

It has been known for some time [2] that the natural way to formulate an approach to space-time in string theory is through conformal field theories. A perturbative approximation to the large-radius limit is given by the non-linear σ-model [3] and this should be regarded as the key tool in matching a superconformal field theory (SCFT) to the corresponding space when we are at large radius. If however the SCFT does not correspond to large radius then we lose any direct correspondence and we shall argue below that it is not possible to fix even the topology of any space that the SCFT is meant to be describing.

The aim of deriving space-time from string theory is a little too ambitious for the current status of the theory but what we can do is to fix the flat (3+1)-dimensional space-time we observe and model the rest of space with a SCFT. If all of space-time were flat, string theory suggests that it should be 10-dimensional and so the SCFT we are considering should give a space of 6 real dimensions in the large radius limit. Of course, for a physically realistic theory, the characteristic size of this space must not get too big since it must be invisible at every-day energies.

What is being alleged is that string theory possesses as ground state solutions a set of field configurations with the property that some of the fields *under certain conditions* can be

---

[1] This leads to potential confusion about the definition of geometry. Either one can say that geometry is governed by the physics of space in which case conventional geometry is only approximately geometrical or one can define geometry as conventional geometry and then say that the physics of space is not geometrical at small scales.

interpreted as parametrizing (coordinatizing) 9+1 dimensional semi-compact manifolds of the type (4-dimensional Minkowski) × (6-dimensional Calabi-Yau). This is the vacuum state of the theory, i.e. the "space-time arena" on which the more conventional particle physics will be played out, the "particles" being excitations of the fields around their background values. In this way the string is believed to circumvent the traditional dichotomy between space and matter; they are inseparable and complementary aspects of the 2-dimensional quantum fields which constitute all there is in this model of Nature, thus providing a true unification of the two fundamental categories into which we organize our physical concepts.

The algebra underlying the SCFT's generating our model is the $N = 2$ superconformal algebra [6]. The $N = 2$ world-sheet supersymmetry is necessary for obtaining space-time supersymmetry, which is believed to be required in order to solve several of the outstanding problems in elementary particle physics. It is well known that the target space of $N = 2\sigma$-models automatically are complex, Kähler, and that those which arise in string theory also have vanishing first Chern class ($k = 0$) [3] and therefore are Calabi-Yau manifolds[2]. Clearly the precise nature of these string ground state solutions; the conditions under which the identification with conventional Riemannian manifolds can be made (and to which extent it is unique), as well as the way in which the relation changes as the vacuum parameters (moduli) are varied, are among the most interesting problems in string theory. We shall review some of our work which addresses these questions [1, 4, 5]. Briefly, the situation is as follows.

A conformal field theory is uniquely characterized by its spectrum of primary fields, together with the operator product algebra, or equivalently the three-point couplings between these fields. This bears a striking resemblance to Wall's classification of six-dimensional spaces [7], which in the case of Calabi-Yau manifolds amounts to specifying the spectrum of harmonic forms together with their triple intersections (Yukawa couplings). (One also requires $c_2(M)$. This is closely related to the triple self-intersections but the real string interpretation of this has yet to be fully elucidated.) Thus it would seem that the classification of conformal field theories requires the same kind of data as those needed to classify three-dimensional Calabi-Yau

---

[2]Physicists do not normally impose $b_1 = 0$ in the definition of a Calabi-Yau manifold.

manifolds. However, we shall see that extra information is required if an arbitrary SCFT is to be identified with a Calabi-Yau manifold. This information will be which target direction in the moduli space away from our starting SCFT will be identified with taking a large radius limit.

The moduli space of a SCFT is very difficult to determine and so it is tricky to take the approach above directly. However, the recently discovered mirror manifold property allows us to circumvent this problem.

2. Large Radius Limit

When considering a space such as a Calabi-Yau manifold, the term "large-radius limit" is clearly ambiguous and we need to give this a more precise definition if we want to do some explicit calculations. This is done most simply by looking at the regime where the non-linear σ-model breaks down badly. This happens when non-perturbative effects (instantons) become important in calculation of physical quantities such as the superpotential. Instantons arise in the σ-model because of classical solutions of maps of the string world-sheet into the target space which are not homotopic to a point. The equations of motion of the string are satisfied when the map of the world-sheet into the target space is holomorphic [8]. If the world-sheet is smooth and this is a smooth map we thus see that instantons correspond to algebraic curves. Also in this case it has been shown [9] that the superpotential is only affected by world-sheets of genus 0. Therefore the instantons of physical importance are rational curves on the Calabi-Yau.

Consider families of maps of the world-sheet (a sphere) into the Calabi-Yau, i.e., we consider $\pi_2(M)$. If M is simply connected then the homotopy group $\pi_2(M)$ is equivalent to the homology group $H_2(M)$. If M is a Calabi-Yau manifold, we also have $h^{2,0} = h^{0,2} = 0$, so $H_2(M)$ can be generated by algebraic curves. It was shown in [8] that a map $S^2 \to M$ given by $X^\mu(\sigma)$ contributes $\exp(-I)$ to the path integral, where I is the Euclidean instanton action with

$$I \geq \frac{1}{4\pi\alpha'} \left| \int d^2\sigma J_{\mu\bar{v}}\epsilon^{\alpha\beta} \frac{\partial X^\mu}{\partial\sigma^\alpha} \frac{\partial \overline{X}^v}{\partial\sigma^\beta} \right| \tag{1}$$

and $J_{\mu\bar{v}}$ is the Kähler form. By definition, the instanton is the field configuration (map) $X^\mu(\sigma)$ which minimizes the instanton action. This occurs when $X^\mu(\sigma)$ is holomorphic, i.e., we have an algebraic curve. An algebraic curve will intersect the generic hyperplane $\mathbb{P}^{N-1}$ of the ambient $\mathbb{P}^N$ at a finite set of points. The number of points is known as the *degree* of the curve. It is known, for example, that the quintic hypersurface in $\mathbb{P}^4$ (sometimes called $Y_{4;5}$) has isolated rational algebraic curves of an infinite number of positive degrees and so has an infinite number of such curves.

If $X^\mu(\sigma)$ is an element of $H_2(M, \mathbb{Z})$, we can write (1) as

$$I \sim \int_X J. \tag{2}$$

It is therefore seen that to have small instanton contributions, the cohomology class of the Kähler form must be such that (2) is large for any $X \in H_2(M, \mathbb{Z})$. This then can be taken as the condition for a geometric Calabi-Yau compactification to work or, in other words, this is the classical region of quantum algebraic geometry.

The ideas above are best expressed if we use the **concept** of the *Kähler cone*. The Kähler form is a real (1,1)-form and so its cohomology class can be considered a point in $\mathbb{R}^{h^{1,1}}$. There are, however, further conditions to be placed upon J for a smooth manifold, M. Namely

$$\begin{aligned}
\int_M J \wedge J \wedge J &> 0 \\
\int_S J \wedge J &> 0 \qquad \forall S \subset M \\
\int_X J &> 0 \qquad \forall X \subset M.
\end{aligned} \tag{3}$$

Here S is any surface and X is any curve embedded in M. If J satisfies the above conditions, it is clear that λJ (where λ is a real positive number) will also provide a satisfactory Kähler form. One therefore sees that the region of space described by (3) is the interior of a cone with its vertex at the origin. This is the Kähler cone. The content of (2) can then be stated as the fact that *one obtains the results of classical algebraic geometry when the cohomology class of the Kähler form lies well within the Kähler cone.* This is the generalization of the expression "large radius limit".

3. An Explicit Example

In order to fix ideas and make the discussion less abstract let us work out the details for the simplest example we have been able to locate which is sufficiently complicated to exhibit at least some of the subtleties associated with the large radius limit. The following calculation first appeared in [4].

From a geometric point of view, the simplest Calabi-Yau manifold to study is $Y_{4;5}$. From a conformal field theory point of view, it has been established [10, 11] that this is similar to 5 copies of the $k = 3$ minimal model quotiented out by a $\mathbf{Z}_5$ group to obtain space-time supersymmetry [12]. This resultant theory has many symmetries. Of particular interest are the symmetries of the type

$$g : \Phi_i \mapsto \Phi_i e^{\frac{2\pi i}{2(k_i+2)}(q_i+\bar{q}_i)}; \qquad q_i, \bar{q}_i = 0, \dots, k - 1. \tag{4}$$

(See [5] for the notation used here.) Thus, in this model we should have a $(\mathbf{Z}_5)^5$ group of symmetries. We can represent an element of this group by a list of five numbers defined modulo 5: $(f_0, f_1, \dots, f_4)$. Actually, the $\mathbf{Z}_5$ group we used to obtain the space-time supersymmetry was $(1, 1, 1, 1, 1)$ and so we have only a $(\mathbf{Z}_5)^4$ group of symmetries of this type remaining in this model. We shall refer these symmetries that act diagonally on the chiral primary fields (Φ_i) as "*phase*" symmetries. Other symmetries of the theory can be obtained by permuting the five $k = 3$ models and so we have for this symmetry group simply the permutation group over five elements P_5. This group of permutations acts by automorphisms on the group of phase symmetries. Hence the total symmetry group of the conformal field theory is a semidirect produdct $P_5 \ltimes (\mathbf{Z}_5)^4$.

These symmetries can now be used to generate a list of mirror pairs of models. See the paper by B.R. Greene elsewhere in this volume for full details. The key point is that pairs can be obtained by dividing the SCFT by groups of phase symmetries satisfying

$$\sum_{i=0}^{4} f_i = 0 \quad (\text{mod } 5). \tag{5}$$

These groups are isomorphic to $(\mathbb{Z}_5)^3$ and its subgroups. The interesting part is that if these SCFT's can be associated to Calabi-Yau manifolds we should be able to obtain the Hodge numbers purely by geometric means. Our starting manifold $Y_{4;5}$ is defined to be a hypersurface of degree 5 in the complex projective space $\mathbb{P}^4$. We know more than this however. The manifold that corresponds to the above conformal field theory must actually be based on the specific defining polynomial

$$p_0 = \frac{1}{5} \sum_{i=0}^{4} z_i^5 = 0, \tag{6}$$

where the z_i here are homogeneous coordinates of $\mathbb{P}^4$. (We should be careful with our language here; the extent to which this SCFT can be associated with this specific manifold should become clearer later in this paper.) Also, since we can think of the z_i's as corresponding to the chiral primary fields we can also identify the way that our symmetry group acts on the manifold. We should therefore be able to quotient out $Y_{4;5}$ geometrically by all the groups above and reproduce the same Hodge numbers. Some of the groups act with fixed points and so the resulting space after quotienting can be singular. In order to determine the Hodge numbers we deform the space a bit to obtain a smooth manifold of vanishing first Chern class. Note that the condition (5) allows this to be possible. This procedure, usually referred to as *"blowing up"*, leads to unique numbers for $h^{1,1}$ and $h^{2,1}$. This procedure can be done for all of the models using the method of toric resolutions and the results do indeed agree with the conformal field theory calculation.

We have not finished with $Y_{4;5}$ however, as we are also free to quotient out by all the permutation symmetries which preserve vanishing first Chern class. Here we consider the group $C_5 \simeq \mathbb{Z}_5$ generated by

$$g_5 : (z_0, z_1, \ldots, z_4) \mapsto (z_4, z_0, z_1, z_2, z_3) \tag{7}$$

and the group $C_3 \simeq \mathbb{Z}_3$ generated by

$$g_3 : (z_0, z_1, \ldots, z_4) \mapsto (z_2, z_0, z_1, z_3, z_4). \tag{8}$$

There is also a permutation group $C_2 \simeq \mathbb{Z}_2$ but for the sake of brevity we will not consider it here. One can now quotient out by all the subgroups of the total symmetry group generated by the phase symmetries and the two permutation symmetries (7) and (8). It is known that it is always possible to blow-up a quotient singularity [13] but it has not yet been shown that a resolution exists that can preserve the $K = 0$ condition. It goes without saying therefore that there is no known general way of explicitly performing the resolution. In the case of points fixed by an abelian group, the method of toric resolutions is guaranteed to work but some of the groups we are now considering can fix points by nonabelian actions. In these cases we need to resort to the Euler characteristic formula of [14] (see also [15]) and use the fact that the blow-up of a point has no deformations of complex structure associated to it. The results of these calculations is shown in Table 1. If we allow ourselves to admit that SCFT's do indeed describe geometry, one can take this as a proof that the resolutions of these non-abelian quotients preserving $K = 0$ are unobstructed.

It is clearly seen that the table gives the mirror manifolds in pairs. It is not immediately obvious how a manifold is related to its mirror but on closer examination a pattern is revealed (see also [16]).

If string compactification on a manifold and compactification on its mirror are supposed to lead to the same physics it must be that the discrete symmetry group of these two manifolds are the same. This discrete symmetry is given by the normalizer of the quotienting group within the total symmetry group $P_5 \ltimes (\mathbb{Z}_5)^4$. That is, when quotienting out by a subgroup G of the total symmetry group S_0, the resultant symmetry is the group $H \subseteq S_0$ containing the elements h such that

$$h^{-1}gh = g', \quad g' \in G, \quad \forall g \in G. \tag{9}$$

The column labeled N in table 1 shows which subgroups of the permutation group P_5 are left in the normalizer when quotienting our by the *phase* part of G. If we only consider the entries in the table where G is generated by phase symmetries alone, it can be seen that the column N is enough to uniquely pair up the mirror

Group	Generators	N	$h^{1,1}$	$h^{2,1}$	χ
1		P_5	1	101	-200
$(\mathbb{Z}_5)^3$	$(0,0,0,1,4)$ $(0,0,1,0,4)$ $(0,1,0,0,4)$	P_5	101	1	200
$\mathbb{Z}_5$	$(0,0,0,1,4)$	C_3	5	49	-88
$(\mathbb{Z}_5)^2$	$(0,1,4,0,0)$ $(0,1,1,4,4)$	C_3	49	5	88
$\mathbb{Z}_3$	g_3	P_5	5	49	-88
$\mathbb{Z}_3 \ltimes (\mathbb{Z}_5)^3$	g_3 $(0,0,0,1,4)$ $(0,0,1,0,4)$ $(0,1,0,0,4)$	P_5	49	5	88
$\mathbb{Z}_5$	$(0,1,2,3,4)$	$\{C_5, C_2\}$	1	21	-40
$(\mathbb{Z}_5)^2$	$(0,1,2,3,4)$ $(0,1,4,4,1)$	$\{C_5, C_2\}$	21	1	40
$\mathbb{Z}_5$	g_5	P_5	1	21	-40
$\mathbb{Z}_5 \ltimes (\mathbb{Z}_5)^3$	g_5 $(0,0,0,1,4)$ $(0,0,1,0,4)$ $(0,1,0,0,4)$	P_5	21	1	40
$(\mathbb{Z}_5)^2$	$(0,1,1,4,4)$ $(0,1,2,3,4)$	C_2	17	21	-8
$\mathbb{Z}_5$	$(0,1,1,4,4)$	C_2	21	17	8
$\mathbb{Z}_{15}$	g_3 $(0,0,0,1,4)$	C_3	17	21	-8
$\mathbb{Z}_3 \ltimes (\mathbb{Z}_5)^2$	g_3 $(0,1,4,0,0)$ $(0,1,1,4,4)$	C_3	21	17	8
$(\mathbb{Z}_5)^2$	g_5 $(0,1,2,3,4)$	$\{C_5, C_2\}$	1	5	-8
$\mathbb{Z}_5 \ltimes (\mathbb{Z}_5)^2$	g_5 $(0,1,2,3,4)$ $(0,1,4,4,1)$	$\{C_5, C_2\}$	5	1	8

TABLE 1. Some quotients of the manifold $Y_{4;5}$.

manifold with its partner. Thus when only quotienting out by phase symmetries we have enough information from the discrete symmetries of the resulting theory to establish which theory is its mirror partner. In order to obtain the rest of the table we now take each pair of theories we have already generated and quotient each table we now take each pair of theories we have already generated and quotient each of these out by the same permutation group (making sure that the resulting total group G is indeed closed). This leads to the other mirror pairs. We see that the phase symmetries and the permutation symmetries appear to play quite different rôles in building up the table.

We now wish to use this mirror property to show how string theory can affect geometrical ideas. First note that the mirror transformation exchanges $(1,1)$-forms and $(2,1)$-forms. In Calabi-Yau manifolds, $(1,1)$-forms can be considered as deformations of the Kähler form whereas $(2,1)$-forms can be regarded as deformations of the complex structure. The mirror property can thus be used to translate between these two types of deformations. The first apparent problem is that $h^{1,1}$ counts *real* deformations of the Calabi-Yau by motion in the Kähler cone but $h^{2,1}$ counts the *complex* dimension of the moduli space of complex structures. This anomaly is cured naturally in string theory by complexifying the Kähler cone [17, 18]. This amounts to introducing an element in $H^2(M,\mathbb{C})$ whose imaginary part is the usual Kähler form and whose real part corresponds to torsion in the metric of the Calabi-Yau manifold.

A natural trilinear product exists in both the $(1,1)$ and the $(2,1)$ sector of the theory. These products correspond to Yukawa couplings in string theory. The coupling of three 2-forms $p, q, r \in H^{1,1}$ is a genuinely topological quantity (on a *three-dimensional* variety), namely the *triple intersection numbers* [19]:

$$p \circ q \circ r = \int_M p \wedge q \wedge r, \tag{10}$$

which are straight forwardly computable integers if a topological basis is chosen. However, these topological contributions to the $(1,1)^3$ couplings are only true to leading order in σ-model perturbation theory, and are expected to be corrected by instanton effects in the full string theory.

The three-point couplings of $(2,1)$-forms are apparently of a completely different character. Three fields $a, b, c \in H^{2,1}$ are uniquely related to three fields $A^{\alpha}, B^{\beta}, C^{\gamma} \in H^1(M, T)$ by the group isomorphism which is induced by the unique holomorphic 3-form Ω on Calabi-Yau manifolds, and their Yukawa coupling is given by

$$a \circ b \circ c = \int_M \Omega \wedge A^{\alpha} \wedge B^{\beta} \wedge C^{\gamma} \Omega_{\alpha\beta\gamma}. \tag{11}$$

One should note that these couplings are not topological invariants but depend on the complex structure. However, one can show that these couplings are not affected by higher loop σ-model corrections or instanton effects [20]. It is this fact which will allow us to calculate the exact SCFT couplings of the $(1,1)$-sector – we need only to calculate the corresponding $(2,1)$-couplings in the mirror using (11).

In the simplest case where the Calabi-Yau manifold is constructed as a polynomial constraint in $\mathbb{P}^n$, the complex structure moduli can be chosen to be the independent coefficients that can appear in the most general defining polynomial (obstructions may sometimes exists to carrying out this identification for all moduli, but this does not happen in the cases considered here). Hence the question we wish to investigate in the remainder of this section is most naively formulated as follows: Does the calculation of $(2,1)$ intersecton numbers of M using deformation theory agree with the calculation of $(1,1)$ intersection numbers of $\widetilde{M}$ using algebraic topology when M is in the large complex structure limit? By large complex structure limit, we mean the limit of M corresponding to the mirror of the above defined large radius limit of $\widetilde{M}$.

We have investigated this question in detail for the simplest mirror pair we could find. These are the manifolds with $\chi = \pm 40$ which appear in Table 1 as the seventh and eighth entry. Firstly, since $M = M_{-40}$ is a simple quotient of a complete intersection Calabi-Yau manifold, deformation theory can be applied straight forwardly to its covering space, a quintic in $\mathbb{P}^4$, which is the simplest of all algebraic Calabi-Yau manifolds. Secondly, since the mirror manifold $\widetilde{M} = M_{40}$ is obtained by blowing up a quotient variety with only isolated singularities, its intersection cubic of 2-forms is also readily obtained.

We need to calculate the coupling of $(1,1)$-forms on M_{40}. This is topological problem since the couplings can be thought of as the intersection numbers of various algebraic surfaces (2 complex-dimensional subspaces) within our manifold. What is a convenient basis for these $(1,1)$-forms or, equivalently, homology surfaces? To answer this question we need to analyze the explicit structure of the way we resolved the singularities of the quotient. The group $G \simeq (\mathbf{Z}_5)^2$ acts fixing 50 points of $Y_{4;5}$, each fixed by a subgroup isomorphic to $\mathbf{Z}_5$. These fixed points are each identified in groups of 5 when quotienting such that we have 10 singular points on the resultant space which need to be blown up. The process of blowing up is rather technical and we will not go into it here (see for example [21, 22]). The space that replaces each of these 10 points upon blowing up contains two algebraic surfaces, A which is a rational surface and $B \simeq \mathbf{P}^2$. We can therefore label the basis of $(1,1)$-forms by $\{A_i, B_i\}$, $i = 1, \dots, 10$ and H, the $(1,1)$-form inherited from the original $\mathbf{P}^4$, i.e., corresponding to the hyperplane section through the manifold. There are no intersections between surfaces arising from different fixed points and so we need only look at one blown-up point to obtain the couplings. There are also no intersections between H and any other $(1,1)$-form. To calculate the nonvanishing intersection numbers we use toric geometry as in [21] to obtain the following

$$A_i \circ A_i \circ A_i = 8$$

$$A_i \circ A_i \circ B_i = 1$$

$$A_i \circ B_i \circ B_i = -3 \tag{12}$$

$$B_i \circ B_i \circ B_i = 9.$$

The self intersection of H can also be calculated as

$$H \circ H \circ H = 3125. \tag{13}$$

All other intersection numbers are zero.

We should also note that this space has a $\mathbf{Z}_5 \simeq C_5$ symmetry which cyclically permutes the 10 fixed points in 2 groups of 5. The resultant Kähler form of the manifold can be expanded in terms of the 21 $(1,1)$-forms A_i, B_i, H. The coefficients in front of the A_i and B_i terms give the size of each of these surfaces within the blow-ups, and the size of the H term gives the size of the ambient $\mathbb{P}^4$ and therefore the distance between the blow-ups. This gives us a clear geometric picture of the meaning of these $(1,1)$-forms which will be of great help when looking at the mirror manifold.

The ring of deformations of a quintic p in $\mathbb{P}^4$ can be described as follows. (See [23] for further details.) All possible deformations of the polynomial p chosen to define the hypersurface in $\mathbb{P}^4$ represent closed forms on the manifold. Deformations that can be obtained by a linear change of coordinates on the ambient $\mathbb{P}^4$ represent exact forms, and these deformations belong to the "ideal of trivial deformations", $\mathfrak{z}$. This ideal is generated by the partial derivatives of the defining polynomial, $\mathfrak{z} = (\partial_i p)$. The deformations which are not in $\mathfrak{z}$ represent the independent harmonic forms, which carry the topological information about the manifold M. Similar results obtain for polynomials of degree different from that of p.

If we call $A = \mathbb{C}[z_0, \ldots, z_4]$ the ring of all polynomials in the homogeneous coordinates of the ambient $\mathbb{P}^4$, then the ring $A/\mathfrak{z}$ is an algebraic characterization of all global properties of M related to the complex structure. This ring has been christened "the chiral ring" in the wake of the realization that the coordinates of $\mathbb{P}^4$ can be regarded as Landau-Ginzburg fields of tensor representations of conformal field theory. If we view $p = 0$ as the equation of an affine variety X in $\mathbb{C}^5$, we see that $A/\mathfrak{z}$ is the ring of regular functions $\mathcal{O}(S)$ on the variety S defined by $\partial_i p = 0$. Thus S is the singular set of X. If M is to be smooth, then X can be singular only at the origin, i.e., $\dim S = 0$. This tells us that the Krull dimension of the chiral ring is zero if M is smooth and thus the dimension of the vector space spanned by the elements of the chiral ring is finite. This ring is also graded if we attach a grade of one to each of the z_i's. We denote the subgroup of elements of $A/\mathfrak{z}$ of grade n by R_n. The following facts can then be shown [27]

$$\dim R_n = \dim R_{15-n}$$
$$\dim R_0 = 1 \tag{14}$$
$$\dim R_5 = h^{2,1}(M).$$

Since $(2,1)$ forms are represented by the elements of R_5, their triple product is a map into R_{15}, which is one dimensional. This top element of the chiral ring represents the $(3,3)$ volume form of M, and can be represented by the Hessian of p, $\det(\partial_i \partial_j p)$.

This means that in order to compute the *unnormalized* Yukawa coupling of three $(2,1)$ modes, all we have to do is to multiply together their algebraic representatives. The resulting polynomial of degree 15 is then necessarily equivalent (mod 3) to the Hessian times a constant. The constant is the value of the coupling in some unknown and arbitrary normalization, and only normalization independent combinations of the couplings will have any invariant physical or mathematical meaning, unless the deformation basis can somehow be identified.

The general problem of reducing a polynomial modulo an ideal to a canonical form is computationally very hard[3], but we are not considering general polynomials and two crucial observations enables us to solve this problem. Firstly, it is clear that each ring A_n of degree n also has the structure of a linear vector space, where we can choose the monomials to define an orthonormal basis. Furthermore, two polynomials are equivalent modulo the ideal iff they differ only by a linear combination of elements in the ideal. Thus reduction of a polynomial q modulo the ideal can be implemented by enforcing the constraints that the members of the ideal vanish in q. This means that we must solve $\dim A_{15} - 1$ equations in $\dim A_{15}$ unknown couplings. Unfortunately the problem is hard – in the case of $Y_{4;5}$ we have to solve 3875 equations algebraically.

The second crucial observation is that we can restrict attention to manifolds with a high degree of symmetry. The chiral ring must also respect the symmetries of p, so that it is sufficient to consider a subring of $A/3$ which is closed under the desired symmetries. The dimensions of the chiral rings are reduced if we impose first the simple $\mathbb{Z}_5$ phase symmetry $(0,1,2,3,4)$ needed to get to M_{-40}, and then in addition the permutation symmetries $z_1 \leftrightarrow z_4$, $x_2 \leftrightarrow z_3$. Call the total symmetry $\mathfrak{G}$. The end result is a tiny $\mathfrak{G}$-invariant subring containing less than 1% of the total chiral ring of $Y_{4;5}$. This ring is just small enough to be exhaustively studied.

---

[3]Many algebraic packages are available at present which use concepts such as the Gröbner basis for reduction of polynomials but these take an inordinately long time to solve problems of the type we are considering. For information about such algorithms see [24, 25, 26].

There are only five $(2, 1)$ modes which respect all these symmetries, namely

$$H = z_0 z_1 z_2 z_3 z_4$$

$$A = z_0^3 z_1 z_4$$

$$B = z_0 z_1^2 z_4^2 \tag{15}$$

$$A' = z_0^3 z_2 z_3$$

$$B' = z_0 z_2^2 z_3^2,$$

so the most general $\mathfrak{G}$-invariant defining polynomial is:

$$p = \frac{1}{5} \sum_{i=0}^{4} z_i^5 + tH + aA + bB + a'A' + b'B'. \tag{16}$$

The ideal-defining equations must be constructed from the $\partial_i p$'s in a way that also respects $\mathfrak{G}$. This is achieved by multiplying the three equations:

$$z_0^4 = -\frac{1}{z_0}(tH + 3aA + bB + 3a'A' + b'B')$$

$$z_1^4 z_4^4 = \frac{1}{z_1 z_4}(tH + aA + 2bB)^2 \tag{17}$$

$$z_2^4 z_3^4 = \frac{1}{z_2 z_3}(tH + a'A' + 2b'B')^2,$$

which are manifestly $\mathfrak{G}$-invariant, with all possible $\mathfrak{G}$-invariant monomials and keeping 35 independent equations. Solving these equations gives all possible nonvanishing couplings on this manifold in terms of H^3. They are all complicated rational functions with polynomials containing over one hundred terms in both the numerator and denominator. We do not record them here for lack of space, and it is sufficient for our purposes to study them only in a special limit which is motivated as follows.

As explained above, two of the blow-ups of $\widetilde{M}$ which are paired by the symmetries still present on the quotient variety, give rise to two pairs of $(1, 1)$ modes on the mirror manifold which are entirely analogous to the $A(A')$ and $B(B')$ modes on M. The symmetry prohibits them from mixing with any other modes except the 2-form inherited from $\mathbb{P}^4$, whose mirror on M we conjecture to be the H mode.

Since the large radius limit should mean that all components of the Kähler form are made large, so that there are no small curves left on the manifold that can give rise to instanton corrections, all $(1,1)$ forms are large in the large radius limit. The mirror limit is that all complex structures are forced to take large values. This is not possible if p is restricted to be given by (16) as we only appear to be blowing up two of the fixed points. What would seem natural, however, is that if we can separate the fixed points by a large distance, then when looking at couplings around one fixed point, instantons from other fixed points should have negligible affect. This can be achieved by letting t go to infinity much faster than a and b.

$$t, a, b, \ldots \to \infty$$
$$a, b, \ldots \ll t. \tag{18}$$

For simplicity we also switch off one of the "mirror blow-ups", i.e., $a' = b' = 0$. The couplings between the A and B modes then reduce to (an overall normalization factor has been suppressed):

$$
\begin{array}{ccccc}
A^3 & \to & (8ab^3 + 9a^2 - b)t^3 H^3 & \to & 8ab^3 t^3 H^3 \\
A^2 B & \to & (a^2 b^2 - 4b^3 - 3a)t^3 H^3 & \to & a^2 b^2 t^3 H^3 \\
AB^2 & \to & (-3a^3 b + 4ab^2 + 1)t^3 H^3 & \to & -3a^3 b t^3 H^3 \\
B^3 & \to & (9a^4 - 13a^2 b + 4b^2)t^3 H^3 & \to & 9a^4 t^3 H^3,
\end{array}
\tag{19}
$$

while the couplings of A, B to H, A', B' vanish in the large t limit.

In general we should allow for arbitrary mixing and normalization of the fields, but in the limit which we are considering, $t \gg a \sim b \gg 1$, the mixing between fields in zero because in this limit the phase symmetries of (16) with $a = b = 0$ become exact. Remarkably, allowing for the remaining multiplicative normalization of the fields A and B, we find that (12) and (19) agree. This observation follows immediately from the fact that the combinations $A^3 B^3 / \{(A^2 B)(AB^2)\}$ and $\{A^3 (AB^2)^3\}/B^3 (A^2 B)^3\}$ are the same (i.e. -24) when calculated using both (19) and (12).

We have established that the couplings of $(2,1)$-forms are the same as the mirror couplings of $(1,1)$-forms. By itself this is a surprising result since the latter rely on purely topological information whereas the former were obtained via deformation theory using apparently quite different information. We can use this analysis to extract further information about $(1,1)$ couplings beyond what is available from purely topological considerations by using the deformation theory results for their mirror partners. As an example, consider the leading corrections to the zeroth order zeros for the couplings between the fields A, B and H as given by deformation theory:

$$HA^2 \sim -125b^3 a^7 t^{-3} H^3$$

$$HAB \sim 75b^3 a^6 t^{-3} H^3$$

$$HB^2 \sim 25b^6 a^4 t^{-4} H^3 \tag{20}$$

$$H^2 A \sim 175b^3 a^8 t^{-4} H^3$$

$$H^2 B \sim 25b^7 a^4 t^{-4} H^3.$$

(We also find the subdominant terms for the couplings of (19) but we do not exhibit them here.)

We expect that the multiplicative normalizations of the fields will change in non-leading order from those determined above but this will not lead to the off-diagonal terms of (20). However, we may also expect mixing between fields which will generate off-diagonal couplings. In this example, to non-leading order, there still remain exact $\mathbb{Z}_5$ phase symmetries which only allow mixing between the A, B and H fields. Even allowing for this there remains non-zero mixing between the rotated fields. Interpreted as $(1,1)$-couplings in the mirror manifold, these correspond to deviations from the large radius (geometric) limit and presumably to world-sheet instanton corrections. To calculate these corrections explicitly it is necessary to obtain the normalizations which does not appear to be possible algebraically. Recently, transcendental methods have been used [28] to solve this problem in the case of the one-dimensional Kähler-form sector of the quintic 3-fold and the deformations of complex structure of its mirror. Unfortunately this model is too simple to exhibit the multiple large radius limits described in the next section.

4. Modifications of the Moduli Space

In the previous section we appear to have shown how, given any SCFT, one can extract topological information required to classify the corresponding Calabi-Yau manifold, thus solving the problem of identifying a geometrical interpretation to a conformal field theory. However, in general this calculation will not be so simple. If we try to repeat the above analysis with any pair of mirror models from Table 1 such that $h^{2,1} > 1$ for both models, as well the pair consisting of the quintic cubic itself and its mirror, then we have problems when we perform the blow-ups. This is because the blow-up procedure is not unique in these cases due to the occurrence of rational curves with normal bundle $\mathcal{O}(-1) \oplus \mathcal{O}(-1)$ appearing in the resolutions. This can be seen from the toric geometry when one performs the star subdivision of the cone required to obtain a smooth manifold. If a slice is taken through the cone one obtains convex quadrilaterals that can be divided into triangles in two ways as shown in Figure 1 (see [22] and references therein). This process is known as a flop (see, for example [29]). There was no problem when calculating the Hodge numbers in Table 1 as the different blow-ups of a specific singularity give identical contributions to the Hodge diamond. What does differ between these resolutions however are the intersection numbers of (10). What's more, these differences cannot, in general, be accounted for by a change of basis of $H^{1,1}(M)$ and so the resultant manifolds are genuinely topologically distinct. The manifolds are however, birationally equivalent and they can be converted into each other by a sequence of flops.

If we look at the structure of the moduli space of the Kähler form around the singular quotient manifold we thus appear to be on the edge of more than one Kähler cone. This is shown schematically in Figure 2. That is, we make a choice when we do the blow-up about which manifold's Kähler cone we wish to enter.

Now, if the SCFT's obtained by taking quotients of Gepner's models are to be considered truly isomorphic to these singular Calabi-Yau manifolds before any blow-ups are performed, we should expect to see this same structure in the moduli space of the SCFT's. We don't. The SCFT's have a set of truly marginal operators which are all mutually marginal. To reproduce the local geometry of the touching Kähler cones we would require sets of marginal operators which destroyed each others marginality such as can happen in the space of conformal field theories with

$c = 1$ (see [30] for a good account of this). Actually the SCFT's of Table 1 all correspond to orbifold points in their local moduli space and this corresponds to a fixed point of the generalization of the "$R \to 1/R$" duality of string theory. The important point is that locally

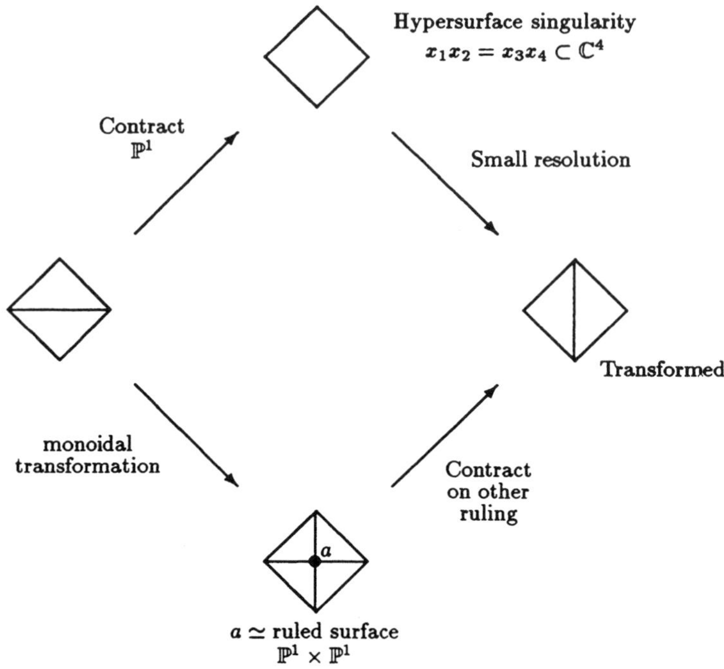

Figure 1. Flop on a $(-1, -1)$-curve.

around the SCFT we do not appear to be making a choice about which blow-up we perform when we deform the SCFT away from the orbifold point.

The last section showed us that the correspondence with geometry is only made clear in some limit however. It is clear that a simple interpretation of the moduli space of SCFT's is to say that there are many "large radius limits" for a specific SCFT and that each of these can give a different Calabi-Yau manifold. Thus we appear to fit Calabi-Yau manifolds differing by flops into the same smooth (up to quotient singularities) moduli space.

The calculation of the last section shows how this is possible. In order to obtain the

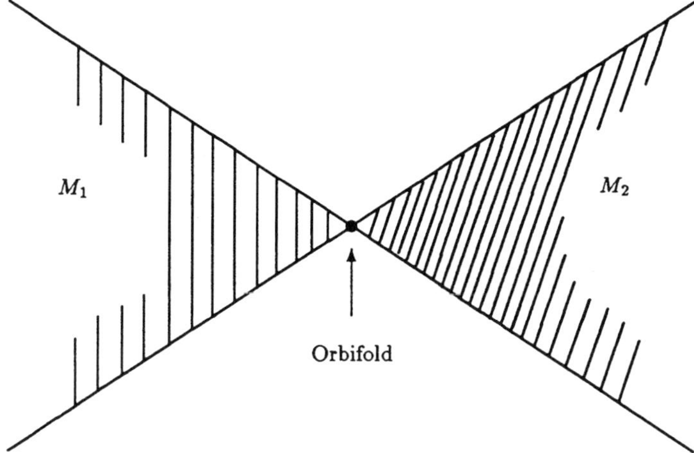

Figure 2. A slice through the Kähler cones of two Calabi-Yau
manifolds that can both be obtained from the same orbifold.

topological intersection numbers of the Calabi-Yau, we had to interpret the connection between
the SCFT and geometry before we could take the large complex structure limit. In general
there are probably many large complex structure limits of any manifold and thus many large
radius limits of the mirror. The meaning of a large complex structure limit was investigated in
[1] where it was suggested that the limit corresponds to the direction of $H^{3,0}(M)$ lining up in a
direction of integral cohomology within $H^3(M,\mathbb{C})$. This means that a large complex structure
limit is not smooth. We therefore have the result that *the mirror of a large radius Calabi-Yau
manifold is almost singular.* Correspondingly, the mirror of a smooth manifold is not at a large
radius and will probably have ambiguous topology.

We are not finished yet however. An interesting generalization of these ambiguous blow-
ups was described by an example in [31]. Here we have again a singular space to be resolved
but this time we can blow it up *or* resolve it by deforming the complex structure. Again the

Hodge diamond is unaffected and again the supposedly corresponding SCFT does not lie on a "branching point" in its local moduli space. Again we see that the string smooths out the global moduli space but this time we smooth out joins between the Kähler cone and the moduli space of complex structures. Thus we see that globally the moduli space is not a product of the moduli space of Kähler form deformations and complex structure deformations.

It should be noted that birational geometry, i.e., deformations where one is allowed to go out to the edge of the Kähler cone, allows for more severs changes in the topology of a manifold. Such an example due to Hirzebruch [32] (see also [33] and references therein) allows for changes in the Hodge numbers. This change cannot be smoothed out by quantum algebraic geometry since the Hodge numbers must always be integers because they count the number of primary chiral fields. What may happen is that the moduli space contains multicritical points which allow for a sudden change in the spectrum of marginal operators. The properties of these points in moduli space has yet to be fully elucidated but analysis has been started in such papers as [34].

In conclusion we see that Calabi-Yau manifolds can be successfully modeled by $N = 2$ superconformal field theories in some limit. When the conformal field theories are not in this limit they appear to be able to interpolate smoothly between manifolds of differing topology. This is possible because the geometrical interpolation would require passing through intermediate spaces where some algebraic curve has been shrunk down to zero volume. It is precisely where rational curves have small volume that conformal field theory deviates strongly from normal geometrical ideas. It is clear that much work needs to be done in exploring more fully the conformal field theory description of transformations such as flops, but it does not seem too much to hope that quantum algebraic geometry may become a powerful tool in algebraic geometry as well as providing a model of space-time produced by superstring theory.

Acknowledgements

This work benefited greatly from conversations with P. Candelas and G.G. Ross. CAL wishes to thank the Norwegian Research Council for Science and the Humanities (NAVF) for financial support.

  P. S. ASPINWALL AND C. A. LÜTKEN

References

[1] P.S. Aspinwall, C.A. Lütken, Nucl. Phys. **B355** (1991) 482.

[2] D. Friedan, E. Martinec, S. Shenker, Nucl. Phys. **B271** (1986) 93.

[3] P. Candelas, G. Horowitz, A. Strominger, E. Witten, Nucl. Phys. **B258** (1985) 46.

[4] P.S. Aspinwall, C.A. Lütken, G.G. Ross, Phys. Lett. **241B** (1990) 373.

[5] P.S. Aspinwall, C.A. Lütken, Nucl. Phys. **B353** (1991) 427.

[6] M. Ademollo et al, Phys. Lett. **62B** (1976) 105; Nucl. Phys. **B111** (1976) 77; P.Di Vecchia, J.L. Petersen, H.B. Zheng, Phys. Lett. **162B** (1985) 327; P. Di Vecchia, J.L. Petersen, M. Yu, Phys. Lett. **172B** (1986) 211; P.Di Vecchia, J.L. Petersen, M. Yu, H.B. Zheng, Phys. Lett. **174B** (1986) 280; S. Nam, Phys. Lett. **172B** (1986) 323.

[7] C.T.C. Wall, Invent. Math.1 (1966) 355.

[8] X.G. Wen, E. Witten, Phys. Lett. **166B** (1986) 397.

[9] E. Martinec, Phys. Lett. **171B** (1986) 189.

[10] B. Greene, C. Vafa, N. Warner, Nucl. Phys. **B324** (1989) 371.

[11] E. Martinec, in V.G. Knizhnik Memorial Volume (ed. L. Brink et al), Phys. Lett. **217B** (1989) 431.

[12] D. Gepner, Phys. Lett. **199B** (1987) 380.

[13] D. Burns, Math. Ann. **211** (1974) 237.

[14] L. Dixon, J. Harvey, C. Vafa, E. Witten, Nucl. Phys. **B261** (1985) 678.

[15] M.F. Atiyah, G.Segal, J. Geom.Phys. **6** (1989) 671.

[16] B.R. Greene, M.R. Plesser, Nucl. Phys. **B338** (1990) 15.

[17] M. Dine, N. Seiberg, X.G. Wen, E. Witten, Nucl. Phys. **B278** (1986) 769.

[18] P. Candelas, X.C. De La Ossa, Nucl. Phys. **B355** (1991) 455.

[19] A Strominger, E. Witten, Commun. Math. Phys. **101** (1985) 341.

[20] E. Witten, Nucl. Phys. **B268** (1986) 79.

[21] D. Markushevich, M. Olshanetsky, A. Perelomov, Commun. Math. Phys. **111** (1987) 247.

[22] P.S. Aspinwall, Commun. Math. Phys. **128** (1990) 593.

[23] P. Candelas, Nucl. Phys. **B298** (1988) 458.

[24] G. Hermann, *The question of finitely many steps in polynomial ideal theory* (in german), Math. Ann. **95** (1926) 736.

[25] B. Buchberger, *gröbner basis: An algorithmic method of polynomial ideal theory*, in *Multidimensional Systems Theory*, ed. N. Bose, Reidel 1985, 184.

[26] *Effective Methods in Algebraic Geometry* eds. T. Mora, C. Traverso, proceedings of the "MEGA-90" symposium held in Castglioncello (Liverno, Italy), in April 1990, published in Progress in mathematics vol. 94, Birkhäuer, Boston 1991.

[27] H. Bass, *On the ubiquity of Gorenstein rings*, Math. Zeitschr. **82** (1963) 8.

[28] P. Candelas, X.C. de la Ossa, P.S. Green, L. Parkes, *A pair of Calabi-Yau Manifolds as an Exactly Soluble Superconformal Theory*, University of Texas Report UTTG-25-1990.

[29] J. Kollár, Bull. Am. Math. Soc. **17** (1987) 211.

[30] P. Ginsparg, *Applied Conformal Field Theory*, Les Houches, Session XLIX, 1988, *Fields, Strings and Critical Phenomena*, eds. E.Brézin and J. Zinn-Justin, North Holland, Elsevier Science Publishers B.V., 1990.

[31] P.S. Aspinwall, *On the Equivalence of Some Three-Generation Manifolds*, Oxford preprint OUTP-91-01P.

[32] F. Hirzebruch, *Some Examples of Threefolds with Trivial Canonical Bundle*, Lecture notes SFB/MPI 85-58, Max-Planck-Institut für Mathematik, Bonn 1985.

[33] P. Candelas, P. Green and T. Hübsch, Nucl. Phys. **B330** (1990) 49.

[34] T. Hubsch, Mod. Phys. Lett. **A6** (1991) 207.

AMS/IP Studies in Advanced Mathematics
Volume **9**, 1998

The Geometry of Calabi–Yau Orbifolds

Shi-Shyr Roan

Max-Planck-Institut für Mathematik

Gottfried-Claren-Strasse 26

5300, Boon 3, Germany

Abstract

In this lecture, I would like to discuss some developments on the topological properties of Calabi–Yau (CY) orbifolds. We shall consider only those related to the study of the $c = 9$ $(2,2)$ conformal field theories (CFT) in the Landau–Ginzburg models. The results we shall describe here are derived from a rigorous mathematical argument, even though the problems stem from the string theory.

© 1998 American Mathematical Society
and International Press

1. Orbifold Euler Characteristic

Let G be a finite group acting on a n-dimensional complex manifold M. The correct Euler characteristic of the quotient M/G for string theory is the expression [4]:

$$\chi(M, G) = \frac{1}{|G|} \sum_{gh=hg} \chi(M^{\langle g,h \rangle})$$

where the summation runs over all pairs of commuting elements in $G \times G$, and $M^{\langle g,h \rangle}$ denotes the common fixed point set of g and h. Using the classical formula for $\chi(M/G)$, one can write $\chi(M, G)$ as a sum over the conjugacy classes of G:

$$\chi(M, G) = \sum_{[g]} \chi(M^g/C(g))$$

here $C(g)$ is the centralizer of g. For a free action we get $\chi(M, G) = \chi(M/G)$. It was noted by Atiyah and Segal [1] that the above "orbifold Euler number" $\chi(M, G)$ is the Euler characteristic of equivalent K-theory $K_G(M)$. For the physicist's interest, the situation we consider here is the case where G preserves the holomorphic volume form of M, or equivalently the canonical sheaf of M/G is locally free $\mathcal{O}_{M/G}$ – module, e.g. $M = \mathbb{C}^n$ and $G = $ a finite subgroup of $SU(n)$. It was pointed out in [4] that in some situations where M/G has a resolution of singularities $\widehat{M/G}$ with trivial canonical bundle, $\chi(M, G)$ is just the Euler characteristic of this resolution. The following cases have been mathematically justified:

(i) When $\dim_{\mathbb{C}} M = 2$, by the Mckay correspondence of Klein group Hirzebruch and Höfer have shown that the equality

$$\chi(M, G) = \chi(\widehat{M/G})$$

holds for $\widehat{M/G} = $ the minimal resolution of M/G [7].

(ii) When $\dim_{\mathbb{C}} M = 3$ and G is abelian, $\chi(M, G)$ is identified with $\chi(\widehat{M/G})$ for $\widehat{M/G} = $ the "minimal toroidal" resolution of M/G. (Note that $\widehat{M/G}$ is unique up to the relation of "small resolution") [8]. In particular for a finite abelian subgroup G of $SU(3)$, one has

$$\chi(\widehat{\mathbb{C}^3/G}) = |G| .$$

For the purpose of study of CY orbifolds in weighted projective space, the above result of (ii) is sufficient for our discussion.

2. Vafa's Formula

Let X be a quasismooth hypersurface in weighted space $W\mathbb{P}^4_{(n_1,\ldots,n_5)}$ defined by a degree d quasihomogeous polynomial $f(Z)$ $(= f(Z_1,\ldots,Z_5))$, and its zero locus in $\mathbb{C}^5$ has the only isolated singularity at the origin. Assume that $\gcd(n_i|i \neq j) = 1$ for all j, and $d = \sum_{j=1}^{5} n_i$. By [3], X is a V-manifold with trivial canonical sheaf. For an element x of X, let $I = \{i|$ the i-th coordinate of $x \neq 0\}$ and $c_I = \gcd(n_i|i \in I)$. In fact, one can easily determine the analytic structure of X near x:

$$(X, x) \simeq (\mathbb{C}^3/\mathbb{Z}_{c_I}, 0)$$

where $\mathbb{Z}_{c_I}$ is some order c_I cyclic subgroup of $SU(3)$. Patching the local resolution of X together, we obtain a resolution of X:

$$\pi : \hat{X} \to X$$

such that $\hat{X}$ is a CY manifold (i.e. a Kahler 3-fold with $c_1 = 0, b_1 = 0$). We have $\chi(\pi^{-1}(x)) = c_I$ for each x in X. Suggested by Vafa's result on the CFT of Landau–Ginzburg model for the potential $f(Z)$ [12], Vafa's formula of Witten index of CFT can be identified with the Euler characteristic of $\hat{X}$ by the topological method:

Theorem [9]. *Let* $q_i = \dfrac{n_i}{d}$ *for* $i = 1,\ldots,5$. *For* $\ell = 0,\ldots,(d-1)$, *denote* $\beta_\ell = \dfrac{1}{d}\sum_{r=0}^{d-1}\prod\left\{\left(1 - \dfrac{1}{q_i}\right)\Big|\ell q_i, rq_i \in \mathbb{Z}\right\}$, *and* $s_\ell = |\{i|1 \leq i \leq 5, \ell q_i \in \mathbb{Z}\}|$. *Then*

$$\chi(\hat{X}) = \sum_{\ell=0}^{d-1} \beta_\ell ,$$

$$b_3(\hat{X}) = -\sum_{s_\ell \geq 3} \beta_\ell ,$$

$$b_2(\hat{X}) = -1 + \frac{1}{2}\sum_{s_\ell < 3} \beta_\ell .$$

We may also consider the case when X is quotiented by a certain group of symmetry, a construction for the mirror CY spaces. Let $d_i = \dfrac{d}{\text{g.c.d.}(d, n_i)}$ for each i. Denote

D_q = the subgroup of $GL_5(\mathbb{C})$ consisting of all the diagonal elements dia.$[t_1, \ldots, t_5]$, $t_i^{d_i} = 1$ for all i.

$SD_q = D_q \cap SL_5(\mathbb{C})$, $q = \text{dia.}[q_1, \ldots, q_5]$.

If G is a subgroup of SD_q such that every element in G preserves the defining polynomial $f(Z)$ of X, Vafa's formula can be generalized to the CY resolution $\widehat{X/G}$ of X/G, and the previous theorem is the case when G = the cyclic group generated by q.

Theorem [10]. *For $g \in G$, define*

$$\beta_g = \frac{1}{|G|} \sum_{h \in G} \prod \left\{ \left(1 - \frac{1}{q_i}\right) \, \middle| \, g(e^i) = h(e^i) = e^i \right\} ,$$

$$s_g = |\{i | 1 \leq i \leq 5 , \; g(e^i) = e^i\}| ,$$

here $\{e^i\}_{1 \leq i \leq 5}$ is the standard base of $\mathbb{C}^5$. Then

$$\chi(\widehat{X/G}) = \sum_{g \in G} \beta_g ,$$

$$b_3(\widehat{X/G}) = - \sum_{s_g \geq 3} \beta_g ,$$

$$b_2(\widehat{X/G}) = -1 + \sum_{s_g < 3} \beta_g .$$

3. Mirror Pairs

In the following discussion, we always assume that d is divisible by n_i and $d_i \left(= \dfrac{d}{n_i}\right) \geq 3$ for all i. By the study of CFT, Greene and Plesser [5] have assigned each subgroup G of SD_q to another group G' of this kind such that

$$h^{1,1}(\widehat{X/G}) = h^{2,1}(\widehat{X/G'}) ,$$

$$h^{2,1}(\widehat{X/G}) = h^{1,1}(\widehat{X/G'}) ,$$

hence

$$\chi(\widehat{X/G}) = -\chi(\widehat{X/G'}) \ .$$

Their construction can also be described through the toroidal geometry. By the expression of teh Betti numbers of $\widehat{X/G}$ in the previous section, we may assume that X is the Fermat hypersurface in $W\mathbb{P}^4_{(n_i)}$:

$$Z_1^{d_1} + Z_2^{d_2} + Z_3^{d_3} + Z_4^{d_4} + Z_5^{d_5} = 0 \ .$$

Denote

$$T = \text{the algebraic torus } (\mathbb{C}^*)^5 \ ,$$

$$exp_q : \mathbb{R}^5 \to T, \ exp_q(x) = \begin{bmatrix} exp(2\pi i x_1 q_1) \\ \vdots \\ exp(2\pi i x_5 q_5) \end{bmatrix} \ ,$$

$$tr_q : \mathbb{R}^5 \to \mathbb{R}, \ tr_q(x) = \sum_j x_j q_j \ .$$

For a subgroup G of D_q, let N_G, M_G be the group of 1-parameter subgroups, characters respectively, of the algebraic torus T/G:

$$N_G = Hom_{\text{alg.group}}(\mathbb{C}^*, T/G) \ ,$$

$$M_G = Hom_{\text{alg.group}}(T/G, \mathbb{C}^*) \ .$$

Then we have the identification of the lattices: $N_G = exp_q^{-1}(G)$, $M_G = \left\{ G - \text{invariant monomial } \prod_{i=1}^{5} Z_i^{k_i} \right\}$. The dual pair G, G' of [5] are characterized by either one of the following equivalent conditions:

$$N_G = M_{G'} \ ,$$

$$M_G = N_{G'} \ .$$

By the Hodge structure of hypersurfaces in weighted projective spaces [3] [11] and the structure of exceptional divisors of the toroidal resolution [6] [8], $H^{2,1}(\widehat{X/G})$ can be described by the "data in M_G", and $H^{1,1}(\widehat{X/G})$ by the "data in N_G", and they are some subsets contained in the part with $tr_q = 1$. And the following result was shown in [10]:

Theorem. *There exists the isomorphisms between the Hodge spaces of* $\widehat{X/G}$ *and* $\widehat{X/G'}$:

$$H^{1,1}(\widehat{X/G}) \simeq H^{2,1}(\widehat{X/G'}) \,,$$
$$H^{2,1}(\widehat{X/G}) \simeq H^{1,1}(\widehat{X/G'}) \,.$$

Reference

[1] M. Atiyah, G. Segal, *On equivariant Euler characteristics*, Journal of Geometry and Physics, Vol. 6, No. 4 (1989) 671-677.

[2] P. Candelas, M. Lynker, R. Schimmrigk, *Calabi–Yau manifolds in weighted* $\mathbb{P}_4$.

[3] I. Dolgachev, *Weighted projective varieties*, Springer-Verlag Lecture Notes in Mathematics 956.

[4] L. Dixon, J. Harvey, C. Vafa, E. Witten, *Strings on orbifolds*, Nuclear Physics **B261** (1985) 678-686.

[5] B.R. Greene, M. R. Plesser, *Duality in Calabi–Yau moduli space*, Nucl. Phys. **B338** (1990) 15.

[6] B.R. Greene, S. S. Roan, S. T. Yau, *Geometric singularities and spectra of Landau–Ginzburg models*, to appear in Communications of Mathematical Physics.

[7] F. Hirzebruch, T. Höfer, *On the Euler number of an orbifold*, Mathematische Annalen **286** (1990) 255-260.

[8] S.S. Roan, *On the generalization of Kummer surfaces*, Journal of Differential Geometry **30** (1989) 523-537.

[9] S.S. Roan, *On the Calabi–Yau orbifolds in weighted projective spaces*, International Journal of Mathematics, Vol. 1, No. 2 (1990) 211-232.

[10] S.S. Roan, *The mirror of Calabi–Yau orbifold*, Max-Planck-Institut Preprint (1991), to appear in International Journal of Mathematics.

[11] J. Steenbrink, *Intersection form for quasi-homogeneous singularities*, Compositio Mathematics, Vol. 34, Fasc. 2 (1977) 211-223.

[12] C. Vafa, *String vacua and orbifoldized LG model*, Mod. Phys. Lett. **A4** (1989) 1169.

AMS/IP Studies in Advanced Mathematics
Volume **9**, 1998

Properties of Superstring Vacua from (Topological) Landau–Ginzburg Models[1]

Amit Giveon[2]

Department of Physics

University of California

Berkeley, CA 94720

Dirk-Jan Smit

Theoretical Physics Group

Lawrence Berkeley Laboratory

1 Cyclotron Rd. Berkeley, CA 94720[3]

Abstract

Duality symmetries of the complex structure moduli of $(2,2)$ string vacua associated with $N = 2$ LG models are realized by LG-field redefinitions. Combined with a mirror symmetry, duality is realized on the Kähler structure moduli of the mirror manifold, thus leading to $R \to 1/R$ type of symmetries. The cubic superpotential in the effective action is manifestly

---

[1] To appear in *Mirror Symmetry* ed. S.-T Yau, MSRI Berkeley, May 1991.

[2] Address after September 1, 1991: School of Natural Sciences, Institute for Advanced Study, Princeton, NJ 08540.

[3] email address: GIVEON@LBL.bitnet, SMIT@LBL.bitnet

© 1998 American Mathematical Society
and International Press

invariant under duality transformations acting on the moduli, combined with the appropriate matter fields redefinitions. The superpotential for the matter fields associated with untwisted (c, c) moduli (and their mirror partners) is found exactly as a function of the moduli from topological LG models. Finally, We discuss a differential equation for the periods of the $(d, 0)$-form of a d-dimensional Calabi–Yau manifold, which determines the prepotential (the generating functional for the Yukawa couplings).

Table of Contents

1. Introduction.
2. Target Space Duality from Landau-Ginzburg Models.
3. Mirror Symmetry.
4. Duality Invariant Superpotential in Effective Actions.
5. Exact Yukawa Couplings from Topological Landau-Ginzburg Models.
6. Differential Equations for Periods and Duality Groups.

1. Introduction

In space-time supersymmetric compactifications of superstring theory (for review see [1]) one considers the string propagating in backgrounds described by two-dimensional conformal field theories which admit at least two supersymmetries on the world-sheet of the string [2, 3]. One refers to such string backgrounds as string *vacua*.[4] For space-time supersymmetric string theories it turns out that the vacua described by (2, 2) superconformal field theories (SCFTs) have the geometry of a Ricci-flat Kähler manifold.

A property of the space of possible string vacua is the possibility to deform a given vacuum continuously. One calls the coupling parameters of such continuous deformations the *moduli* of

---

[4]The (2, 2) vacua that will be considered in this paper, are those for which both left- and right-handed SCFTs have $N = 2$ world sheet SUSY.

the string vacuum. Thus, in general, to each SCFT a moduli space of continuous deformations is attached. By construction, each deformation of the vacuum will lead to a geometrically different target space. For example, in a string compactification on a circle one may change continuously the radius, R, corresponding to just one real modulus.

An important aspect of the moduli spaces of string vacua is that they respect a number of surprising symmetries. An example of such a symmetry is the well known duality symmetry in the theory describing the circle compactification, which relates a circle with redius R to one with radius $1/R$ [4]. This duality is an automorphism of the conformal field theory algebra, which gives rise to an isomorphism of two models with *distinct* underlying geometry. The target space duality symmetry was generalized to d-dimensional toroidal compactifications [5, 6, 7, 8, 9], toroidal orbifolds [10] and particular types of curved backgrounds [11]. Such duality transformations are important in constructing an effective low energy field theory [7, 12, 13, 14] and possible applications to cosmology [15] and black holes in string theory [16].

So far, however, there is no systematic way to classify these symmetry groups for general $(2,2)$ string vacua. In [17, 18] an approach to this problem is discussed using a description of $(2,2)$ vacua in terms of 'orbifoldized' $N = 2$ Landau-Ginzburg (LG) models.[5] It turns out that such models can be used to describe $(2,2)$ vacua corresponding to target spaces corresponding to Kähler manifolds with vanishing first Chern class [19, 20, 21, 22, 23, 24, 25, 26]. These LG models are characterized by superpotentials which are weighted homogeneous polynomials with isolated singularities. Using the mathematical theory of isolated singularities one may obtain insight in the structure of the symmetry groups mentioned above.

The LG models that we will consider are characterized by superpotentials $W(X_i)$ that are quasi-homogeneous functions with isolated singularities. The weight q_i of the scalar chiral superfield X_i is its $U(1)$-charge at the critical point. Examples of such superpotentials are

$$W = \sum_{i=1}^{n} X_i^{l_i} \, , \tag{1.1}$$

---

[5]The study of the symmetries in the $c = 3$ cases was done already in [27, 28], although in an opposite way: the known symmetries in the $(2,2)$ $c = 3$ toroidal-orbifolds were used in order to find the symmetries of the $c = 3$ LG moduli.

where $l_i = 1/q_i$. Such superpotentials correspond to Gepner models of the type $\prod_{i=1}^{n} k_i$, where $k_i = l_i - 2$ is the level of a minimal $N = 2$ SCFT using the A-type modular invariant [3].

Let us next briefly recall some of the properties an $N = 2$ SCFT should have in order to serve as a vacuum of string theory. Essential for a string-like interpretation are two conditions which come from the relation between the holomorphic (the 'left-handed') and anti-holomorphic (the 'right-handed') properties of string states and their $U(1)$-charges. Because of the super-symmetry the string states have in addition chiral and anti-chiral parts. The chiral holomorphic and chiral anti-holomorphic states are denoted by (c, c). Their anti-chiral partners, which are obtained by complex conjugation in *field space*, are denoted by (a, a). In addition there are (a, c) states, and their conjugated partners (c, a). In order to serve as a vacuum of the superstring the $U(1)$ charges in the NS-sector must be integral. Furthermore, for a space-time interpretation the theory should have a central charge which is a multiple of three.

The condition that the central charge be a multiple of three permits one to derive an algebraic equation describing the target space. It arises in the path integral of the LG theory as a delta-function constraint, $\delta(W)$, evaluated in a weighted projective space, WCP^{n-1}.[6] This procedure is described in detail for the deformations of the models (1.1) in [20, 22], where it is claimed that this hypersurface is a candidate target space in which the string propagates. A more mathematical treatment of this technique is described in [29].

2. Target Space Duality from Landau-Ginzburg Models

Some continuous deformations of a given LG superpotential lead to perturbations by marginal operators of the original LG model and hence to a moduli space of SCFTs. For simplicity we discuss the deformations of a model of the type (1.1), described by the superpotential

$$W(X, a) = \sum_{i=1}^{n} X_i^{l_i} + \sum_{I=1}^{m} a_I \phi_I(X) , \qquad (2.1)$$

---

[6]The identification $X_i = e^{2\pi i q_i} X_i$ which give rise to the weighted projective space amounts to a twisting or 'orbifoldizing' of the original LG model by a product of cyclic groups. This twisting turns out to correspond to the generalized GSO-projection [25].

where ϕ_I is a (c, c) primary field of charge $(1, 1)$ (and therefore of super-dimension $(1/2, 1/2)$); m is the number of such fields in the chiral ring (for a definition and review see for example [22, 26, 17]) and a_I is a complex parameter. The space of marginal couplings $a = (a_1, \cdots, a_m)$ is called 'the a-moduli space'. At each point of this space we have a LG model which at the conformal point defines an $N = 2$ SCFT. Thus the a-moduli space is identified with the moduli space of geometrically different target spaces of the string theory. (There are, however, more moduli attached to a target space, as we will discuss in the next section, that cannot be described as deformations of the superpotential.)

An important property of the a-moduli space is that different points representing geometrically different target manifolds may be related by a symmetry of the SCFT. Such an apparent symmetry on the space of string vacua is referred to as a *target space duality* symmetry. We will briefly indicate how these symmetries may be found from simple considerations of the path integral of for LG models. The idea is to consider these symmetries as a result of certain field redefinitions of the chiral superfields X_i for which the kinetic term in the $N = 2$ LG action remains a Kähler potential, and for which the effect on the superpotential can be expressed as a transformation involving only the parameters a. For example consider the transformation

$$X_i'^{n_i} = \sum_j U_{ij} X_j^{n_j} \ . \tag{2.2}$$

The $U(1)$ symmetry in the $N = 2$ superconformal algebra restricts the power n_i to be such that the charges $q(X_j^{n_j})$ are equal for all j. Any non-singular linear transformation on the fields X_i, mixing only fields of the same charge, changes K to a new Kähler potential. The universality class of the flow of the kinetic term is preserved since the operator corresponding to the difference $K' - K$ is an irrelevant perturbation.

Thus, we are interested in transformations of the form

$$X_i' = \sum_j U_{ij} X_j \ ; \quad \overline{X}_i' = \sum_j \overline{U}_{ij} \overline{X}_j \ , \tag{2.3}$$

where U is a non-singular matrix, and the indices i, j run over all chiral scalar superfields with the same $U(1)$ charge.[7] In some simple cases U can be a diagonal matrix consisting just of phases or a permutation matrix, permuting different superfields X_i. The property of the phase and permutation transformation is that the kinetic term and the part of W independent of the moduli are invariant. Only the moduli parameters, a, are transformed into a'. In the cases where U is diagonal the parameters a' are related to a by phase factors. The point $a = 0$ is a fixed point of such U.

More generally, if (2.3) is a symmetry of the SCFT we require that it describes a symmetry of the path integral of the LG model. In view of the remark on the kinetic term made above only the piece in the path integral depending on the superpotential is important for us. It is then easily seen that in order for U to describe a symmetry the transformed superpotential, $W'(X, a')$ must have the same form as the original W, i.e.

$$W(X', a) = W'(X, a') = \sum_i X_i^{l_i} + \sum_I a'_I \phi_I \,, \qquad (2.4)$$

where an a-dependent rescaling of the superfields is understood. This ensures that the only effect in the physical theory is a duality transformation $a \to a'$. In other words: the points a and a' in the moduli space are physically equivalent.

Equation (2.4) gives conditions on the entries of the transformations U. In [17] it was shown that there will be at most a discrete set possibilities for the matrices U thus giving rise to at most discrete duality groups acting on the a-moduli space. Furthermore, it was shown that this program gives a systematic way to find explicit nontrivial solutions for superpotentials giving rise to torus ($c = 3$), K3 ($c = 6$) and Calabi–Yau ($c = 9$) compactifications. Subgroups of the automorphism group of the hypersurface defined by the superpotential plus these stabilized marginal operators.

3. Mirror Symmetry

---

[7]Non-linear field redefinitions are also possible in general if one orbifoldizes the theory properly in order to achieve a coordinate transformation. We will not discuss this possibility here.

As was discussed earlier, the full conformal field theory describing a string vacuum consists of a tensoring of two copies of the symmetry algebra acting separately in the left-handed and right-handed sectors, thus combining into four different rings denoted by (c, c), (a, c), (c, a), (a, a), where the first (second) entry denoting the left-handed (right-handed) sector.

An important target space duality symmetry of a $(2, 2)$ SCFT is the so-called 'mirror symmetry'. It is related to the operation of interchanging chiral and anti-chiral states in only one sector.[8] In the $N = 2$ SCFT this interchanging can be trivially realized by flipping the sign of the $U(1)$ charge in just one sector. This operation is a Z_2 automorphism of the SCFT, however, it has a non-trivial consequence for the target space. This has to do with the fact that the marginal deformations in the (c, c) ring and the (c, a) ring have *geometrically* different interpretations, although in the SCFT they appear on an equal footing. More precisely, the moduli of a target space describing the complex structure deformations are, by convention, given by the marginal operators of the (c, c) ring. These marginal deformations in the (c, c) ring come from the $(1, 1)$ operators, where the notation denotes the left- respectively right-handed $U(1)$-charge, and describe perturbations of the superpotential of the kind discussed above.

The Kähler structure deformations of the target space are represented by the moduli coming from the (a, c) ring. These operators always come from the twisted sectors of the orbifoldized LG theory.[9] Although the mirror transformation is a symmetry of the SCFT, in general, interchanging of Kähler and complex structure moduli changes the geometry and (for $c = 9$) also the topology of the target space. Stated differently, superstring compactifications described by $N = 2$ SCFT have a tendency to come in pairs of geometrically distinct target manifolds related to each other by interchanging the complex structure moduli and the Kähler moduli [30, 31, 32].[10]

---

[8]The SCFT is invariant under a simultaneous interchange in the left and right sectors of chiral (c) and anit-chiral (a) states. This operation leaves both the SCFT as well as the target space strictly invariant. This symmetry could be seen as the 'square' of the mirror symmetry discussed in the text.

[9]Some of the moduli in the (c, c) ring come from the untwisted sector, and correspond to a perturbation of the superpotential, while others come from the twisted sector of the orbifoldized LG model [25].

[10]In [32, 17] it is explicitly shown that interchanging the role of Kähler structure moduli and the complex

In [32] it was shown that the mirror manifold can be obtained in geometrical terms by dividing out a subgroup of the phase symmetries of the original manifold. This is useful if the target space manifold is given by the locus of a superpotential in some weighted projective space as the phase symmetries are simply read of from the superpotential.

The mirror symmetry plays an important role in the study of symmetries on the moduli space of $(2, 2)$-superstring vacua arising from orbifoldized LG models. One immediate consequence of the mirror phenomenon is that the duality groups following from the techniques discussed in the previous section act not only on the complex structure moduli space (for which the duality symmetries are geometrically understood as changing the complex structure of the target space) but also on the Kähler structure moduli of the mirror of the manifold defined by the LG superpotential. Such transformations are 'quantum' symmetries peculiar to string theory, which are analogous to the interchange of momentum states with winding modes of the string for the $R \to 1/R$ duality of the circle compactification. In fact we find 'quantum' symmetries of the Kähler structure moduli which relates target manifolds with different sizes corresponding to different Kähler structures for which the SCFT is however the same. Thus we find a generalization of the '$R \to 1/R$' symmetry known in circle compactifications. Furthermore, the mirror phenomenon is also useful in discussing the Yukawa couplings in the SCFT as we will discuss in the following sections.

4. Duality Invariant Superpotential in Effective Actions

In this section we use the field redefinition technique of section 2 to show that the cubic superpotential in the effective action of Calabi–Yau string vacua described by an $N = 2$ Landau–Ginzburg theory, is manifestly invariant under target space duality. This problem was addressed already in the context of orbifold compactifications in [33].

The cubic term of the effective actions superpotential in the Landau–Ginzburg basis is

---

structure moduli indeed corresponds to a symmetry of the SCFT corresponding to a sign change of the $U(1)$ charge in only one sector.

given (for the untwisted (c,c) moduli and its mirror partners) by

$$W = C_{IJK}(a)M^I M^J M^K \,, \tag{4.1}$$

where C_{IJK} are the Yukawa couplings in the Landau–Ginzburg basis, namely, the structure constants of the deformed chiral ring algebra (which will be computed in the next section); M^I are the matter fields in the 27 representation of E_6 and $I = 1,\dots,m$, where m is the number of marginal operators in the untwisted (c,c) ring (see section (2)). In the Landau–Ginzburg basis one expects the superpotential W to be duality invariant. (This is different in the supergravity basis [33] in which the superpotential is a non-trivial 'duality form', which transforms as a modular form of weight -1 in case the duality group is isomorphic to the modular group $SL(2,Z)$.)

Each matter field M^I in (4.1) is associated with a marginal superfield $\phi^I(X_i)$. Under a field redefinition (2.3), the superfield $\phi^I(X_i)$ is transformed to

$$\phi^I(X_i) \to (\phi')^I(X_i) = U_J^I(a)\phi^J(X_i) = \phi^I(U_i^j(a)X_j) \quad \text{modulo} \quad (\partial_i W = 0) \,. \tag{4.2}$$

It now follows that the superpotential in (4.1) is invariant under the simultaneous transformation

$$M^I \to (M')^I = U_J^I(a)M^J \;; \quad a_I \to (a')_I \,. \tag{4.3}$$

Here the linear transformation U_J^I is defined in (4.2), and $(a')_I$ are the duality transformed moduli parameters as discussed in section 2. The cubic superpotential in the effective action is, therefore, manifestly duality invariant.

To conclude this section, we remark that the superpotential (4.1) is invariant under a group isomorphic to $Sp(2m+2, Z)$ (see for example [34], and references therein). The symplectic group acts naturally on the periods of the holomorphic 3-form defining the Calabi–Yau target space. It thus follows that the target space dualities of the untwisted (c,c) sector form a sub-group of $Sp(2m + 2, Z)$. Now, using mirror symmetry, we conclude that the dualities discussed in this work generate a sub-group of $Sp(2m_2 + 2, Z) \times Sp(2m_1 + 2, Z)$, where m_1 (m_2) is the number of untwisted (c,c) moduli of the mirror (original) manifold.

5. Exact Yukawa Couplings from Topological Landau–Ginzburg Models

Following ref. [35], in this section we present the exact 27^3 Yukawa couplings C_{IJK}, appearing in eq. (4.1), as functions of the $(1,2)$ moduli parameters a^I, using $d = 3$ topological Landau–Ginzburg models (or, using mirror duality, the $\overline{27}^3$ Yakawa couplings as functions of the $(1,1)$ moduli parameters). As before, we will consider compactifications on CY manifolds, which can be obtained from an orbifoldized $N = 2$ Landau–Ginzburg (LG) theory at the renormalization group fixed point. The Yukawa couplings are given as the three point functions of two space-time fermions and one scalar or, equivalently, as the chiral algebra of (c, c) superfields with $U(1)$-charge $q = \bar{q} = 1$. The Yukawa couplings we will discuss correspond to the untwisted (c, c) moduli, namely, to states in the untwisted sector of the orbifoldized LG model.

An $N = 2$ superconformal field theory (SCFT) with central charge c can be 'twisted' to a topological conformal field theory (TCFT) with $U(1)$-anomaly $d = c/3$ [36]. In particular, an $N = 2$ LG theory can be twisted to a topological LG model. The correlators as functions of the untwisted (c, c) moduli parameters in the topological model can be computed exactly [37, 38].

The Yukawa couplings of the matter fields, corresponding to untwisted (c, c) marginal operators, are given by the three point functions of the $q = \bar{q} = 1$ chiral superfields of the form

$$C_{IJK}(a) = \langle \phi_I(X_i)\phi_J(X_i)\phi_K(X_i) \rangle_a \ . \tag{5.1}$$

The $\langle \ \rangle$ in (5.1) includes the insertion of a charge $q = \bar{q} = -3$ operator $e^{-i\sqrt{3}(\varphi+\bar{\varphi})}$, where φ is the free boson which corresponds to the $U(1)$-current of the $N = 2$ algebra, $J = i\sqrt{3}\partial\varphi$.

Define the monomial

$$F_{IJK}(X_i) \equiv \phi_I(X_i)\phi_J(X_i)\phi_K(X_i) \ , \tag{5.2}$$

which has a $U(1)$-charge $q = 3$. The state corresponding to this monomial is unique, since a property of the chiral ring ensures that there is only one operator with charge $q = d$ (modulo terms proportional to the first partial derivatives of the superpotential). Therefore we may write

$$F_{IJK}(X) = f_{IJK}(a)\rho(X) + g^i(X,a)\partial_i W(a) , \tag{5.3}$$

where $\partial_i \equiv \dfrac{\partial}{\partial X_i}$, and ρ is the polynomial which corresponds to the unique $q = 3$ state in the chiral ring

$$\rho = \det(\partial_i \partial_j W_0(X)) . \tag{5.4}$$

In eq. (5.3), g^i are polynomials in the superfields X, and f_{IJK} are functions of the moduli parameters a.

Using eqs. (5.1) and (5.3) and the vanishing relations $\partial_i W \equiv 0$ one finds [39] that the functions f_{IJK} are related to the Yukawa couplings by

$$C_{IJK}(a) = f_{IJK}(a)\langle \rho \rangle . \tag{5.5}$$

This concludes the discussion of the a-dependent Yukawa couplings for the superconformal $(2,2)$ string vacua.

Let us turn now to the TCFTs one obtains by a 'twist' of the $N = 2$ SCFTs. In topological LG theories obtained by twisting the $N = 2$ models with superpotential $W(X,a)$ one can follow the same steps described before. Thus one gets for Yukawa couplings

$$C_{IJK}^{Top}(a) \equiv \langle \phi_I(X)\phi_J(X)\phi_K(X) \rangle_a = f_{IJK}(a)\langle \rho \rangle_a^{Top} , \tag{5.6}$$

where f_{IJK} are given in (5.3) (notice that they are *necessarily* the same functions as the ones in the SCFT), and $\langle \rho \rangle_a^{Top}$ is the expectation value of ρ computed in the topological case. Thus, after rescaling the marginal fields

$$\phi_I \to \left(\frac{\langle \rho \rangle^{N=2}}{\langle \rho \rangle_a^{Top}} \right)^{\frac{1}{3}} \phi_I \tag{5.7}$$

we recover the Yukawa couplings of the ordinary $(2,2)$ string vacua in (5.5). (That is, (5.7) maps the structure constants of the deformed chiral algebra of the topological model to those of the untwisted $N = 2$ SCFT.)

The Yukawa couplings as functions of the complex (or Kähler) structure moduli of the Calabi–Yau Manifold can, therefore, be computed in the topological LG theory (as long as there is a LG superpotential for the CY manifold). The advantage of working in the framework of topological LG models, is that we know how to compute th correlation functions *exactly*! The three point functions $C_{IJK}^{Top}(a)$ are given by [37]

$$C_{IJK}^{Top}(a) \equiv \langle F_{IJK}(X_i) \rangle_a^{Top} = \sum_{dW=0} F_{IJK} H^{-1}(a) , \tag{5.8}$$

where the Hessian H is defined by

$$H(a) = \det(\partial_i \partial_j W(a)) , \tag{5.9}$$

which coincides with ρ at $a = 0$.[11] The sum in (5.8) is over the singular points of W, namely, the roots of the vanishing relations $\partial_i W = 0$; the polynomials F_{IJK} and H on the right hand side of (5.8) are evaluated at the critical points. (To be more precise, one should first resolve the singularity of the superpotential by relevant operators, and in the end of the computation take all the relevant coupling to zero.)

Let us discuss some other properties of the ordinary effective theory one can study in the topological framework. It is known [34, 31] (and references therein) that one can make a holomorphic redefinition of moduli fields and a Kähler transform that together yield the following form of the Yukawa couplings

$$C_{IJK}^{1,2}(a) = \partial_I \partial_J \partial_K \mathcal{F}^{1,2}(a) , \tag{5.10}$$

where $\partial_I \equiv \dfrac{\partial}{\partial a^I}$, and $\mathcal{F}^{1,2}$ are holomorphic functions of the $(1,1)$, $(1,2)$ moduli fields, respectively. These holomorphic functions determine not only the Yukawa couplings, but also the Kähler potential for the metric on the moduli spaces of respectively $(1,1)$-forms and $(1,2)$-forms of the CY manifold. (This information is necessary to find the normalization of the states.)

---

[11] Because the topological theory is conformal, the correlator (5.8) only gets a worldsheet contribution from the sphere.

For TCFTs, it is known [40] that there is a unique function (up to second order terms) $F(a)$ of all couplings (relevant and marginal) such that $C_{IJK}(a) = \partial_I \partial_J \partial_K F(a)$ are the structure constants of the perturbed chiral ring.[12] This result is given for a specific a-dependent basis of the chiral operators, for which the metric is a-independent. By projecting F on the moduli couplings (that is, taking the relevant couplings to zero), one recovers the holomorphic function $\mathcal{F}$ of the ordinary vacua. As will be discussed in the next section, the function F can be computed exactly for topological LG models and, as a result, also the kinetic term in the low-energy effective theory.

Finally, we mention that the results above could be seen as a particular consequence of the non-renormalization theorems familiar in sigma model perturbation theory [41]. It would be interesting to study how TCFTs in combination with the mirror hypothesis for CY manifolds [42] could be used to obtain more general exact results for $(2,2)$ string vacua. The first step of such a program is the subject of the next section.

6. Differential Equations for Periods and Duality Groups

In the previous section we showed that the moduli dependence of the Yukawa couplings in the $N = 2$ SCFT can be computed exactly by relating the 'prepotential' $\mathcal{F}$ of the SCFT with the tree level free energy of a topological conformal field theory associated with the same LG superpotential. This apparent relation between three-point couplings in a SCFT and the free energy of the associated TCFT is rather delicate since it *only* holds in a particular set of coordinates, namely those in which the two-point function of the TCFT is manifestly independent of the moduli.

The fact that such a basis exists in a TCFT is a characterizing feature of such a theory. One would like to have a more fundamental understanding of this relation. That is, *how does the prepotential in a SCFT know about the flatness of the coupling parameter space of the associated*

---

[12]To be more precise, we assume that the space of coupling constants is an affine space and has no cohomology.

TCFT?

It is important to have an answer to this question if we would like extract results in the SCFT from the associated *topological* LG model, for example the invariant Yukawa couplings. Let us recall that in order to find the physical couplings the result of the previous section is not enough, as one needs to normalize the states in the SCFT for which we need the Kähler potential on the moduli space. This Kähler potential can be computed from the prepotential $\mathcal{F}$ of the SCFT via

$$K = -\log\left(i\bar{t}^J \frac{\partial \mathcal{F}}{\partial t^J} - it^J \frac{\partial \overline{\mathcal{F}}}{\partial \bar{t}^J}\right) , \tag{6.1}$$

where K denotes the Kähler potential, and t^J denotes a set of coordinates parameterizing the Kähler moduli of the manifold. Clearly, it would be advantageous if we could use the free energy of a TCFT which is in principal easier to compute. However, if we want to substitute for $\mathcal{F}$ the free energy of the associated TCFT the coordinates $t^J = t^J(a)$ must be the flat coordinate functions of the moduli alluded to above.

In this section we shall argue that such coordinate functions can be found from the invariants of a certain linear differential equation satisfied by the periods of the holomorphic d-form (as function of the moduli) defining the CY manifold. The holomorphic d form $\Omega(a)$ is defined as

$$\Omega(a) = \text{P.R.}(\omega)|_{W(X,a)=0} , \quad \omega(a) = \frac{dX_1 \wedge \cdots \wedge dX_n}{W(X_1,\ldots,X_n;a)} , \tag{6.2}$$

where P.R. denotes the Poincaré residue [43] and $n = d + 2$. The form of ω is useful since it is shows explicitly the dependence on the moduli appearing in the superpotential.

We should emphasize that one is not forced to use topological conformal field theory considerations to find the Yukawa couplings and the prepotential. In fact in particular example discussed in [42] the prepotential was computed using explicitly the 'period map' which gives the coordinate functions $t(a)$ described above. This was done using the monodromy properties of the linear differential equation for $\omega(a)$. In [42] this differential equation was found in an adhoc manner, which does not seem to be practical for general LG superpotentials.

Alternatively, it was argued in [44] that an analogous linear differential equation follows from solving the compatability conditions on the structure constants of the chiral ring of the topological theory. In particular this differential equation implies that the topological two-point function is independent of the coupling parameters. In other words, it specifies the correct coordinates to be used in SCFT. However, for general models it seems a formidable task to solve explicitly the compatability conditions from which one may determine the required differential equation.

In contrast, our approach, based on a mathematical technique known as 'variation of the Hodge structure', immediately leads to the linear differential equation satisfied by $\omega(a)$. It is an efficient tool applicable to all superpotentials defining isolated singularities which furthermore provides a clue to the geometric origin of the differential equation found in [42] and in [44]. Namely, it suggests that the compatability condition between the structure constants C_{IJ}^K of the operator product algebra of the topological field theory and (5.10) can be phrased in geometrical terms as the condition that the two point function in the topological theory be covariant with respect to the monodromy group of the linear differential equation satisfied by the periods, which is closely related to the duality group discussed in earlier sections.

The content of this section resulted from discussions with W. Lerche and N. Warner. Below we shall only outline the general principles of this technique and provide a simple example. A more complete discussion of these ideas, put in a more general context including also relevant perturbations, will appear elsewhere [45].

We consider differential forms generically of the form

$$\omega = \frac{P(X_1,\ldots,X_n)}{W(X_1,\ldots,X_n)}dX_1 \wedge \cdots \wedge dX_n , \qquad (6.3)$$

where P and W are arbitrary polynomial functions in the variables $X_1,\ldots,X_n$. In the case $P = 1$ in (6.3), W is the LG superpotential. (We ignore the moduli parameters for the moment.) For clarity we restrict to the case $n = 1$ for the moment; a generalization to n variables is straightforward. The contour integral $\int_\Gamma \omega$ with Γ a closed curve in the complex plane encircling the poles of ω is called a period of ω.

The differential equation we will derive is in fact a property of the periods ω. We will do this in two steps: first we discuss a reduction property of the differential forms. Then we discuss a variational property of the periods.

A reduction property of rational differential forms.

The integral may formally be expressed as a sum over the residues

$$\int_\Gamma \frac{P(X)}{W(X)} dX = \frac{1}{2\pi i} \left\{ \sum_{j=1}^{N} \text{Res}_{p_j} \left(\frac{P(X)}{W(X)} \right) \right\} \tag{6.4}$$

where p_j denotes the j-th pole of ω on $\mathbf{CP}^n$ (i.e. the complex plane with infinity). The variety $V = \{p_n, \ldots, p_n\}$ is called the polar locus of ω, that is V coincides with the zeros of W plus infinity if $\deg W < \deg P + 2$. We assume that $V \subset \mathbf{CP}^n$ is nonsingular. The contour Γ encircling all the poles can be split into a sum of contours Γ_j encircling p_j. The computation of the integral (6.4) may be reduced t a computation of $\int_{\Gamma_j} \tilde{\omega}$ where $\tilde{\omega}$ has a first order pole at p_j which brings the period into a canonical form [46]. This may be generalized to rational differential forms ω (6.3) on $\mathbf{CP}^n$. It is customary to go over to inhomogeneous coordinates ξ_i on $\mathbf{CP}^n$ and to write

$$\omega = (P(\xi)/W(\xi))\Lambda , \tag{6.5}$$

with $\Lambda = \sum_{j=0}^{n} (-1)^{i+j\xi^j} (d\xi_1 \wedge \cdots d\hat{\xi}_j \cdots \wedge d\xi_n)$. By assumption the polar locus $V == \{W(\xi) = 0 \subset \mathbf{CP}^n\}$ is nonsingular so that we may write $\omega = P(\xi)/(W(\xi))^k$, k a positive integer, and P and W are relatively prime. In [46] a reduction of ω is discussed based on the following consequence of Stoke's theorem:

$$\int_\Gamma \omega = \int_\Gamma \omega + d\phi , \tag{6.6}$$

with ϕ a rational $(n-1)$-form on $\mathbf{CP}^n$ with polar locus V. The reduction is given by two theorems which essentially say the following

1. Given the n-form ω then there exists a ϕ as above such that $\omega + d\phi$ has a pole of order n on $V \subset \mathbf{CP}^n$. So we can always reduce ω to have a pole of order n or less.

2. Suppose we can reduce the pole of ω from k to $k-1$. Then this can be done by a ψ with pole of order $k-1$ on V such that $\omega + d\psi$ has a pole of order $k-1$.

In order to apply this we use the fact [46] that a rational $n-1$ form ϕ can be written as

$$\phi = \frac{1}{W(\xi)^l} \left(\sum_{i<j} (-1)^{i+j} (\xi^i A_j(\xi) - \xi^j A_i(\xi))(d\xi_1 \wedge \cdots d\hat{\xi}^i \cdots d\hat{\xi}^j \cdots \wedge d\xi_n) \right) , \tag{6.7}$$

with $l \cdot \deg W = \deg A_j + n$. Then it follows that

$$d\phi = \frac{P\Lambda}{W^{l+1}} \,, \tag{6.8}$$

with

$$P(\xi) = l \left(\sum_{j=0}^{n} A_j(\xi) \frac{\partial W(\xi)}{\partial \xi^j} \right) - W(\xi) \left(\sum_{j=0}^{n} \frac{\partial A_j(\xi)}{\partial \xi^j} \right) .$$

From this it follows that the pole of ω can be reduced from k to $k-1$ if and only if P belongs to the (prime) ideal generated by the first partial derivatives of W. Subsequently, along similar lines one proves that with ϕ as in (6.7), there exists a rational $(n-2)$-form ψ such that $\phi + d\psi$ has a pole of order $l-1$ on V provided that there exists a skew-symmetric matrix C_{ij} with entries homogeneous polynomials in ξ with

$$A_j(\xi) = \sum_{j=0}^{n} C_{ij}(\xi) \frac{\partial W(\xi)}{\partial \xi^j} \quad \mathrm{mod} \ \ W . \tag{6.9}$$

Finally let $\omega = P\Lambda/W^k$ have a pole of order k on V. Then one can reduce the pole to $k-1$ if

$$k \geq \frac{(q-1)(n+1)+1}{q} \,, \quad q \equiv \deg \ W . \tag{6.10}$$

This concludes the discussion on the reduction property of ω, necessary below.

Variation of the periods.

In order to derive the differential equation we need to know in addition to this reduction property how the period ω varies with W. For that we introduce a (formal) parameter a which in our case will be identified as a moduli parameter. As an example we set $W = W(X,a) = X(X-a)$ and for the sake of the argument we put $P = 1$. Thus the period

$$\pi(a) = \int_\Gamma \frac{dX}{X(X-a)}$$

is a function of a. Now taking the derivative with respect to a (under the integral sign) one finds the linear differential equation

$$a \frac{d\pi(a)}{da} + \pi(a) = 0 \tag{6.11}$$

which has only regular singular point. This phenomenon generalizes to differential forms $\omega(a)$ of the form (6.3) with W depending on a parameter.[13] Generically one finds a linear differential equation of order k:

$$P_k(a)\omega^{(k)} + \cdots + P_1\omega^{(1)} + P_0\omega = d\alpha ,\tag{6.12}$$

with $\omega^{(i)}$ denoting the i-th derivative, and $\alpha(X,a)$ a rational function in X and a given as a linear combination of the ideal generated by the first partial derivatives of $W(X,a)$ with respect to X_i. As shown in [46] the differential equation thus obtained has at worst regular singular points. This implies that the monodromy group of the equation is a *finite group*. We will come back to this fact in relation to the duality group below.

It is now clear that derivatives of order k (exceeding the bound in (6.10)) of the differential form $\omega(a)$ on a Ricci-flat target manifold defined as the Poincaré residue, will satisfy a linear differential equation.[14]

Let us apply these techniques to a simple example. We consider the superpotential

$$W = \frac{1}{2}(X_1^4 + X_2^4) + X_3^2 - aX_1^2X_2^2 ,\tag{6.13}$$

describing a torus with complex structure parameterized by a defining a marginal deformation of a SCFT. We write the differential form in homogeneous coordinates ($P = 1$):

$$\omega(a) = \frac{X_1 dX_2 \wedge dX_3 - X_2 dX_1 \wedge dX_3 + X_3 dX_1 \wedge dX_2}{W(X_1, X_2, X_3, a)} .\tag{6.14}$$

From (6.10) we conclude that the poles in the second order derivative of ω with respect to a can be reduced so that $d^2\omega/da^2$ can be expressed in terms of linear combinations of $d\omega/da$ and ω. This then defines the differential equation. Applying the general techniques sketched above and repeatedly using Stoke's theorem one finds

$$\left((1 - a^2)\frac{d}{da^2} - 2a\frac{d}{da} - \frac{1}{4}\right)\omega = 0 .\tag{6.15}$$

---

[13] We will discuss below the case of only one parameter, however, the phenomenon occurs also for arbitrary parameters.

[14] It should be noted that the technique of [46] is far more general; in particular it will apply also to superpotentials describing arbitrary topological field theories, not just conformal ones. This will be discussed in [45].

This linear differential equation is a lower dimensional analogue of the fourth order equation found in [42].[15]

Since the differential equation has only three regular singular points, it is possible to recast it in a hypergeometric differential equation from which one can read off what the monodromy group is. Standard techniques (see e.g. [47]) involving the Schwarzian differential of the ratio t of two independent solutions allow one to determine this hypergeometric equation. One thus finds for the monodromy group the modulary group $\Gamma/\Gamma(2)$. It is amusing to observe that this is also the duality group of the target manifold defined by Λ. This is true also for theother two superpotentials describing $c = 3$ SCFTs. The precise relation between the monodromy group of the linear differential equation satisfied by the periods of Ω and the duality group of the associated LG model is at present unclear. In [45] we hope to come back to this.

The solution of this equation is a Schearz-triangle function $s\left(\frac{1}{2}, \frac{1}{2}, \frac{1}{3}; J\right)$, (with $J(a(\tau)) = \frac{1}{27}\frac{(a^2 + 3)^3}{(a^2 - 1)^2}$, which is the result found in [27]. As was also pointed out in [44] the rational function J of a determines the basis in which the two-point function of the TCFT becomes a independent. That is, the structure constants expressed in the coordinate $t(a) \equiv J(a)$ defines the prepotential $\mathcal{F}$.

Now that we can associate to any CY compactification relatively straight-forwardly the linear differential equation satisfied by the periods of its defining holomorphic d-form (and in principle the monodromy group) one can proceed with computing the Kähler potential and the normalized Yukawa couplings for the associated $N = 2$ SCFT as outlined in the beginning of this section.

Acknowledgements

This work is partially supported by the Director, Office of Energy Research, Office of High Energy and Nuclear Physics, Division of High Energy Physics of the U.S. Department

---

[15]One can derive the equation for the holomorphic three-form of the quintic discussed in [42] in the same way.

of Energy under Contract DE-AC03-76SF00098. The work of A.G. is supported in part by a Chaim Weizmann fellowship. The work of D.-J.S. is supported in part by NSF grant PHY85-15857.

References

[1] M.B. Green, J.H. Schwarz and E. Witten, *Superstring theory*, Cambridge Univ. Press, Cambridge, 1987.

[2] T. Banks, L.J. Dixon, D. Friedan and E. Martinex, Nucl. Phys. **B299** (1988) 613; A. Sen Nucl. Phys. **B278** (1986) 289; Nucl. Phys. **B284** (1987) 423.

[3] D. Gepner, Phys. Lett. **199B** (1987) 380, Nucl. Phys. **B285** (1988) 732, Nucl. Phys. **B296** (1988) 757.

[4] K. Kikkawa and M. Yamasaki, Phys. Lett. **149B** (1984) 357; N. Sakai and I. Senda, Prog. theor. Phys. **75** (1986) 692.

[5] V.P. Nair, A. Shapere, A. Strominger and F. Wilczek, Nucl. Phys. **B287** (1987) 402.

[6] R. Dijkgraaf, E. Verlinde and H. Verlinde, in 'Perspectives in String Theory', P. Di Vecchia and J.L. Petersen eds., (World scientific, 1988).

[7] a. Giveon, E. Rabinovici and G. Veneziano, Nucl. Phys. **B322** (1989) 167.

[8] A. Shapere and F. Wilzcek, Nucl. Phys. **B320** (1989) 669.

[9] A. Giveon, N. Malkin and E. Rabinovici, Phys. Lett. **220B** (1989) 551.

[10] J. Lauer, J. Mas and H.P. Nilles, Phys. Lett. **226B** (1989) 251; M. Dine, P. Huet and N. Seiberg, Nucl. Phys. **B322** (1989) 301; J.M. Molera and B.A. Ovrut, preprint UPR-0404T, December 1989.

[11] T.H. Buscher, Phys. Lett. **B201** (1988) 466; E. Smith and J. Polchinski, preprint UTTG-07-91, January 1991; A.A. Tseytlin, *Duality and the Dilaton*, John Hopkins preprint, JHU-TIPAC-91008, May 1991; E.B. Kiritsis, *Duality in gauged WZW Models*, Berkeley preprint, LBL-30747, May 1991.

[12] S. Ferrara, D. Lust, A. Shapere and S. Theisen, Phys. Lett. **225B** (1989) 363.

[13] A. Giveon, N. Malkin and E. Rabinovici, Phys. Lett. **238B** (1990) 57.

[14] A. Giveon and M. Porrati, Phys. Lett. **B246** (1990) 54; A. Giveon and M. Porrati, Nucl. Phys. **B355** (1991) 422.

[15] R. Brandenberger and C. Vafa, Nucl. Phys. **B316** (1989) 301; B. Greene, A. Shapere, C. Vafa and S.T. Yau, Harvard Preprint HUTP-89/A047 and IASSNS-HEP-89/47.

[16] A. Giveon, *Target Space Duality and Stringy Black Holes*, Berkeley preprint, LBL-30671, April 1991; R. Dijkgraaf, E. Verlinde and H. Verlinde, *String Propagation in Black Hole Geometry*, IAS preprint, IASSNS-HEP-91/22, May 1991.

[17] A. Giveon and D.-J. Smit, Nucl. Phys. **B349** (1991) 168; A. Giveon and D.-J. Smit, preprint LBL-29470, UCB-PTH-90/37, August 1990, in the proceedings of the Yukawa International Seminar on *Common trends in Mathematics and Quantum Field Theory*, May 1990 Kyoto.

[18] M. Cvetic, J.M. Molera and B.A. Ovrut, Phys. Lett. **B248** (1990) 83.

[19] C. Vafa and N.P. Warner, Phys. Lett. **218B** (1989) 51.

[20] B. Greene, C. Vafa and N.P. Warner, Nucl. Phys. **B324** (1989) 371.

[21] W. Lerche, C. Vafa and N.P. Warner, Nucl. Phys. **B324** (1989) 427.

[22] N.P. Warner, Lectures given at ICTP Spring Workshop, 1989.

[23] E. Martinec, Phys. Lett. **217B** (1989) 431.

[24] E. Martinec, *Criticality, Catastrophes and Compactifications*, in the V.G. Knizhnik memorial volume, 1989.

[25] C. Vafa, Mod. Phys. Lett. **A4** (1989) 1169.

[26] C. Vafa, *Superstring Vacua*, preprint HUTP-89/A057, 12/89.

[27] W. Lerche, D. Lust and N.P. Warner, Phys. Lett. **231B** (1989) 417.

[28] E.J. Chun, J. Mas, J. Lauer and H.P. Nilles, Phys. Lett. **226B** (1989) 251.

[29] B.R. Greene, S. Roan and S.-T. Yau, *Geometric singularities and spectra of LG models*, Cornell preprint, CLNS-91-1045, February 1991.

[30] L. Dixon, In *Proceedings of the 1987 ICTP Summer Workshop in High Energy Physics and Cosmology*, Trieste, Italy, edited by G. Furlan, J.C. Pati, D.W. Sciama, E. Sezgin and Q. Shafi.

[31] L. Dixon, V.S. Kaplunovsky and J. Louis, Nucl. Phys. **B329** (1990) 27.

[32] B.R. Greene and M.R. Plesser, Nucl. Phys. **B338** (1990) 15.

[33] S. Ferrara, D. Lust and S. Theisen, Phys. Lett. **B233** (1989) 147; Phys. Lett. **B242** (1990) 39.

[34] S. Ferrara, *Superstring Effective Field Theories on (2,2) Vacua*, preprint CERN-TH-

5873/90, September 1990.

[35] A. Giveon and D.-J. Smit, *Exact Yukawa Couplings from Topological Landau-Ginzburg Models*, Berkeley preprint, LBL-30388, UCB-PTH-91/10, March 1991, to appear in Mod. Phys. Lett. **A**.

[36] E. Witten, Comm. Math. Phys. **117** (1988) 353; E. Witten, Nucl, Phys. **B340** (1990) 281; T. Eguchi and S.-K. Yang, Mod. Phys. Lett. **A5** (1990) 1693.

[37] C. Vafa, Mod. Phys. Lett. **A6** (1991) 337.

[38] A. Giveon and D.-J. Smit, preprint LBL-30342, UCB-PTH-91/8, February 1991, to appear in Int. J. Mod. Phys. A.

[39] D. Gepner, Nucl. Phys. **B322** (1989) 65.

[40] R. Dijkgraaf, E. Verlinde and H. Verlinde, Princeton preprint PUPT-1204, IASSNS-HEP-90/71, October 1990; R. Dijkgraaf, E. Verlinde and H. Verlinde, Princeton preprint PUPT-1217, IASSNS-HEP-90/80, November 1990.

[41] E. Witten, Nucl. Phys. **B268** (1986) 79; J. Distler and B.R. Greene, Nucl. Phys. **B309** (1988) 295.

[42] P. Candelas, X.C. de la Ossa, P.S. Green and L. Parkes, University of Texas preprint UTTG-25, October 1990.

[43] P. Griffiths and J. Harris, *Principles of Algebraic Geometry*, Wiley Interscience, New-York (1978).

[44] E. Verlinde and N. Warner, *Topological Landau-Ginzburg Matter at $c = 3$*, Princeton preprint IASSNS-HEP-91/16, March 1991.

[45] W. Lerche, D.-J. Smit and N. Warner, preprint to appear.

[46] P. Griffiths, *On the periods of certain rational integral I*, Ann. Math. **90** (1969), 460.

[47] A. Forsyth, *Theory of Differential Equations*, Vol. 4 pp. 174, Dover Publications, New-York (1959).

AMS/IP Studies in Advanced Mathematics
Volume **9**, 1998

An $SL(2,\mathbb{C})$ Action on Certain Jacobian Rings and the Mirror Map

Tristan Hübsch[♣]

Departments of Mathematics and Physics

Harvard University, Cambridge, MA 02138

hubsch@ zariski.harvard.edu, @ huhepl.bitnet

and

Shing-Tung Yau[◇]

Department of Mathematics

The National Tsing-Hua University

Hsinchu, Taiwan, R.O.C.

styau @ math.nthu.edu.tw

[♣]On leave form the Institute "Rudjer Bokova", Zagreb, roatia.
[◇]On leave from the Harvard University Math. Dept., Cambridge, MA 02138, U.S.A.

© 1998 American Mathematical Society
and International Press

T. HÜBSCH AND S.-T. YAU

Abstract

We consider the Jacobian ring of quasi-homogeneous hypersurfaces in weighted projective N-spaces. For a generic hypersurfaces $f^{-1}(0)$ with trivial canonical class, there exists an $SL(2,\mathbb{C})$ action and an associated Lefschetz-type decomposition of the middle-dimensional cohomology and also of the degree-k $\deg(f)$ subring of the Jacobian ring. This is also a corollary of the so-called 'mirror map', which has attracted considerable interest.

1. The Main Result and its Background

Recent studies of complex 3-folds with trivial canonical class, predominantly motivated by their application in superstring models [1], indicate that such varieties occur in so-called mirror pairs [2,3,4,5]. For complex 3-folds with trivial canonical class and finite structure group, $h^{p,q} = 0$ unless $p = \pm q \pmod{3}$, $h^{p,q} = 1$ if $p = q \pmod{3}$. The cohomology ring then naturally splits into $\oplus_1 H^{q,q}$ and $\oplus_q H^{3-q,q}$, both of which can be given a ring structure, denoted $\mathcal{R}_Q(H^{q,q})$ and $\mathcal{R}_Q(H^{3-q,q})$, respectively (see below).

Two 3-folds with trivial canonical class, $\mathcal{M}$ and $\mathcal{W}$, are said to be each other's mirror model if their (quantum) cohomology rings satisfy the isomorphisms

$$\mathcal{R}_Q\left(H^{q,q}(\mathcal{M})\right) \approx \mathcal{R}_Q\left(H^{3-q,q}(\mathcal{W})\right) , \tag{1.1a}$$

$$\mathcal{R}_Q\left(H^{3-q,q}(\mathcal{M})\right) \approx \mathcal{R}_Q\left(H^{q,q}(\mathcal{W})\right) . \tag{1.1b}$$

In specifying the ring structure of $\mathcal{R}_Q(H^{q,q})$, the choice of the Kähler class, J, is essential: in the $J \to \infty$ limit, $\mathcal{R}_Q(H^{q,q})$ becomes the standard ring $(\oplus_q H^{q,q}, \wedge)$. In this sense, $\mathcal{R}_Q(H^{q,q})$ is a quantum deformation of $(\oplus_q H^{q,q}, \wedge)$. Now recall that multiplication with the Kähler class $J \in H^{1,1}$ and the $*$ isomorphism

$$* : H^{p,q} \longrightarrow H^{3-p,3-q} , \qquad 0 \le p, q \le 3 , \tag{1.2}$$

which is of course related to the standard duality

$$H^{3-p,3-q} \times H^{p,q} \longrightarrow H^{3,3} \approx \mathbb{C} , \qquad 0 \le p,\ q \le 3 , \tag{1.3}$$

induce the well known $SL(2)$ action and the Lefschetz decomposition. No such preferred element of $\oplus_q H^{3-q,q}$ appears to be chosen by conventional cohomology theory, whereas the isomorphisms (1.1) suggest that such a structure in $\mathcal{R}(H^{3-q,q})$ does exist.

The following two theorems ensure the existence of such a structure:

Lemma A. *Let $f(x)$ be a transversal quasi-homogeneous degree-d polynomial in $x_0, \ldots, x_N$ and let $\mathcal{R}(f)$ be the Jacobian ring of $\mathcal{M} \stackrel{\text{def}}{=} f^{-1}(0) \hookrightarrow \mathbb{P}^N_{(\mathbf{w})}$ and Q its element of top degree. If also $c_1(\mathcal{M}) = 0$, then*

1. *there exists a $\vartheta \in \mathcal{R}^{\deg f}(f)$ such that $\vartheta^{N-1} \equiv Q \pmod{\Im[\partial f]}$;*

2. *for a generic choice of $f(x)$, the monomial $\vartheta \overset{\mathrm{def}}{=} \left(\prod_{i=0}^{N} x_i \right)$ is a universally valid choice, regardless of the weight $\mathbf{w}$ and dimension N.*

It appears reasonable to call ϑ the fundamental monomial of $\mathcal{R}(f)$.

Theorem B. *Let $f(x)$, $\mathcal{M}$, $\mathcal{R}(f)$ and ϑ be as defined in Theorem A and also define $\mathcal{R}(f) \overset{\mathrm{def}}{=}$ $\bigoplus_{k=0(\bmod \deg f)} \mathcal{R}^k(f)$. Then*

1. *multiplication with ϑ generates an $SL(2,\mathbb{C})$ action on $\mathcal{R}(f)$;*

2. *this induces a Lefschetz type decomposition of $\overset{\circ}{\mathcal{R}}(f)$;*

3. *corresponding to $\vartheta \in \mathcal{R}^{\deg f}(f)$, there exists a $\varphi \in H^{N-2,1}(\mathcal{M})$ along with the $SL(2)$ action and the Lefschetz type decomposition of $\bigoplus_q H^{N-1-q,q}(\mathcal{M})$.*

Two remarks

At least when $f(x)$ is a Fermat polynomial or a relatively simple deformation thereof, the mirror model $\mathcal{W}$ is obtained as a quotient of $\mathcal{M} = f^{-1}(0)$, typically harboring cyclic quotient singularities [3, 4, 5]. The $(1,1)$-cohomology of $\mathcal{W}$ is then generated by J_0, the pull-back of the Kähler form of the embedding $\mathbb{P}^N_{(\mathbf{w})}$, and the vanishing cocycles associated to the resolution of the quotient singularities. When $\mathcal{W}$ is an orbifold, suitable limits of the vanishing cocycles (defined by the 'minimal' blow-up which preserves the triviality of the canonical class) must be retained to match the string theory spectrum. Since $\mathcal{W}$ is singular, $J_0 \in H^{1,1}(\mathcal{W})$ and the associated Lefschetz decomposition are ill defined. Indeed, mirroring this in $H^{2,1}(\mathcal{M})$ and the Jacobian ring of $\mathcal{M}$, multiplication by $\vartheta = (x_1 \cdots x_N)$ also develops as kernel and our Lefschetz type decomposition of $\overset{\circ}{\mathcal{R}}(f)$ is also ill defined, as expected.

In the case $\dim \mathcal{M} = 3$, it appears that the fundamental class $\vartheta \in H^{2,1}(\mathcal{M})$ (defined by the fundamental monomial $\vartheta \in \mathcal{R}(f)$) is the (complexified) preimage of $J_0 \in H^{1,1}(\mathcal{W})$ under the mirror map $\mathcal{M} \to \mathcal{W}$. This identification has been made by Aspinwall and Lütken for some Fermat polynomials [12], based on their high degree of symmetry. The present results are rather more general in that no symmetry is required of $f(x)$.

1.1   *Yukawa couplings and rings*

The main motivation for the work presented here comes from the existence of the "mirror map", which follows from 2-dimensional conformal field theory. For the benefit of the mathematically oriented reader, we here describe in some detail the background in which to discuss the mirror map, leaving aside the conformal field theory reasons for its existence.

The (q,q)-cohomology has a (classical) ring structure, $\mathcal{R}(H^{q,q})$, provided by the wedge product which fits in the bottom row of the commutative diagram

$$
\begin{array}{ccc}
\mathrm{Sym}^3 H^{1,1} & \xrightarrow{\text{Yukawa coupling}} & H^{3,3} \approx \mathbb{C} \\
\Big\uparrow{\scriptstyle\mathrm{Sym}} & & \Big\uparrow{\scriptstyle\mathrm{dual}} \\
\left(H^{1,1} \times H^{1,1}\right) \times H^{1,1} & \xrightarrow{\ \wedge\ } & H^{2,2} \times H^{1,1}
\end{array}
\tag{1.4}
$$

The map in the top row, evaluated by integration over the manifold, is known as the classical $\overline{27}^3$ Yukawa coupling; in (super)string models, it accounts for the constant maps of the (super)string into the target space $\mathcal{M}$. That is, the (super)string is mapped into points of $\mathcal{M}$, whence integration over $\mathcal{M}$. Quantum corrections then involved holomorphic maps, the image of which are (at genuse-0) rational curves $L \subset \mathcal{M}$; there are no other (finite genus) corrections to the Yukawa coupling in models with $(2,2)$-supersymmetry. The resulting Yukawa coupling is specified as [10]

$$
\kappa(A,B,C) \overset{\text{def}}{=} \left(\int_{\mathcal{M}} A \wedge B \wedge C \right) + \sum_{\{L\}} \frac{e^{-(\int_L J)}}{1 - e^{-(\int_L J)}} \left(\int_L A \right)\left(\int_L B \right)\left(\int_L C \right) .
\tag{1.5}
$$

for $A,B,C \in H^{1,1}(\mathcal{M})$; summation over $\{L\}$ of course implies also integration over the relevant moduli space of rational curves $L \subset \mathcal{M}$. Re-inserting $\hbar$, the exponential factor becomes $\exp\left\{-\left(\frac{1}{\hbar}\int_L J\right)\right\}$, so that the classical limit $\hbar \to 0$ is identifiable as the 'large size limit' $J \to \infty$. In this limit, only the first (classical) contribution to the Yukawa coupling (1.5) remains. The motivates the interpretation of $\kappa(A,B,C)$ as a quantum deformation of $\left(\int_{\mathcal{M}} ABC \right)$ and so also $\mathcal{R}_Q(H^{q,q})$ as a quantum deformation of the classical ring $(\oplus_q H^{q,q}, \wedge)$ (up to completely unknown 'infinite genus' corrections).

This deformation of the product map and the deformation of the duality map are *not* specified separately. In point of fact, while the wedge product, the star operation and the classical Yukawa coupling ($\int_{\mathcal{M}} ABC$) in (1.4) are canonical[1], their quantum deformations depend crucially on the choice of $J \in H^{1,1}$ and the web of rational curves $\{L\} \subset \mathcal{M}$.

Owing to the triviality of the canonical class, the (multiplication) maps

$$\Omega : H^q(\mathcal{M}, \wedge^p \mathcal{T}_{\mathcal{M}}) \quad \longrightarrow \quad H^{3-p,q}(\mathcal{M}) , \qquad 0 \le p,\ q \le 3 , \tag{1.6}$$

are all isomorphisms, depending only (and holomorphically, moreover linearly) on the holomorphic 3-form Ω, that is, on the choice of the complex structure of $\mathcal{M}$. This permits the definition of a multiplication[2], $\Diamond$, which fits in the commutative diagram

$$
\begin{array}{ccc}
\mathrm{Sym}^3 H^{2,1} \overset{\Omega}{\approx} \mathrm{Sym}^3 H^1(\mathcal{T}_{\mathcal{M}}) & \xrightarrow{\text{Yukawa coupling}} & \left(H^3(\mathcal{M}, \mathcal{K}_{\mathcal{M}\cdot})\right)^2 \approx \mathbb{C} \\
\uparrow{\scriptstyle \mathrm{Sym}} & & \uparrow{\scriptstyle \mathrm{dual}} \\
\left(H^1(\mathcal{T}_{\mathcal{M}}) \times H^1(\mathcal{T}_{\mathcal{M}})\right) \times H^1(\mathcal{T}_{\mathcal{M}}) & \xrightarrow{\Diamond} & H^2(\wedge^2 \mathcal{T}_{\mathcal{M}}) \times H^1(\mathcal{T}_{\mathcal{M}})
\end{array}
\tag{1.7}
$$

The right hand side vertical map can be identified with the dual pairing (1.3), since

$$H^2(\wedge^2 \mathcal{T}) \times H^1(\mathcal{T}) \overset{\sim}{\longrightarrow} H^{1,2} \times H^{2,1} , \tag{1.8}$$

in view of (1.6). The map in the top row of (1.7), known as the $\mathbf{27}^3$ Yukawa coupling, is obtained by straightforward evaluation [6,7,8] and receives no quantum correction in superstring models [9]. It can be specified as

$$\kappa(\alpha,\beta,\gamma) \overset{\text{def}}{=} \int_{\mathcal{M}} \Omega \wedge \left((\alpha \wedge \beta \wedge \gamma) \cdot \Omega\right) , \tag{1.9}$$

where $\alpha, \beta, \gamma \in H^1(\mathcal{T}_{\mathcal{M}})$ and $\Omega \in H^{3,0}$. Let $\mathcal{R}_Q(H^{3-q,q})$ denote the ring obtained by equipping the (middle) cohomology group $\oplus_q H^{3-q,q}$ with this multiplication.

---

[1] Since $H^{2,0} = H^{0,2} = H^{3,1} = H^{1,3} = 0$ for 3-folds with trivial canonical class and finite structure group, $H^{q,q}$ may be identified with the integral even-dimensional cohomology and the Yukawa coupling is in fact topological.

[2] Essentially, $\Diamond$ is the wedge product combined with the skew-symmetric product in $\mathcal{T}_{\mathcal{M}}$.

1.2   Hypersurfaces and Jacobian rings

We now remind the reader of some basic facts and notation usually related to the study of (the various form of) the Torelli problem; see for instance Refs. [13, 14]. Let $(\mathbf{w}) = (w_0 : \cdots : w_N)$ and let $\mathbb{P}^N_{(\mathbf{w})} = \{\mathbb{C}^{N+1} - 0\}/\varpi$ be the weighted complex projective space with

$$\varpi : (x_0 : \cdots : x_N) \longmapsto (\lambda^{w_0} x_0 : \cdots : \lambda^{w_N} x_N) \,. \tag{1.10}$$

Let $f(x)$ be a section of $\mathcal{O}_{\mathbb{P}^N_{(\mathbf{w})}}(d)$, this is, $f(x)$ is a degree-d quasi-homogeneous polynomial in $x_0, \ldots, x_N$. We define $\mathcal{M} \hookrightarrow \mathbb{P}^N_{(\mathbf{w})}$ as the zero-set of $f(x)$. Then there are natural isomorphism $\left(F^k \overset{\text{def}}{=} \oplus_{i=0}^k H^{N-1-k,k}\right)$

$$\begin{aligned} \mathcal{R}^{(N-1-k)d}(f) &\overset{\sim}{\longrightarrow} \{F^{N-1-k}/F^{N-2-k}\} \,, \qquad k = 0, \ldots, (N-2) \,, \\ \mathcal{R}^0(f) &\overset{\sim}{\longrightarrow} F^0 = H^{N-1,0} \,, \end{aligned} \tag{1.11}$$

depend holomorphically on f. $\mathcal{R}^k(f)$ denotes the degree-k piece of the Jacobian ring

$$\mathcal{R}(f) \overset{\text{def}}{=} \mathbb{C}[x_0, \ldots, x_N]/\Im[\partial f] \,, \tag{1.12}$$

where $\mathbb{C}[x_0, \ldots, x_N] \overset{\text{def}}{=} \operatorname{Sym}^* H^0(\mathbb{P}^4, \mathcal{O}_{\mathbb{P}^4_{\mathbf{w}}}(w_i))$ is the ring of complex homogeneous polynomials in $x_0, \ldots, x_N$ and $\Im[\partial f] \subset \mathbb{C}[x_0, \ldots, x_N]$ is the Jacobian ideal generated by the partials[3] of $f(x)$. Of course, $\mathcal{R}(f)$ is naturally graded in terms of $\mathbf{w}$.

We record the following more or less known results:

Theorem 0. *If the system of partials* $\left\{\dfrac{\partial f}{\partial x_0}, \cdots, \dfrac{\partial f}{\partial x_N}\right\}$ *is non-degenerate.*

1. *$f^{-1}(0) \hookrightarrow \mathbb{P}^N_{(\mathbf{w})}$ is transversal and thus quasi-smooth[4];*

2. *$\dim \mathcal{R}(f) < \infty$;*

3. *the highest degree occurring in $\mathcal{R}(f)$ is $\sigma \overset{\text{def}}{=} \sum_{i=0}^N (d - 2w_i) = (N+1) \cdot d - 2\left(\sum_{i=0}^N w_i\right)$;*

4. *$\dim \mathcal{R}^\sigma(f) = 1$;*

---

[3] Recall that, at the zero-set of $f(x)$, the partials are independent of any connection and $(\partial f/\partial x_i)$ are in fact covariant derivatives on $\mathcal{M}$.

[4] A hypersurface in a weighted projective space is quasi-smooth if its singularities stem solely from intersecting the singular set of the weighted projective space.

5. *the duality (multiplication) map*

$$* : \mathcal{R}^k \times \mathcal{R}^{\sigma-k} \to \mathcal{R}^\sigma \approx \mathbb{C} \ , \tag{1.13}$$

is perfect for any $k = 0, \dots, \sigma$;

6. *the representative*

$$Q \stackrel{\text{def}}{=} \det \big[Hess(f) \big] = \det \Big[\frac{\partial^2 f}{\partial x_i \partial x_j} \Big] \in \mathcal{R}(f)\big|_{\deg=\sigma} \tag{1.14}$$

is natural, in that it depends only (and holomorphically) on the choice of f.

Sketch of proof

The statement 1. is a standard application of the inverse function theorem. For a proof of 2.-5., see p.656–662 of Ref.[11] and also Theorem 2.2 in [14]. The last statement, 6. follows from 4., the observation that Q is coordinate independent, being a determinant, and that $Hess(f)$ and so also Q are independent of any connection used to define them:

$$(\partial_i + \Gamma_i)(\partial_j + \Gamma_j)f = \text{Hess}_{ij}(f) + (\Gamma_i\partial_j f + \Gamma_j\partial_i f) + (\partial_i\Gamma_j)f \ , \tag{1.15}$$

by quasi-homogeneity, $f = (1/\deg f)\sum_i x_i(\partial_i f)$ and so all the lower order terms belong to $\mathfrak{S}[\partial f]$. $\checkmark$

The above duality has a straightforward consequence:

Corollary 0. *The quadratic form (inner product)* $\langle \cdot, \cdot \rangle$ *defined by*

$$\langle \varphi, \varphi' \rangle \cdot Q \stackrel{\text{def}}{=} -(*\varphi) \cdot \varphi' \pmod{\mathfrak{S}[\partial f]} \ , \qquad \varphi, \varphi' \in \mathcal{R}^k(f) \ , \tag{1.16}$$

is non-degenerated for all $k = 0, \dots, \sigma$.

(The negative sign in the definition comes from $** = -1$, as appropriate for 3-folds.)

2. The Fundamental Monomial

We now prove Theorem A. For a $\vartheta \in \mathcal{R}(f)$ to exist and satisfy $\vartheta^{N-1} \equiv Q(\mathrm{mod}\,\Im[\partial f])$, it must be that $\deg Q = (N-1)\deg\vartheta$, that is, that $\sigma = (N-1)d$. Therefore,

$$\sum_{i=0}^{N}(d - 2w_i) = (N-1)d \Longrightarrow d - \Big(\sum_{i=0}^{N} w_i\Big) = 0 \ . \tag{2.1}$$

On the other hand, $c(\mathbb{P}^N_{(\mathbf{w})}) = \prod_i(1 + w_i\xi)$, where ξ generates the $(1,1)$-cohomology of $\mathbb{P}^N_{(\mathbf{w})}$. Also, since $f(x)$ is a section of $\mathcal{O}_{\mathbb{P}^N_{(\mathbf{w})}}(d)$,

$$c(\mathcal{M}) = \frac{\prod_{i=0}^{N}(1 + w_i\xi)}{(1 + d\xi)} = 1 - \Big(d - \big(\sum_{i=0}^{N} w_i\big)\Big)\xi + \cdots \tag{2.2}$$

by the Adjunction formula for $\mathcal{M} = f^{-1}(0)$. Therefore, $c_1\big(f^{-1}(0)\big) = 0$ is necessary for the existence of ϑ.

Once the degree has been cleared, the existence of ϑ is generically unhindered. Since $\deg\vartheta = \deg f = d$, the deformation space of ϑ may be identified with the deformation space of f. For hypersurfaces in $\mathbb{P}^N_{(\mathbf{w})}$, this space may be identified with the space of complex structures for $f^{-1}(0)$, i.e., $H^1(\mathcal{T}) \stackrel{\Omega}{\approx} H^{2,1}$. Thus, the variations of the fundamental monomial take form as $\vartheta + \sum_a t^\alpha \varphi_\alpha$, where $\varphi_\alpha \in \mathcal{R}^d(f)$ and t^α are complex parameters.

Note that $\vartheta = \big(\prod_{i=0}^{N} x_i\big)$ trivially satisfies $\deg\vartheta = \big(\sum_{i=0}^{N} w_i\big)$. Clearly, ϑ is a section of $\otimes_i \mathcal{O}_{\mathbb{P}^N_{(\mathbf{w})}}(w_i)$, which equals to anticanonical bundle of $\mathbb{P}^N_{(\mathbf{w})}$, as is easily seen from the (determinant of the) Euler sequence. Now, $\oplus_{i=1}^{N} \mathcal{O}_{\mathbb{P}^N_{(\mathbf{w})}}(w_i)$ has N algebraically independent sections, $x_1,\dots,x_N$, one for each w_i. Therefore, a choice of ϑ determines a basis $\{x_i\}$ (up to a permutation among coordinates of equal weight) in which ϑ is a monomial.

Finally, we must exclude the cases when some power of the fundamental monomial (of $\mathbb{P}^N_{(\mathbf{w})}$) is in the ideal $\Im[\partial f]$ and is therefore trivial in $\mathcal{R}(f)$. These conditions are clearly closed (they specify a subset of the space of choices of strictly smaller complex dimension) and the statements follow.

We remark that requiring genericity *is* essential: there exist even smooth models for which some power of the fundamental monomial is trivial in $\mathcal{R}(f)$. Consider

$$f(x) = \frac{1}{8}\Big(\sum_{i=1}^{4} x_i{}^8\Big) + \frac{1}{2}x_5{}^2 \,, \qquad (2.3)$$

which provides a degree-8 quasi-smooth 3-fold with trivial canonical class embedded in $\mathbb{P}^4_{(1:1:1:1:4)}$. Clearly, $(\partial f/\partial x_5) = x_5$, whence $\vartheta \in \Im[\partial f]$. Similarly, in the degree-6 hypersurface

$$f(x) = \frac{1}{6}\Big(\sum_{i=1}^{4} x_i{}^6\Big) + \frac{1}{3}x_5{}^3 \qquad (2.4)$$

in $\mathbb{P}^4_{(1:1:1:1:2)}$, $(\partial f/\partial x_5) = x_5{}^2$, whence $\vartheta \notin \Im[\partial f]$ but $\vartheta^2 \in \Im[\partial f]$.

Finally, owing to Eq. (2.1), the duality (1.13) and the inner product (1.6) of the Jacobian ring $\mathcal{R}(f)$ descend also to its subring $\mathcal{R}(f)$.

3. The $SL(2)$ Action and the Lefschetz Decomposition

Next, we turn to Theorem B and restrict $\overset{o}{\mathcal{R}}(f) \overset{\text{def}}{=} \oplus_{k=0(\text{mod deg }f)}\mathcal{R}^k(f)$. The collection $\{[1],[\vartheta],[\vartheta^2],\dots,[\vartheta^{N-1}]\}$ clearly forms a subring of $\overset{o}{\mathcal{R}}(f)$. This corresponds to a subring of $\oplus_q H^{N-q,q}(\mathcal{M})$ which is the (complexified) preimage of the primitive cohomology subring in $\oplus_q H^{q,q}(\mathcal{W})$ under the mirror map. The quotient $\{\overset{o}{\mathcal{R}}(f)/\oplus_k [\vartheta^k]\}$ remains to be identified with the subset of $\oplus_q H^{N-q,q}(\mathcal{M})$ which should be the preimage of the primitive cohomology in $\oplus_q H^{q,q}(\mathcal{W})$ under the mirror map. In the so stratified ring $\overset{o}{\mathcal{R}}(f)$, we find a $*$-operator and introduce the operations

$$L(\varphi) \overset{\text{def}}{=} \vartheta \cdot \varphi(\operatorname{mod}\Im[\partial f]) \,, \qquad (3.1a)$$

$$\Lambda(\varphi) \overset{\text{def}}{=} *\big(L(*\varphi)\big) \qquad (3.1b)$$

$$h(\varphi) \overset{\text{def}}{=} [L,\Lambda](\varphi) \,. \qquad (3.1c)$$

as usual and need to show that the action of these operators in indeed the standard one.

This can easily be shown by explicit computation. Rather then attempting to devise a universal notation in which to prove the general case, we present a simplex example; the extension to any particular case is straightforward. In any case, the statement 3 follows directly from the existence and holomorphicity of the isomorphisms (1.11).

Consider the pencil of quintics $\mathcal{M}_\psi = P_\psi^{-1}(0)$, with

$$P_\psi \stackrel{\text{def}}{=} \frac{1}{5} \sum_{i=1}^{5} x_i{}^5 - \psi \prod_{i=1}^{5} x_i \, , \tag{3.2}$$

where $(x_1 : x_2 : x_3 : x_4 : x_5) \in \mathbb{P}^4$ and $\psi \in \mathbb{P}^1$. It is convenient to use the basis of monomials for $\overset{\circ}{\mathcal{R}}(P_\psi)$ as given in Table 1. We compute $Q_\psi = 2^{10}(1 - \psi^5)\left(\prod_{i=1}^{5} x_i{}^3\right)$ and easily check that $\vartheta^3 \equiv Q_\psi(\text{mod}\,\Im[\partial f])$ for $\psi^5 \neq 1$ and $\psi \neq \infty$; the factor $2^{10}(1 - \psi^5)$ may then be absorbed by an overall rescaling of the defining polynomial $f(x)$ and will be ignored henceforth.

By Theorems 0 and A, there exists a perfect pairing amongst elements of $\overset{\circ}{\mathcal{R}}(P_\psi)$ except when $\psi^5 = 1$ or $\psi = \infty$, where the system of partials of P_ψ degenerates. We find a $*$-operator by writing $* : \phi \to \alpha_\phi \phi^*$ for $\phi \in \mathcal{R}^{\deg \phi}(f)$ and $\phi^* \in \mathcal{R}^{\sigma - \deg \phi}(f)$, such that $\phi \cdot \phi^* \equiv Q(\text{ mod } \Im[\partial f])$.[5] To fix the coefficients $\alpha(\phi)$, we require that $** = -1$ (as appropriate for odd-dimensional varieties) and the standard normalizations of the eigenvalues of $h = [\Lambda, L]$. Table 2 presents the results for the particular basis in Table 1. Note also that this basis is orthogonal with respect to the quadratic form (1.16). $\qquad\qquad \checkmark$

Concluding remarks: Just as the $\overline{27}^3$ Yukawa coupling (1.5) can be evaluated in the classical $(J \to \infty)$ limit, the 27^3 Yukawa coupling (1.9) can likewise be evaluated in the $\vartheta \to \infty$ limit. The latter then defines the mirror image of the topological coupling $\int_{\mathcal{M}} ABC$ and has been checked in a number of examples by Aspinwall and Lütken [contribution in this volume].

One may have wished to obtain a canonical Lefschetz type decomposition of $\overset{\circ}{R}(f)$. However, as we have noted, by having deformed the ring structure of $\oplus_q H^{q,q}$ so as to obtain (1.5) for

---

[5] At this point, the construction is by far simplest if one chooses a basis of monomials to span $\mathcal{R}(f)$. For each element, ϕ, the dual ϕ^* is required to be the simplest appropriate element from $\mathcal{R}^{\sigma - d}(f)$. Also, the so determined $\alpha(\phi)\phi^*$ must be independent to span $\mathcal{R}^{\sigma - d}(f)$.

the Yukawa coupling, the standard Lefschetz decomposition of $\oplus_q H^{q,q}$ has also been deformed. Since the mirror isomorphisms (1.1) involve the quantum cohomology rings $\mathcal{R}_Q(H^{q,q})$ and $\mathcal{R}_Q(H^{3-q,q})$, the Lefschetz type decomposition of $\overset{\circ}{\mathcal{R}}(f)$ and so of $\oplus_q H^{3-q,q}$ is the mirror of the *deformed* Defschetz decomposition of $\oplus_q H^{q,q}$ and is therefore not canonical.

Finally, the basis in Table 1 makes it manifest that the fifty monomials $(x_i{}^3 x_j{}^2)$ and $(x_i{}^3 x_j x_k)$ are in the kernel of both L and A and also of $*$ when $\psi = 0$. Of course, they also become null vectors of the quadratic form $\langle \cdot, \cdot \rangle$. This obstructs the extension of the $SL(2, \mathbb{C})$ action, as defined in Table 2, from $\overset{\circ}{\mathcal{R}}(P_{\psi \neq 0})$ to $\overset{\circ}{\mathcal{R}}(P_0)$.

Related to this is the fact that $\mathcal{W}_\psi$, the *mirror* of $\mathcal{M}_\psi$, has cyclic quotient singularities when $\psi = 0$ and the standard Lefschetz $SL(2)$ action on $H^{1,1}(\mathcal{W}_\psi)$ becomes ill-defined. At $\psi = 0$, $\mathcal{W}_\psi$ is the quotient $\mathcal{M}_0 / \overline{\omega}$, where $\overline{\omega}$ is a diagonal $\mathbb{Z}_5 \times \mathbb{Z}_5 \times \mathbb{Z}_5$ action on $(x_1 : x_2 : x_3 : x_4 : x_5) \in \mathbb{P}^4$ which leaves invariant $\mathcal{M}_0$ and its holomorphic 3-form

$$\Omega \overset{\text{def}}{=} \text{Res}_{\mathcal{M}_0}\left[\frac{d\mu}{P_0}\right] \,, \quad d\mu \overset{\text{def}}{=} \frac{1}{5!} \sum_\sigma \text{sign}(\sigma) x_{\sigma_1} dx_{\sigma_2} dx_{\sigma_3} dx_{\sigma_4} dx_{\sigma_5} \,, \tag{3.3}$$

that is, the top element $Q \in \mathcal{R}^\sigma(P_0)$.

Acknowledgements

We thank Joe Harris, Paul Green and Brian Greene for helpful discussions. T.H. was supported by the DOE grant DE-FG02-88ER-25065 and would like to thank the Mathematics Department of the National Tsing-Hua University at Hsinchu, Taiwan, for the warm hospitably while part of this research was being completed.

Tables

$\begin{array}{l}[\mathbf{1}]\\ [\vartheta]\\ [\vartheta^2]\\ [\vartheta^3]\end{array}$	$\begin{array}{l}[x_i{}^3 x_j{}^2]\\ [x_j x_k{}^3 x_m{}^3 x_n{}^3]\end{array}$	$\begin{array}{l}[x_i{}^3 x_j x_k]\\ [x_j{}^2 x_k{}^2 x_m{}^3 x_n{}^3]\end{array}$	$\begin{array}{l}[x_i{}^2 x_j{}^2 x_k]\\ [x_i x_j x_k{}^2 x_m{}^3 x_n{}^3]\end{array}$	$\begin{array}{l}[x_i{}^2 x_j x_k x_m]\\ [x_i x_j{}^2 x_k{}^2 x_m{}^2 x^3]\end{array}$

Table 1: A suitable basis of monomials for the degree-0 (mod5) subring of $\overset{\circ}{\mathcal{R}}(P_\psi)$. Note: $\vartheta \overset{\text{def}}{=} (x_1 x_2 x_3 x_4 x_5)$ and $i, j, k, m, n = 1 \ldots 5$ are all different.

	$*$	L	Λ	h
$[\mathbf{1}]$	$\dfrac{i}{6}[\vartheta^3]$	$[\vartheta]$	0	-3
$[\vartheta]$	$\dfrac{i}{2}[\vartheta^2]$	$[\vartheta^2]$	$-3[\mathbf{1}]$	-1
$[\vartheta^2]$	$2i[\vartheta]$	$[\vartheta^3]$	$-4[\vartheta]$	$+1$
$[\vartheta^3]$	$6i[\mathbf{1}]$	0	$-3[\vartheta^2]$	$+3$
$[x_i{}^3 x_j{}^2]$	$i\psi^2[x_j x_k{}^3 x_m{}^3 x_n{}^3]$	$\psi^2[x_i x_k{}^3 x_m{}^3 x_n{}^3]$	0	-1
$[x_i{}^3 x_j x_k]$	$i\psi[x_j{}^2 x_k{}^2 x_m{}^3 x_n{}^3]$	$\psi[x_i{}^3 x_k{}^3 x_m{}^2 x_n{}^2]$	0	-1
$[x_i{}^2 x_j{}^2 x_k]$	$i[x_i x_j x_k{}^2 x_m{}^3 x_n{}^3]$	$[x_i{}^3 x_j{}^3 x_k{}^2 x_m x_n]$	0	-1
$[x_i{}^2 x_j x_k x_m]$	$i[x_i x_j{}^2 x_k{}^2 x_m{}^2 x_n{}^3]$	$[x_i{}^3 x_j{}^2 x_k{}^2 x_m{}^2 x_n]$	0	-1
$[x_j x_k{}^3 x_m{}^3 x_n{}^3]$	$i\psi^{-2}[x_i{}^3 x_j{}^2]$	0	$-\psi^{-2}[x_i{}^2 x_j{}^3]$	$+1$
$[x_j{}^2 x_k{}^2 x_m{}^3 x_n{}^3]$	$i\psi^{-1}[x_i{}^3 x_j x_k]$	0	$-\psi^{-1}[x_i{}^3 x_m x_n]$	$+1$
$[x_i x_j x_k{}^2 x_m{}^3 x_n{}^3]$	$i[x_i{}^2 x_j{}^2 x_k]$	0	$-[x_k x_m{}^2 x_n{}^2]$	$+1$
$[x_i x_j{}^2 x_k{}^2 x_m{}^2 x_n{}^3]$	$i[x_i{}^2 x_j x_k x_m]$	0	$-[x_j x_k x_m x_n{}^2]$	$+1$

Table 2: The $*$ and the $SL(2,\mathbb{C})$ action on the degree-$5k$ subring of $\overset{\circ}{\mathcal{R}}(P_\psi)$. Note: in the last column. only the eigenvalues of $h \overset{\text{def}}{=} [L, \Lambda]$ are listed.

References

[1] P. Candelas, G. Horowitz, A. Strominger and E. Witten, Vacuum Configurations for Superstrings, *Nucl. Phys.* **B258** (1985) 46.

[2] For a review and references, see L. Dixon, Some World Sheet Properties of Superstring Compactifications, on Orbifolds and Otherwise. in *Superstrings, Unified Theories and Cosmology 1987*, 67–127, eds G. Furlan et al. (World Scientific, Singapore, 1988).

[3] B.R. Greene and M.R. Plesser, Duality in Calabi–Yau Moduli Space, *Nucl. Phys.* **B338** (1990) 15–37;

P. Candelas, M. Lynker and R. Schimmrigk, Calabi–Yau Manifolds in Weighted $\mathbb{P}^4$. *Nucl. Phys.* **B341** (1990) 383–402.

[4] P.S. Aspinwall, C.A. Lütken and G.G. Ross, Construction and Couplings of Mirror Manifolds, *Phys. Lett.* **241B** (1990) 373–380.

[5] P.S. Aspinwall and C.A. Lütken, Geometry of Mirror Manifolds, *Nucl. Phys.* **B353** (1991) 427–461.

[6] A. Strominger, Yukawa Couplings in Superstring Compactification, *Phys. Rev. Lett.* **55** (1985) 2547.

[7] E. Cremmer, C. Kounnas, A. Van Proeyen, J.-P Derendinger, S. Ferrara, B. de Wit and L. Girardello, Vector Multiplets Coupled to $N = 2$ Supergravity: Super-Higgs Effect, Flat Potentials and Geometric Structure, *Nucl. Phys.* **B250** (1985) 385.

[8] D.R. Morrison, Mirror Symmetry and Rational Curves on Quintic Threefolds: A Guide for Mathematicians, *Duke University report* DUK-M-91-01.

[9] J. Distler and B.R. Greene: Some Exact Results on the Superpotential Calabi–Yau Compactifications, *Nucl. Phys.* **B309** (1988) 295;

for a different proof, see T. Hübsch, How Singular a Space Can Superstrings Thread? *Mod. Phys. Lett.* **A6** (1991) 207–216.

[10] M. Dine, N. Seiberg, X.G. Wen and E. Witten, Non-Perturbative Effects on the String World Sheet, *Nucl. Phys.* **B278** (1986) 769, *ibid.* **B289** (1987) 319.

[11] P. Griffiths and J. Harris, *Principles of Algebraic Geometry* (John Wiley, New York, 1978).

[12] P.A. Aspinwall and C.A. Lütken, Quantum Algebraic Geometry of Superstring Compact-ifications. *Nucl. Phys.* **B355** (1991) 482–510.

[13] R. Donagi, Generic Torelli for Projective Hypersurfaces, *Comp. Math.* **50** (1983) 325–353.

[14] L. Tu, Macaulay's Theorem and Local Torelli for Weighted Hypersurfaces. *Comp. Math.* **60** (1986) 33–44.

AMS/IP Studies in Advanced Mathematics
Volume **9**, 1998

A Generalized Construction of Mirror Manifolds

Per Berglund

CERN, Theory Division

CH-1211 Geneva 23

Switzerland

and

Theory Group, Department of Physics

University of Texas, Austin, TX 78712

berglun@cernvm.bitnet

Tristan Hübsch[♠]

Departments of Mathematics and Physics

Harvard University

Cambridge, MA 02138

hubsch@zariski.harvard.edu, @humal.bitnet

[♠] On leave from the Institute "Rudjer Bokova", Zagreb, roatia.

© 1998 American Mathematical Society
and International Press

                     P. BERGLUND AND T. HÜBSCH

Abstract

We generalize the known method for explicit construction of mirror pairs of (2,2)-superconformal field theories, using the formalism of Landau–Ginzburg orbifolds. Geometrically, these theories are realized as Calabi–Yau hypersurfaces in weighted projective spaces. This generalization makes it possible to construct the mirror partners of many manifolds for which the mirror was not previously known.

1. Introduction

String vacua which lead to $N = 1$ space-time supersymmetry can be described by (2, 0)-superconformal field theories [1]. Another approach is to consider Calabi–Yau manifolds as the classical background in which the string is propagating [2]. Although at first sight very different, it is by now well-known that a large class of Calabi–Yau spaces can be described in terms of (2,2)-superconformal field theories (see for example Refs. [3-5]). In fact, it was conjectured that the Calabi–Yau spaces come in pairs, where for two spaces in such a pair the role of (2,1)-forms and (1,1)-forms respectively are interchanged[1]. Two such theories are said to form a mirror pair. Although the two respective underlying conformal field theories are isomorphic and differ only in the relative sign of the $U(1)$ currents in the (2,2)-superconformal field theory, it is far from straightforward to explicitly construct manifolds which exbibit the above mirror symmetry.

The first construction was given by Greene and Plesser [6], who considered the 3^5 theory, i.e. a tensor product of five A_4 superconformal minimal models. The idea is to use the fact that [7]

$$A_{k+1}/\mathbb{Z}_{k+2} \cong A_{k+1} \; ,$$

the effect of the modding being a change of the relative sign of the left- and right-moving $U(1)$ charge. This procedure can be extended to tensor products of minimal models [6, 8]. In particular,

$$A_4^5/\mathbb{Z}_5 \cong A_4^5 \; .$$

Using the fact that a quintic hypersurface in $\mathbb{P}^4$, which we denote by $\mathbb{P}^4[5]$, can be though of as the conformal field theory corresponding to $A_4^5/\mathbb{Z}_5$, where the quotient is the GSO-type projection, one finds that $\mathbb{P}^4[5]/\mathbb{Z}_5^3$ is the mirror to $\mathbb{P}^4[5]$. The class of theories which can be described by a polynomial (superpotential) of the Fermat type is however quite small. A larger class is formed by those polynomials which can be related to a Fermat type by a non-linear

---

[1] We adopt the *convention* [1] where (2,1)-forms are equivalent to $U(1)$ charge-(1,1) states in the conformal field theory language while (1,1)-forms are analogs of charge-$(-1, 1)$ states.

change of coordinates with a constant Jacobian [9, 10]. Hence the above construction for Fermat polynomials can also be used in this case to obtain the mirror manifold in a straightforward manner.

In this paper we describe a construction of the mirror theory from a (2,2)-superconformal field theory defined in terms of a Landau–Ginzburg field theory. Unlike in the constructions outlined above, we will not restrict (or even relate) the superpotential to the Fermat type polynomial, which corresponds to a tensor product of A_k-type minimal models. Instead, we consider the rather more general class of non-singular polynomials with the number of monomials the same as the number of coordinates. In order to find the mirror, we need to consider quotients of another theory whose defining polynomial is the transpose, in a sense that will be made precise, of the original one.

We will make use of the recently established ties [4, 5, 11] between the geometric point of view and the Landau–Ginzburg orbifold approach. Corresponding to a hypersurface $\mathcal{M}$ in a weighted projective space $\mathbb{P}^4_{(l_1,\ldots,l_5)}$ defined by $P(x_i) = 0$, the (2,2)-superconformal field theory is determined by the superpotential $P(\Phi_i)$. The fact that $P(\Phi_i)$ is the same polynomial as $P(x_i)$ leads to some important identifications. In particular, the function ring of the variety $\mathcal{M}$ [12] and the chiral ring of the corresponding Landau–Ginzburg model [13] are identical. Moreover, the ring structure of the full (p, q)-cohomology on $\mathcal{M}$ can be identified *in complete detail* with the full ring of marginal operators of the Landau–Ginzburg *orbifold*–including untwisted and twisted, (c, c)- and (a, c)-sections [11]. Also, the scaling symmetry and the associated GSO-type projection correspond to the projectivity of the ambient space $\mathcal{M}$. This allows us to freely toggle between Calabi–Yau hypersurfaces and the corresponding Landau–Ginzburg orbifolds, and we make no notational distinction between them.

The paper is organized as follows. We first give the general construction in Section 2. In Section 3, we work out an explicit example and present the general arguments and explicit computations in verification of the mirror pairing of $\mathcal{W}$ with $\mathcal{M}$. Section 4 contains our closing remarks, and some technical details are left for the Appendices.

2. The Construction

Consider a smooth hypersurface $\mathcal{M}$ in a weighted projective space $\mathbb{P}^4_{(l_1,\ldots,l_5)}$ of dimension four. The generalization to other dimensions is straightforward[2]. For $\mathcal{M}$ to be Calabi–Yau, it must be defined as the zero-set of a polynomial of degree $d = \sum_{i=1}^{5} l_i$:

$$P(x_1,\ldots,x_5) = 0 , \quad (x_1,\ldots,x_5) \in \mathbb{P}^4_{(l_1,\ldots,l_5)} , \tag{2.1}$$

and x_i has scaling weight l_i. The corresponding statement for a (2,2)-superconformal field theory leads to a theory with central charge $c = 9$. Let us also define $Q_{\mathcal{M}} = \mathbb{Z}_d$ to be the scaling symmetry $Q_{\mathcal{M}}$ (requiring $\lambda^d = 1$),

$$x_i \mapsto \lambda^{l_i} x_i , \quad P(x_i) \mapsto \lambda^d P(x_i) , \tag{2.2}$$

associated with P. The associated Landau–Ginzburg orbifold is obtained from the Landau–Ginzburg field theory with superpotential P, by implementing the $\mathbb{Z}_4$ GSO-type projection. This Landau–Ginzburg orbifold will also be denoted by $\mathcal{M}$ and we note that this $\mathbb{Z}_d$ becomes the so-called 'quantum symmetry' of $\mathcal{M}$[14]. The group of all phase symmetries of P, excluding $Q_{\mathcal{M}}$, is called the 'geometric symmetry'[3] of $\mathcal{M}$ and is denoted $G_{\mathcal{M}}$.

For the general analysis of Refs. [4, 5, 11] to apply, we must require the system of gradients $\partial_i P$ to vanish only at the origin $x_i = 0$. In this note, we also restrict $P(x_i)$ to be a sum of only as many monomials as there are coordinates (five in the present case), which is clearly the minimal choice. Under this minimality condition it is straightforward to extend the analysis of Chapter 13 in Ref. [15] and we list all 16 minimal non-singular polynomials in Table 1; they are contained in the list obtained previously in Ref. [16]. Of course, by allowing more than the minimal number of monomials, more general superpotentials are obtained and a similar study of such a larger class is under way; we hope to report on these results in a detailed study.

---

[2]$n > 4$ is relevant when considering theories with more than five fields, such as complete intersection Calabi–Yau manifolds and tensor products of models from the ADE series.

[3]The *full* geometric symmetry group will of course also contain permutation symmetrices, but we include these separately as usual.

Given a model $\mathcal{M}$ with one of the superpotentials from Table 1, we now want to find the mirror model $\mathcal{W}$. The idea is to construct another model, $\mathcal{W}$, such that the roles of the quantum symmetry and the geometrical symmetry are interchanged [10], that is,

$$Q_{\mathcal{M}} \cong G_{\mathcal{W}} , \tag{2.3a}$$

$$G_{\mathcal{M}} \cong Q_{\mathcal{W}} . \tag{2.3b}$$

Note that $\mathcal{W}$ will in general be a quotient of a manifold with the fixed points blown up.

Let us now study the different cases at hand. For the first (Fermat-type) polynomial in Table 1, $\mathcal{W}$ is obtained by dividing $\mathcal{M}$ by the action of $\Pi_{\mathcal{M}}$, the group of phase symmetries which leave the (3,0)-form Ω invariant. This is the technique used by Greene and Plesser as mentioned previously [6, 17].

We turn therefore to those polynomials for which dividing by the action of $\Pi_{\mathcal{M}}$ does not yield the mirror. To demonstrate the procedure, let us describe it in detail for one of the polynomials in Table 1. The other cases will then follow easily (see also Section 3 and Tables 1 and 2).

From Table 1, we take for example

$$P = x_1^{a_1} x_2 + x_2^{a_2} x_3 + x_3^{a_3} x_4 + x_4^{a_4} x_5 + x_5^{a_5} . \tag{2.4}$$

To this superpotential, we may associate the matrix of exponents

$$P \simeq \begin{bmatrix} a_1 & 0 & 0 & 0 & 0 \\ 1 & a_2 & 0 & 0 & 0 \\ 0 & 1 & a_3 & 0 & 0 \\ 0 & 0 & 1 & a_4 & 0 \\ 0 & 0 & 0 & 1 & a_5 \end{bmatrix} . \tag{2.5}$$

whose columns are the degree vectors of the respective monomials of P. It is convenient at this point to note that P has a $\mathbb{Z}_{a_1 \cdots a_5}$ phase symmetry. To see this, let the charge of x_5 be $\vartheta_5 = 1/a_5$. For the nomial $x_4^{a_4} x_5$ to transform with an integral (not necessarily unity!) charge, we may choose $\vartheta_4 = \dfrac{-1}{a_4 a_5}$. Thereafter, $\vartheta_3 = \dfrac{+1}{a_3 a_4 a_5}$, $\vartheta_2 = \dfrac{-1}{a_2 a_3 a_4 a_5}$, $\vartheta_1 = \dfrac{+1}{a_1 a_2 a_3 a_4 a_5}$ and we have a manifest $\mathbb{Z}_{a_1 \cdots a_5}$ action.

The new polynomial $\widehat{P}$ is then defined to correspond to the transposed matrix and so is

$$\widehat{P} = x_5^{a_5} x_4 + x_4^{a_4} x_3 + x_3^{a_3} x_2 + x_2^{a_2} x_1 + x_1^{a_1} \,, \qquad \widehat{P} \simeq \begin{bmatrix} a_1 & 1 & 0 & 0 & 0 \\ 0 & a_2 & 1 & 0 & 0 \\ 0 & 0 & a_3 & 1 & 0 \\ 0 & 0 & 0 & a_4 & 1 \\ 0 & 0 & 0 & 0 & a_5 \end{bmatrix}. \qquad (2.6)$$

We say that $\widehat{P}$ is the transpose[4] of P. Of course, transposing again gives back P. Apart from this transposition, $\widehat{P}$ still belongs to the same class of polynomials as P (see Table 1). Most crucially, the total phase symmetry of $\widehat{P}$ is again $\mathbb{Z}_{a_1 \cdots a_5}$, i.e. $Q_\mathcal{M} \times G_\mathcal{M}$.

The zero locus of the transposed polynomial $\widehat{P}$ defines a hypersurface $\widehat{\mathcal{W}} = \mathbb{P}^4_{(\hat{l}_1, \ldots, \hat{l}_5)}[\hat{d}]$, of which the degree $\hat{d}$ and the weights $\hat{l}_i$ are determined by $\widehat{P}$ using (2.2). In general, Eqs. (2.3) will not be fulfilled because $Q_{\widehat{\mathcal{W}}} \subset Q_\mathcal{W}$, but is typically smaller than $Q_\mathcal{W}$, see Section 3 and Table 2 for examples. To enlarge $Q_{\widehat{\mathcal{W}}}$, we must divide $\widehat{\mathcal{W}}$ by a suitable group of phase symmetries H such that

$$Q_\mathcal{W} \stackrel{\text{def}}{=} Q_{\widehat{\mathcal{W}}} \times H \cong G_\mathcal{M} \,, \qquad (2.7a)$$

$$G_\mathcal{W} \stackrel{\text{def}}{=} Q_{\widehat{\mathcal{W}}}/H \cong Q_\mathcal{M} \,. \qquad (2.7b)$$

We then assert that $\mathcal{W} = \widehat{\mathcal{W}}/H$ is the mirror of $\mathcal{M}$ and also check, case by case, that a suitable H does indeed exist. In Table 1, we list the transpose $\widehat{P}$ for each of the defining polynomials (superpotentials) P. We are not aware of a definition of H in closed form and suitable for display with the general classes as in Table 1; see Table 2 for some examples.

We note in passing that all Fermat-type polynomials are self-transposed. From the point of view of our construction, this is precisely the reason why the method of Greene and Plesser works for the Fermat polynomials. More generally, if a polynomial P is self-transposed, whether of the Fermat type or not[5], we find that the mirror of $\mathcal{M}$ is a quotient of P such that Eqs. (2.3) are fulfilled. This accounts for the original class of mirror pairs [6, 17].

---

[4] It appears that this transposition is related to the ($\mathbb{Z}_2$) reversal in the relative sign between the $U(1)$-currents in the underlying (2,2)-superconformal field theory.

[5] There exist special self-transposed examples such as Eq. (2.4), with $l_i = l_{5-i}$ and so $a_i = a_{5-i}$. Most often however, transposition associates weighted hypersurfaces of very different weights.

With regard to the ADE classification of the $N = 2$ superconformal minimal models, it is worth pointing out the following interesting fact. It is easy to check that all models except D_k ($k > 3$) are invariant under transposition. In particular, the E_7 polynomial ($x^3 + xy^3$) is also self-transposed. Since it occurs in the $1 \cdot 16^3$ model, this explains why the mirror of the Schimmrigk manifold [18], $\mathcal{M}$, is just $\mathcal{M}/\Pi_{\mathcal{M}}$.

For D_k we have the following defining polynomial:

$$P(D_k) = x_1^{k-1} + x_1 x_2^2 \, , \tag{2.8}$$

the transpose of which is

$$\widehat{P}(D_k) = x_2^2 + x_2 x_1^{k-1} \, . \tag{2.9}$$

For $k > 3$ we see that $P \neq \widehat{P}$, and so the mirror cannot in general be found by the standard technique. That is, explicit calculation of the (c,c)-and (a,c)-rings of relevant operators shows that unless k is even no quotient of $P(D_k)$ can be identified with the mirror of $P(D_k)$. Instead, there is a $\widehat{P}(D_k)/\mathbb{Z}_{2(k-1)}$ Landau–Ginzburg orbifold which may be identified with the mirror of $P(D_k)$ (the details are presented in Appendix A).[6]

Given our short discussion of the $P(D_k)$ and $\widehat{P}(D_k)/\mathbb{Z}_{2(k-1)}$ Landau–Ginzburg models and the details in Appendix A, it is easy to see that the basic argument for the mirror pairing of $\mathcal{W}$ with $\mathcal{M}$ is the same as in the usual case [6]. There, the mirror map followed from two facts. Firstly,

$$P(A_k)/\mathbb{Z}_k \approx P(A_k) \, , \tag{2.10}$$

for a 1-variable A-type Landau–Ginzburg model. Second, such models could be combined into a $c = 9$ theory, where the isomorphism (2.10) reverses the sign of the Euler characterisitc. Likewise here, we break up the polynomials in Table 1 into irreducible n-variable models ($n = 2,\ldots,5$) and generalizing the above $P(D_k) \approx \widehat{P}(D_k)/\mathbb{Z}_{2(k-1)}$ paradigm, we find that the transposition and quotient by a suitable H provide the generalization of the isomorphism (2.10) for all the irreducible n-variable models used in Table 1. The application of this to the "compound" polynomials in Table 1 — and so also the mirror pairing of $\mathcal{W}$ with $\mathcal{M}$ – is then straightforward.

---

[6]Note that if we think of the D_k model as a $\mathbb{Z}_2$ orbifold of A_{2k-3} and a trivial quadratic piece the mirror of D_k is then given in a straightforward manner as a $\mathbb{Z}_{2(k-1)}$ orbifold of $A_{2k-3} + A_1$.

Greene and Plesser recently described a somewhat larger class of models [10]. They considered mirror pairs $(\mathcal{M}, \mathcal{W})$, where $\mathcal{W}$ is constructed by first performing a fractional but holomorphic change of coordinates to get to a Fermat polynomial. They then deform the latter so as to obtain $\mathcal{W}$, with Eqs. (2.3) satisfied.

Although it is always possible to get from one of the polynomials in Table 1 to a Fermat polynomial, by a fractional holomorphic coordinate transformation, in general it will not be possible to also deform to a $\mathcal{W}$ such that the roles of the quantum and geometric symmetry are interchanged. Thus, the construction described in this paper generalizes the known method for finding the mirror manifold and makes the $\mathbb{Z}_2$ transposition manifest.

3. An Example

To illustrate the method let us consider the following example. The manifold below is listed in Ref. [16] as one without a mirror. Let $\mathcal{M}$ be a degree-75 hypersurface, $\mathcal{M} = \mathbb{P}^4_{(5,8,12,15,35)}[75]^{27,30}_{+6}$, with $\chi_E = +6$, $b_{2,1} = 27$, and $b_{1,1} = 30$, defined by the zero locus of

$$P = x_1^{15} + x_2^5 x_5 + x_1 x_5^2 + x_3^5 x_4 + x_4^5 \; . \tag{3.1}$$

Note that $Q_{\mathcal{M}} = \mathbb{Z}_{75}$ and $G_{\mathcal{M}} = \mathbb{Z}_{25} \times \mathbb{Z}_2$. To P we can associate the following matrix (see Eq. (2.5))

$$p \simeq \begin{bmatrix} 15 & 0 & 0 & 0 & 1 \\ 0 & 5 & 0 & 0 & 0 \\ 0 & 0 & 5 & 0 & 0 \\ 0 & 0 & 1 & 5 & 0 \\ 0 & 1 & 0 & 0 & 2 \end{bmatrix} = \begin{bmatrix} 15 & 0 & 1 \\ 0 & 5 & 0 \\ 0 & 1 & 2 \end{bmatrix} \oplus \begin{bmatrix} 5 & 0 \\ 1 & 5 \end{bmatrix} \; . \tag{3.2}$$

Next, we construct

$$\widehat{P} = x_1^{15} x_5 + x_5^2 x_2 + x_2^5 + x_4^5 x_3 + x_3^5 \; , \tag{3.3}$$

corresponding to the transposed matrix. Using Eq. (2.4), we find that $\widehat{P} = 0$ defines a hypersurface $\widehat{\mathcal{W}} = \mathbb{P}^4_{(1,5,5,4,10)}[25]^{69,9}_{-120}$. Note that $Q_{\widehat{\mathcal{W}}} = \mathbb{Z}_{25}$ and $G_{\widehat{\mathcal{W}}} = \mathbb{Z}_2 \times \mathbb{Z}_{75}$. Thus $Q_{\widehat{\mathcal{W}}}$ is not isomorphic to $G_\mathcal{M}$. However, on dividing by[7]

$$H \overset{\text{def}}{=} (\mathbb{Z}_2 : 1, 0, 0, 0, 1) , \tag{3.4}$$

we find that $\mathcal{W} = \widehat{\mathcal{W}}/H$ satisfies Eqs. (2.3) and identify $\mathcal{W}$ as the mirror of $\mathcal{M}$.

Using the method described in Appendix B, it is now straightforward to find not only χ_E but also $b_{2,1}$ and $b_{1,1}$. For the example in Section 3, we find $\chi_E(\mathcal{W}) = -6$ and that $b_{2,1} = 30$ and $b_{1,1} = 27$. Hence, as expected from the general arguments above, $\mathcal{W}$ satisfies the basic numerical requirements for the mirror of $\mathcal{M}$.

One way wonder if there exists a polynomial whose zero locus in a weighted projective space is the manifold $\mathcal{W}$ as constructed in the above example. After all, taking the quotient by H may be enforced by the following fractional change of coordinates:

$$(x_1, x_2, x_3, x_4, x_5) = (\tilde{x}_1^{1/2}, \tilde{x}_2, \tilde{x}_3, \tilde{x}_4, \tilde{x}_5 \tilde{x}_1^{1/2}) . \tag{3.5}$$

This gives us the polynomial

$$\widetilde{P} = \tilde{x}_1^8 \tilde{x}_5 + \tilde{x}_5^2 \tilde{x}_1 \tilde{x}_2 + \tilde{x}_2^5 + \tilde{x}_4^5 \tilde{x}_3 + \tilde{x}_3^5 . \tag{3.6}$$

It is easy to check that $\widetilde{P} = 0$ describes a degree-25 hypersurface in $\widetilde{\mathcal{W}} = \mathbb{P}^4_{(2,5,5,4,9)}$. However, $\widetilde{P}$ is not transverse at $p^\sharp = (0, 0, 0, 0, 1)$. Hence we cannot expect Vafa's formula for the Euler number [24] (see also Eq. (B.1)) to give us the correct spectrum as it only holds for smooth manifolds, that is non-degenerate Landau–Ginzburg orbifolds. Note however that $p^\sharp$ is in the fixed point set of H and is also a local $\mathbb{Z}_9$-singularity as inherited from the weighted projective space. Interpreting (as usual) the Landau–Ginzburg orbifold as a blow-up at the fixed point set, the singularity may actually be smoothed completely and the situation may not be as bad as it seems. In a local coordinate system with $\tilde{x}_5 = 1$ and $u = (\tilde{x}_4^5 + \tilde{x}_3^4)$ at $p^\sharp$, we have

$$\widetilde{P}\big|_{p^\sharp} = \tilde{x}_1^8 + \tilde{x}_1 \tilde{x}_2 + \tilde{x}_2^5 + u \tilde{x}_3 , \quad \det[\partial^2 \widetilde{P}]\big|_{p^\sharp} = 1 \tag{3.7}$$

The singularity is therefore also a node, the Landau–Ginzburg orbifold is degenerate and we are unable to compute the numerical characteristics reliably [19].

---

[7]We will use the notation $(\mathbb{Z}_k : r_1, r_2, r_3, r_4, r_5)$ for a $\mathbb{Z}_k$ symmetry with the action $(x_1, x_2, x_3, x_4, x_5) \to (\alpha^{r_1} x_1, \alpha^{r_2} x_2, \alpha^{r_3} x_3, \alpha^{r_4} x_4, \alpha^{r_5} x_5)$, where $\alpha^k = 1$.

In spite of the technical problems in computing the correct spectrum for a non-transverse polynomial, we believe that $\widetilde{W}$ indeed is the mirror to $\mathcal{M}$. We base this conjecture on the fact that an appropriate resolution of the singular $\widetilde{W}$ may be identified with the quotient $\widehat{W}/H$. Note, however, that $\widetilde{W}$ and W will correspond in general to two different points in the moduli space. This would lead to a special class of Calabi–Yau conifolds [20], which are non-transverse precisely at the fixed points of the scaling symmetry $Q_{\widetilde{W}}$. It would be interesting to list all non-transverse polynomials of the above type to see if they would complete the list of weighted projective spaces [16] – as far as providing every manifold there with its mirror.

In Table 2, we give several more examples of theories for which the mirror was not previously known. As well as being of potential phenomenological interest, we hope that they will illustrate the ease with which the mirror model is constructed.

4. Discussion and Conclusions

The cautious reader may worry whether there is a hidden caveat to our argument for the mirror pairing of W with $\mathcal{M}$. Indeed, the only completely unambiguous proof would have us compute the (correctly) normalized three-point functions $\langle \phi_{(1,1)}^3 \rangle$ and $\langle \phi_{(-1,1)}^3 \rangle$ in the underlying (2,2)-superconformal field theory. If

$$\langle \phi_{(1,1)}^3 \rangle_{\mathcal{M}} = \langle \phi_{(-1,1)}^3 \rangle_W \ , \quad \langle \phi_{(1,-1)}^3 \rangle_{\mathcal{M}} = \langle \phi_{(1,1)}^3 \rangle_W \ , \tag{4.1}$$

the proof is complete, since the three-point functions completely determine the (2,2)-superconformal field theory.

Unfortunately, it does not seem possible at this time to complete the above calculation, because the (2,2)-superconformal field theories we find are not tensor products of minimal models and the correct normalizations seem to elude us. Nevertheless, using the ring structure of the Landau–Ginzburg orbifolds, specified through the Jacobian ideal generated by the system of gradients of the superpotential, as well as the quantum and geometric symmetries, it is straightforward to show that the general structure of the Yukawa couplings, such as the zeros, is the same.

Without delving into the calculation (which will be presented in full detail elsewhere [21]), for the example discussed earlier we find a one-to-one relation between the (1,1) states in the superconformal field theory corresponding to $\mathcal{M}$ and the $(-1,1)$ states in $\mathcal{W}$ and vice versa.

We have described a technique which generalizes the existing methods of obtaining the mirror manifold to a given hypersurface in a weighted projective space. The method also applies for complete intersection Calabi–Yau manifolds, for which the covering space can be embedded in a (product of) weighted projective space(s) — as long as the superpotential has as many monomials as there are variables. We conjecture that in general the mirror $\mathcal{W}$ to a manifold $\mathcal{M}$ does not have to be described by a transverse polynomial — although $\mathcal{W}$ can be expressed as a quotient of a non-singular covering space $\widehat{\mathcal{W}}$ and the non-transversality occurs at fixed points of $Q_{\mathcal{W}}$. The straightforward way in which the method is applied makes us believe that one ought to be able to mechanize the procedure in terms of a computer code. From the list in Ref. [16], one would then be able to obtain a very large class of mirror pairs. Although there may still remain manifolds without a constructed mirror, we shall be a step closer to verifying that every Calabi–Yau manifold indeed has a mirror.

Acknowledgements

P.B. acknowledges useful discussions with P. Candelas, E. Derrick, X. de la Ossa and J. Louis. P.B. was supported by the Foundation Blanceflor-Boncompagni-Ludovisi née Bildt, the Fulbright Program, a University of Texas Fellowship, and in part by the NSF grant PHY 9009850 and the Robert A. Welch foundation. T.H. was supported by the DOE grant DE-FG02-88ER-25065 and would also like to thank the Department of Mathematics of the National Tsing-Hua University at Hsinchu, Taiwan, for the warm hospitality during the time when part of this research was completed.

Appendix A. The Mirror of D_k

In this appendix, we will show that the mirror to D_k is given by $\widehat{P}(D_k)/\mathbb{Z}_{2(k-1)}$ for $k > 2$. We have that the defining polynomial for D_k is given by

$$P(D_k) = x^{k-1} + xy^2 \ . \tag{A.1}$$

The charges are $q_x = \dfrac{1}{k-1}$ and $q_y = \dfrac{k-2}{2(k-1)}$. The corresponding Landau–Ginzburg theory has the following ring structure

$$(c,c) : \{|0\rangle^{(0)}_{(c,c)},\ x|0\rangle^{(0)}_{(c,c)}, \ldots, x^{k-2}|0\rangle^{(0)}_{(c,c)},\ y|0\rangle^{(0)}_{(c,c)}\}_k$$
$$(a,c) : \{|0\rangle^{(0)}_{(a,c)}\}_1$$

where the subscripts k and 1 indicate the number of states in the (c,c) and (a,c) rings respectively. To find the mirror theory we consider quotients of D_k. Depending on whether k is even or odd, there are two situations.

For k even, the scaling symmetry j of D_k is $(\mathbb{Z}_{k-1} : 1, \dfrac{k-2}{2})$. Using the result in Ref. [22] for the transformation of a given state under a symmtry, we find the states which are invariant under j. This gives rise to the ring structure

$$(c,c) : \{|0\rangle^{(0)}_{(c,c)}\}_1$$
$$(a,c) : \{|0\rangle^{(0)}_{(a,c)},\ |0\rangle^{(k-2)}_{(a,c)}, \ldots, |0\rangle^{(2)}_{(a,c)},\ x^{\frac{k}{2}-1}|0\rangle^{(1)}_{(a,c)},\ y|0\rangle^{(1)}_{(a,c)}\}_k \ .$$

It is easy to check, using Eq. (B.4), that the (a,c) states above have the same charges as the (c,c) states in the D_k theory except that $q_L \to -q_L$. So $D_k/\mathbb{Z}_{k-1}$ is the mirror to D_k for k even.

When k is odd the scaling symmetry j is $(\mathbb{Z}_{2(k-1)} : 2, k-2)$. A calculation similar to the one above shows that, unless $k = 3$, D_k/j is not the mirror; neither is any other quotient. We then turn to the transposed polynomial for D_k,

$$\widehat{P}(D_k) = y^2 + yx^{k-1} \ , \tag{A.2}$$

which has charges $q_1 = \dfrac{1}{2(k-1)}$ and $q_2 = \dfrac{1}{2}$. Note that for all k (even and odd), the scaling symmetry is $\hat{j} = (\mathbb{Z}_{2(k-1)} : 1, k-1)$. It is then straightforward to obtain the ring structure for the LG-orbifold $\widehat{P}(D_k)/\hat{j}$:

$$(c,c) : \{|0\rangle^{(0)}_{(c,c)}\}_1$$

$$(a,c) : \{|0\rangle^{(0)}_{(a,c)},\ |0\rangle^{(2(k-1)-2)}_{(a,c)},\ldots,|0\rangle^{(2)}_{(a,c)},\ y|0\rangle^{(1)}_{(a,c)}\}_k \ .$$

Note that for $\ell = 2p+1$, the states $x^{k-2}|0\rangle^{(1)}_{(a,c)}$ are not projected out, but $\partial_x P(D_k)|_{y=0} = x^{k-2}$ and the state is in the ideal. (We restrict to $y=0$ because $\Theta_y(2p+1) \notin \mathbb{Z}$.) Using Eq. (2.3), one can check that the charges are the same as for the (c,c) states in the D_k model, with a change of sign for q_L. Thus, $\widehat{P}(D_k)/\hat{j}$ is the mirror of D_k for any $k > 2$.

Appendix B. Computation of χ_E, $b_{2,1}$ and $b_{1,1}$

We next turn to compute the χ_E, $b_{2,1}$ and $b_{1,1}$ for $\mathcal{W}$ (see also Ref. [23]). To this end, we need the expression for the Euler number for a weighted projective hypersurface [24, 25]:

$$\chi_E = \frac{1}{d} \sum_{r,\ell=0}^{d-1} (-1)^{(D-N)(r+\ell+r\ell)} \prod_{\ell q_i, r q_i \in \mathbb{Z}} \left(1 - \frac{1}{q_i}\right) , \tag{B.1}$$

where $D = 3$ is the dimension of the Calabi–Yau space, $N = 5$ is the number of homogeneous coordinates, and $q_i = \ell_i/d$. We now want to generalize Eq. (B.1) so that it can also be valid for a quotient. First, rewrite it as

$$\chi_E = \sum_{r=0}^{d-1} S_r , \tag{B.2}$$

where

$$S_r = \frac{1}{d} \sum_{\ell=0}^{d-1} (-1)^{(D-N)(r+\ell+r\ell)} \prod_{\Theta_i(\ell), \Theta_i(r) \in \mathbb{Z}} \left(1 - \frac{1}{q_i}\right) . \tag{B.3}$$

Note that r and ℓ run over all twisted sectors, including the sectors due to dividing by the discrete symmetry; $\Theta_i(\ell)$ is the ith twist charge from the ℓth twisted sector and similarly for $\Theta_i(r)$. This is a generalization of ℓq_i and $r q_i$ in Eq. (B.1) [22, 11]. The point of writing χ_E in the form (B.2) is that it resembles the usual expression for the Euler number. In fact, the S_r will determine the Hodge numbers.

To extract $b_{2,1}$ and $b_{1,1}$, we need to know which sectors contribute to (2,1)- and (1,1)-forms, respectively. The simplest way to see this is by looking at the conformal field theory. Since χ_E is given by Eq. (B.1), it is enough to consider $b_{2,1}$. Recall that we associate charge-(1,1) states to (2,1)-forms. The question is then which sectors contain (1,1) states. To answer this, we need the expression for the charges of the Ramond vacuum in the ℓth twisted sector [24] :

$$\frac{J_0}{\bar{J}_0} |0\rangle_R^{(\ell)} = \left\{ \pm \left[\sum_{\Theta_i(\ell) \notin \mathbb{Z}} \left(\Theta_i(\ell) - [\Theta_i(\ell)] - \frac{1}{2} \right) \right] + \left[\sum_{\Theta_i(\ell) \in \mathbb{Z}} \left(q_i - \frac{1}{2} \right) \right] \right\} |0\rangle_R^{(\ell)} . \tag{B.4}$$

The $|0\rangle_{(c,c)}^{(\ell)}$ vacua are obtained by spectral flow $\mathcal{U}_{(1/2,1/2)}$, of charges $\left(\frac{c}{6}, \frac{c}{6} \right)$. It is then an easy exercise to find those sectors that contain marginal states of charge (1,1). So we find that

$$b_{2,1} = \frac{1}{2}b_3 - 1 \quad = -\frac{1}{2}\sum_{r_+} S_{r_+} - 1 , \qquad (\text{B.5a})$$

$$b_{1,1} = \frac{1}{2}\chi_E - b_{2,1} = \frac{1}{2}\sum_{r \neq r_+} S_r + 1 , \qquad (\text{B.5b})$$

where r_+ runs over the sectors contributing charge-(1,1) states. The reason for the form of the expression for $b_{2,1}$ is that S_{r_+} also contains an equal number of (2,2) states as well as one (0,0) and one (3,3) state respectively from the untwisted sector, $p = 0$. The analogy between 3-forms and integral (c, c)-states is self-evident and well-known.

P	$\widehat{P}$
$x_1^{a_1} + x_2^{a_2} + x_3^{a_3} + x_4^{a_4} + x_5^{a_5}$	$x_1^{a_1} + x_2^{a_2} + x_3^{a_3} + x_4^{a_4} + x_5^{a_5}$
$x_1^{a_1} x_2 + x_2^{a_2} + x_3^{a_3} + x_4^{a_4} + x_5^{a_5}$	$x_2^{a_2} x_1 + x_1^{a_1} + x_3^{a_3} + x_4^{a_4} + x_5^{a_5}$
$x_1^{a_1} x_2 + x_2^{a_2} x_3 + x_3^{a_3} + x_4^{a_4} + x_5^{a_5}$	$x_3^{a_3} x_2 + x_2^{a_2} x_1 + x_1^{a_1} + x_4^{a_4} + x_5^{a_5}$
$x_1^{a_1} x_2 + x_2^{a_2} x_3 + x_3^{a_3} x_4 + x_4^{a_4} + x_5^{a_5}$	$x_4^{a_4} x_3 + x_3^{a_3} x_2 + x_2^{a_2} x_1 + x_1^{a_1} + x_5^{a_5}$
$x_1^{a_1} x_2 + x_2^{a_2} x_3 + x_3^{a_3} x_4 + x_4^{a_4} x_5 + x_5^{a_5}$	$x_5^{a_5} x_4 + x_4^{a_4} x_3 + x_3^{a_3} x_2 + x_2^{a_2} x_1 + x_1^{a_1}$
$x_1^{a_1} x_2 + x_2^{a_2} + x_3^{a_3} x_4 + x_4^{a_4} + x_5^{a_5}$	$x_2^{a_2} x_1 + x_1^{a_1} + x_4^{a_4} x_4 + x_3^{a_3} + x_5^{a_5}$
$x_1^{a_1} x_2 + x_2^{a_2} x_3 + x_3^{a_3} + x_4^{a_4} x_5 + x_5^{a_5}$	$x_3^{a_3} x_2 + x_2^{a_2} x_1 + x_1^{a_1} + x_5^{a_5} + x_5 x_4^{a_4}$
$x_1^{a_1} + x_2^{a_2} + x_3^{a_3} + x_4^{a_4} x_5 + x_5^{a_5} x_4$	$x_1^{a_1} + x_2^{a_2} + x_3^{a_3} + x_4^{a_4} x_5 + x_5^{a_5} x_4$
$x_1^{a_1} + x_2^{a_2} x_3 + x_3^{a_3} + x_4^{a_4} x_5 + x_5^{a_5} x_4$	$x_a^{a_1} + x_3^{a_3} x_2 + x_2^{a_2} + x_4^{a_4} x_5 + x_5^{a_5} x_4$
$x_1^{a_1} x_2 + x_2^{a_2} x_3 + x_3^{a_3} + x_4^{a_4} x_5 + x_5^{a_5} x_4$	$x_3^{a_3} x_2 + x_2^{a_2} x_1 + x_1^{a_1} + x_4^{a_4} x_5 + x_5^{a_5} x_4$
$x_1^{a_1} + x_2^{a_2} x_3 + x_3^{a_3} x_2 + x_4^{a_4} x_5 + x_5^{a_5} x_4$	$x_1^{a_1} + x_2^{a_2} x_3 + x_3^{a_3} x_2 + x_4^{a_4} x_5 + x_5^{a_5} x_4$
$x_1^{a_1} + x_2^{a_2} + x_3^{a_3} x_4 + x_4^{a_4} x_5 + x_5^{a_5} x_3$	$x_1^{a_1} + x_2^{a_2} + x_5^{a_5} x_4 + x_4^{a_4} x_3 + x_3^{a_3} x_5$
$x_1^{a_1} x_2 + x_2^{a_2} + x_3^{a_3} x_4 + x_4^{a_4} x_5 + x_5^{a_5} x_3$	$x_2^{a_2} x_1 + x_1^{a_1} + x_5^{a_5} x_4 + x_4^{a_4} x_3 + x_3^{a_3} x_5$
$x_1^{a_1} x_2 + x_2^{a_2} x_1 + x_3^{a_3} x_4 + x_4^{a_4} x_5 + x_5^{a_5} x_3$	$x_1^{a_1} x_2 + x_2^{a_2} x_1 + x_5^{a_5} x_4 + x_4^{a_4} x_3 + x_3^{a_3} x_5$
$x_1^{a_1} + x_2^{a_2} x_3 + x_3^{a_3} x_4 + x_4^{a_4} x_5 + x_5^{a_5} x_2$	$x_1^{a_1} + x_5^{a_5} x_4 + x_4^{a_4} x_3 + x_3^{a_3} x_2 + x_2^{a_2} x_5$
$x_1^{a_1} x_2 + x_2^{a_2} x_3 + x_3^{a_3} x_4 + x_4^{a_4} x_5 + x_5^{a_5} x_1$	$x_5^{a_5} x_4 + x_4^{a_4} x_3 + x_3^{a_3} x_2 + x_2^{a_2} x_1 + x_1^{a_1} x_5$

Table 1: $P = 0$ defines a hypersurface in a weighted projective 4-space. $\widehat{P} = 0$ gives the convering space, $\widehat{\mathcal{W}}$, of $\mathcal{W}$, the mirror of $\mathcal{M}$. $\mathcal{W}$ will in general be a quotient of $\mathcal{W}$.

$\mathcal{M}$, $\widehat{\mathcal{W}}$	P, $\widehat{P}$	$\mathcal{W}$
$\mathbb{P}^4_{(2,8,29,49,59)}[147]^{51,48}_{-6}$	$x_2^{11}x_5 + x_5^2x_3 + x_3^5x_1 + x_1^{49}x_4 + x_4^3$	
$\mathbb{P}^4_{(1,5,6,18,25)}[55]^{97,28}_{-138}$	$x_4^3x_1 + x_1^{49}x_3 + x_3^5x_5 + x_5^2x_2 + x_2^{11}$	$\widehat{\mathcal{W}}/(\mathbb{Z}_2 : 1,0,1,1,1)$
$\mathbb{P}^4_{(3,4,14,21,21)}[63]^{35,32}_{-6}$	$x_2^{15}x_1 + x_1^{21} + x_3^3x_4 + x_4^3 + x_5^3$	
$\mathbb{P}^4_{(2,3,10,15,15)}[45]^{49,22}_{-54}$	$x_1^{21}x_2 + x_2^{15} + x_4^3x_3 + x_3^3 + x_5^3$	$\widehat{\mathcal{W}}/(\mathbb{Z}_3 : 1,0,1,1,0)$
$\mathbb{P}^4_{(3,5,8,24,35)}[75]^{43,40}_{-6}$	$x_1^{25} + x_2^{15} + x_3^5x_5 + x_4^3x_1 + x_5^2x_2$	
$\mathbb{P}^4_{(2,3,15,25,30)}[75]^{75,27}_{-96}$	$x_1^{25}x_4 + x_4^3 + x_2^{15}x_5 + x_5^2x_3 + x_3^5$	$\widehat{\mathcal{W}}/(\mathbb{Z}_2 : 0,1,0,0,1)$
$\mathbb{P}^4_{(4,5,26,65,95)}[195]^{67,70}_{+6}$	$x_1^{25}x_5 + x_5^2x_2 + x_2^{39} + x_3^5x_4 + x_4^3$	
$\mathbb{P}^4_{(1,3,15,20,36)}[75]^{145,31}_{-228}$	$x_2^{39}x_5 + x_5^2x_1 + x_1^{25} + x_4^3x_3 + x_3^5$	$\widehat{\mathcal{W}}/(\mathbb{Z}_2 : 1,0,0,0,1)$
$\mathbb{P}^4_{(5,6,14,45,65)}[135]^{42,45}_{+6}$	$x_3^5x_5 + x_5^2x_1 + x_1^{27} + x_2^{15}x_4 + x_4^3$	
$\mathbb{P}^4_{(1,3,9,14,18)}[45]^{95,23}_{-144}$	$x_1^{27}x_5 + x_5^2x_3 + x_3^5 + x_4^3x_2 + x_2^{15}$	$\widehat{\mathcal{W}}/(\mathbb{Z}_2 : 1,0,0,0,1)$
$\mathbb{P}^4_{(5,8,12,15,35)}[75]^{27,30}_{+6}$	$x_2^5x_5 + x_5^2x_1 + x_1^{15} + x_3^5x_4 + x_4^5$	
$\mathbb{P}^4_{(1,5,5,4,10)}[25]^{69,9}_{-120}$	$x_1^{15}x_5 + x_5^2x_2 + x_2^5 + x_4^5x_3 + x_3^5$	$\widehat{\mathcal{W}}/(\mathbb{Z}_2 : 1,0,0,0,1)$
$\mathbb{P}^4_{(2,6,9,17,17)}[51]^{31,34}_{+6}$	$x_4^3 + x_1^{17}x_5 + x_5^3 + x_2^7x_3 + x_3^5x_2$	
$\mathbb{P}^4_{(3,6,9,17,16)}[51]^{66,51}_{-102}$	$x_4^3 + x_5^3x_1 + x_1^{17} + x_3^5x_2 + x_2^7x_3$	$\widehat{\mathcal{W}}/(\mathbb{Z}_2 : 0,1,1,0,0)$
$\mathbb{P}^4_{(4,4,11,17,19)}[55]^{24,21}_{-6}$	$x_1^{11}x_3 + x_3^5 + x_2^9x_5 + x_5^2x_4 + x_4^3x_2$	
$\mathbb{P}^4_{(1,1,2,2,5)}[11]^{109,4}_{-210}$	$x_3^5x_1 + x_1^{11} + x_4^3x_5 + x_5^2x_2 + x_2^9x_4$	$\widehat{\mathcal{W}}/(\mathbb{Z}_5 : 0,1,1,1,2)$

Table 2: A list of $\mathcal{M}$ and their mirrors $\mathcal{W}$. $\widehat{\mathcal{W}}$ is defined by $\widehat{P} = 0$. A GSO-type $\mathbb{Z}_d$-projection is implicitly understood, where d is the degree of the defining polynomial. $\mathcal{M}$ are manifolds for which no mirror was listed in Ref. [16].

References

[1] For a review and references, see L. Dixon: in *Superstrings, Unified Theories and Cosmology 1987*, eds. G. Furlan et al. (World Scientific, Singapore, 1988) p. 67-127.

[2] P. Candelas, G. Horowitz, A. Strominger and E. Witten, Nucl. Phys. **B258** (1985) 46.

[3] D. Gepner, Phys. Lett. **199B** (1987) 380, *String theory on Calabi–Yau manifolds: the three generations case*, Princeton University report (December 1987, unpublished).

[4] B.R. Green, C. Vafa and N.P. Warner, Nucl. Phys. **B324** (1989) 371;
J.I. Latorre and C.A. Lütken, Phys. Lett. **222B** (1989) 55;
S.J. Gates and T. Hübsch, Phys. Lett. **226** (1989) 100, Nucl. Phys. **B343** (1990) 741;
B.R. Greene, Commun. Math. Phys. **130** (1990) 335.

[5] S. Cecotti, L. Girardello and A. Pasquinucci, Nucl. Phys. **B328** (1989) 701, Int. J. Mod. Phys. **A6** (1991) 2427;
S. Cecotti, Int. J. Mod. Phys. **A6** (1991) 1749, Nucl. Phys. **B355** (1991) 755.

[6] B.R. Greene and M.R. Plesser, Nucl. Phys. **B338** (1990) 15.

[7] D. Gepner and Z. Qiu, Nucl. Phys. **B285** (1987) 423.

[8] D. Gepner, Nucl. Phys. **B296** (1988) 757.

[9] M. Lynker and R. Schimmrigk, Phys. Lett. **249B** (1990) 237;
R. Schimmrigk, *Mirror Symmetry in String Theory and Fractional Transformtions*, to appear in the proceedings of the PASCOS-91 Symposium, Boston 1991.

[10] B.R. Greene and M.R. Plesser, Cornell and Yale University preprints CLNS 91-1109, YCTP-P32-91.

[11] P. Berglund, B. Greene and T. Hübsch, *Classical vs. Quantum Geometry of Compactification*, University of Texas report UTTG-21-91 (1991).

[12] P. Griffiths and J. Harris, *Principles of Algebraic Geometry* (John Wiley, New York, 1978).

[13] W. Lerche, C. Vafa and N. Warner, Nucl. Phys. **B324** (1989) 427.

[14] C. Vafa, Mod. Phys. Lett. **A4** (1989) 1615.

[15] V.I. Arnold, S.M. Gusein-Zade and A.N. Varchenko, *Singularities of Differentiable Maps*, Vol. I (Birkhäuser, Boston, 1985).

[16] P. Candelas, M. Lynker and R. Schimmrigk, Nucl. Phys. **B341** (1990) 383.

[17] P.S. Aspinwall, C.A. Lütken and G.G. Ross, Phys. Lett. **241B** (1990) 373.

[18] R. Schimmrigk, Phys. Lett. **193B** (1987) 175.

[19] T. Hübsch, Mod. Phys. Lett. **A6** (1991) 207, Class. Quant. Grav. 8 (1991) L31.

[20] P. Candelas, P.S. Green and T. Hübsch, Nucl. Phys. **B330** (1990) 49.

[21] P. Berglund and T. Hübsch, in **preparation**.

[22] K. Intrilligator and C. Vafa, Nucl. Phys. **B339** (1990) 95.

[23] S.-S. Roan, Int. J. Math. **2** (1991) 439.

[24] C. Vafa, Mod. Phys. Lett. **A4** (1989) 1169.

[25] S.-S. Roan, Int. J. Math. **1** (1990) 211.

AMS/IP Studies in Advanced Mathematics
Volume **9**, 1998

New Constructions of Mirror Manifolds:
Probing Moduli Space far from Fermat Points

B.R. Greene

F.R. Newman Laboratory of Nuclear Studies

Cornell University

Ithaca, NY 14853

M.R. Plesser

Department of Physics

Yale University, New Haven, CT 06511

S.-S. Roan

Max-Planck-Institut für Mathematik

D-5300 Bonn 3, Germany

© 1998 American Mathematical Society
and International Press

347

     B. R. GREENE, M. R. PLESSER, AND S.-S. ROAN

Abstract

We describe recent work which indicates that explicit pairs of mirror manifolds can be constructed far from the Fermat points emphasized by the initial procedure based on minimal model conformal theories. This new approach utilizes certain holomorphic nonlinear field transformations on Landau–Ginzburg representations of conformal field theories. In mathematical terms, singularity theory and toric geometry provide the natural setting for our constructions. In addition to classical/quantum symmetry arguments and the study of hypergeometric differential equations for period maps, the latter mathematical formalism provides additional support for the veracity of our constructions although no complete proof has been found.

1. Introduction

Mirror manifolds are pairs of Calabi–Yau manifolds which, as target spaces space for string propagation, determine identical physical theories. In particular, the identification of geometrical objects with the ground states of the $N = 2$ superconformal field theory for the two manifolds differs by a reversal of the left-moving $U(1)$ charges. This means, among other properties, that the Hodge diamonds for the two manifolds differ by a reflection about a diagonal axis (i.e. $h_M^{p,q} = h_{\widetilde{M}}^{d-p,q}$ where M and $\widetilde{M}$ are a mirror pair of Calabi–Yau spaces each of complex dimension d).[1] The existence of mirror pairs has been conjectured by physicists as a resolution of certain naturality issues in the association of superconformal field theories with Calabi–Yau spaces [2] [3]. A partial proof of this conjecutre was given in [4] by direct construction of the mirror manifold in a limited class of examples. The existence of mirror manifolds implies some rather strong and sometimes startling conclusions in physics and mathematics [4] [5] [6], and it is thus an important problem to delineate the precise class of Calabi–Yau spaces which possess a mirror manifold.

In this work we do not present a complete solution to this problem. Rather, we initiate what we hope will be a sequence of generalizations of presently known results.[2] In particular, while neither the methods of [4] nor those used in most of this work are restricted to this class, we will focus on Calabi–Yau manifolds represented as hypersurfaces in weighted projective spaces. Specialized to this case, the arguments of [4] are valid for manifolds lying in a neighborhood (in moduli space) of a Fermat hypersurface, for which the associated superconformal field theory is exactly solvable [8]. Strong evidence that mirror manifolds exist for a larger class of hypersurfaces was found by Candelas, Lynker and Schimmrigk [9] who completed a thorough computer search of

---

[1] For a review see [1].

[2] A more detailed presentation of this work will appear in [7].

Calabi–Yau hypersurfaces in weighted projective four space initiated in [10].[3] They found that the set of all such constructions is almost invariant under the interchange of $h^{1,1}$ and $h^{2,1}$. Having two Calabi–Yau manifolds whose Hodge numbers obey this relation is a requisite property for having a mirror pair but proving that they correspond to the same conformal field theory (thus giving rise to identities between n-point functions on each manifold) is a far more formidable task. In particular, the superconformal field theories associated to these more general examples are not exactly solvable, and the direct methods of [4] are not applicable here, so a new formulation is required. This paper is concerned with steps in this direction.

A useful framework for describing this more general class of superconformal models was developed in [12] [13]. This approach utilizes nonrenormalization theorems to encode the properties of the theory in a Landau–Ginzburg superpotential. This offers the possibility of applying the methods of Lagrangian field theories (and in particular, using path-integral arguments) as well as a partial description of the moduli space. In this work we will use this formulation to extend the construction of mirror manifolds to a somewhat larger class of hypersurfaces than those admitting a Fermat point. Specifically, we will show that the quotient construction of [4] may be realized by certain nonlinear changes of variables. In fact, this notion of inducing the orbifold operation via nonlinear field transformations has been known and studied by a number of groups [14] [15] [5] for some time. We will describe some subtleties of the argument, and show that resolving these leads to a consistent understanding of the procedure which generaates surprising new conclusions. This approach will allow us to extend the class of mirror manifolds (as we shall see) to cover all Landau–Ginzburg theories for which there exists a (nonlinear) field redefinition taking the theory (up to global

---

[3]More recently, this search has been extended to quotients of such hypersurfaces in [11] with similar results.

identifications) to Fermat form. These ideas will become more clear shortly, and will be illustrated with explicit examples.

From the point of view of mathematics, the procedure we follow here has a very natural description in the language of toric geometry. As we shall see in section six, our procedure for generating candidate mirror pairs corresponds to a particular kind of basis change in the toric lattice associated with one of the manifolds of the original mirror pair. By using the mathematical framework for mirror symmetry advocated in [16], this induces a change of basis in the lattice associated with the function data on the other member of the original mirror pair. This has the effect of moving us to a new distinct point in the moduli space.

Mathematically, then, we are performing certain birational transformations on the original mirror pair to build new explicit mirror manifolds. From a more physical vantage point one can interpret the procedure we study as a systematic way of generating points in moduli space which respect enhanced quantum symmetry groups. This in turn picks out corresponding points in the moduli space of the mirror with enhanced classical (geometrical) symmetries.

There is one important point that we emphasize at the outset. At present the only Calabi–Yau manifolds definitively demonstrated to constitute mirror pairs are Fermat hypersurfaces in weighted projective space and their mirrors,[4] and local deformations thereof [4].[5] In this paper we make use of these known constructions to present strong evidence for new explicit mirror manifold pairs. Although we hope to convince the reader that our evidence is quite compelling, we note that we do not have a definitive

---

[4]To be more precise, the only class of Calabi–Yau manifolds proven to constitute mirror pairs are those associated with minimal model conformal field theories. The majority of these correspond to Calabi–Yau hypersurfaces in weighted projective space, but some such conformal theories give rise to higher codimension examples as well.

[5]As discussed in [1] there is evidence that one can relax the restriction to *local* deformations.

proof that our constructions yield mirror manifolds in the full sense of conformal field theory.

In section two, we review the essential elements of the construction of [4], and the Landau–Ginzburg description of Calabi–Yau hypersurfaces. We introduce the notion of quantum symmetries which will play an important role in the sequel. In section three we describe the realization of quotients by nonlinear field redefinitions, and the use of quantum symmetry to obtain an understanding of the procedure. In particular, we show that these transformations yield explicit mirror manifolds at naively unexpected points in moduli space (not Fermat points). In section four we provide further evidence for the validity of our conclusions: we compute the differential equations satisfied by the periods and show that they are precisely of the form predicted by mirror symmetry. In both of these sections we emphasize the rich and ubiquitous example of the quintic hypersurface and its mirror. In section five we apply singularity theory to organize our analysis. This will yield a rather rich construction of mirror pairs that we explicitly illustrate with, again, the quintic hypersurface. To indicate the flavor of the type of results we find we remind the reader that the mirror of the fermat quintic in [4] was constructed by performing a $(\mathbb{Z}_5)^3$ orbifold. Here we find that, for example, we can generate the mirror by a single $\mathbb{Z}_{41}$ or $\mathbb{Z}_{51}$ orbifold on particular quintic hypersurfaces with complex structures differing from the Fermat form. In section six we rephrase our analysis in the mathematical formalism of toric geometry. We will see that what appears to be a correct mathematical setting for mirror manifolds [16] accomodates our construction in a natural way, adding further weight to the validity of our analysis. In section seven we shall discuss the application of our results in the more general setting and reach the conclusion regarding the domain of validity of mirror symmetry announced earlier. Section eight will present our conclusions and outline some open problems.

2. Calabi–Yau Hypersurfaces, Landau–Ginzburg Models, and Mirrors

In this section we will review a representation of Calabi–Yau compactification given in [12] [13], utilizing a representation of $N = 2$ superconformal models as fixed points in the renormalization group flows of $N = 2$ Landau–Ginzburg models.[6]

An $N = 2$ Landau–Ginzburg model in two dimensions [19] is constructed from chiral scalar superfields X_i, $i = 1, \ldots, r$, with superspace action

$$S = \int d^4\theta d^2 z\, K(X_i, \overline{X}_i) + \left(\int d^2\theta d^2 z\, W(X_i) + \text{c.c.} \right) . \qquad (2.1)$$

The bosonic potential is given by

$$V = \left| \frac{\partial W}{\partial X_i} \right|^2 , \qquad (2.2)$$

so that the vacuum states corresponds to critical points of the superpotential W. We will shift so that the critical point we deal with is at the origin of field space, and restrict ourselves to isolated critical points, i.e. those for which a small enough neighborhood of the origin contains no other critical points. This action does not generically describe a conformal model. However, through the renormalization group flow, it represents a universality class of theories which flow to an infrared fixed point at which the theory is invariant under the $N = 2$ superconformal algebra.

The crucial observation of [19] is that the renormalization group flow is completely determined by W, with virations of the Kähler potential K comprising irrelevant perturbations. Furthermore, the superpotential is not renormalized along the flow, to within field renormalizations, the effect of which is that at the fixed point the remaining part of W is quasihomogeneous, i.e. under $X_i \to \lambda^{\omega_i} X_i$ for some set of numbers ω_i,

---

[6]A more rigorous derivation of this correspondence was later given in [17] [18].

we have $W \to \lambda W$. Thus, we can specify an $N = 2$ superconformal theory by a quasi-homogeneous analytic function W.[7] Under the $U(1)$ subalgebra of the superconformal algebra W must clearly have unit charge, identifying the charge of X_i as ω_i.

We are now in a position to review the connection to Calabi–Yau vacua found in [12]. We will find that a quotient of the superconformal theory defined by (2.1) is equivalent to propagation on a Calabi–Yau manifold given by the vanishing locus of the superpotential W, considered as a function on an appropriate weighted projective space. Since the Kähler potential is irrelevant we may choose it at will, and in particular we may choose it very small, and to first approximation ignore it. The path integral representing the partition function of the theory now become

$$\int [dX_1] \cdots [dX_r]\ e^{i \int d^2 z\, d^2 \theta (W(X_1,\ldots,X_r) + \text{c.c.})} \ . \tag{2.3}$$

In a patch of field space in which $X_1 \neq 0$, we can rewrite the path integral (2.3) in terms of new variables

$$\xi_1^{\omega_1} = X_1 \ ; \quad \xi_i = \frac{X_i}{\xi_1^{\omega_i}} \ . \tag{2.4}$$

Since

$$W(X_1,\ldots,X_r) \equiv W'(\xi_1,\ldots,\xi_r) = \xi_1 W'(1,\xi_2,\ldots,\xi_r) \ , \tag{2.5}$$

(2.3) becomes

$$\int [d\xi_1] \cdots [d\xi_r]\ J\ e^{i \int d^2 z\, d^2 \theta \xi_1 (W'(1,\xi_2,\ldots,\xi_r) + \text{c.c.})} \ , \tag{2.6}$$

where J is the Jacobian for the change of variables, and is given by $J = X_1^{\nu}$, with

$$\nu = \frac{1}{\omega_1} \left(1 - \sum_{i=1}^{r} \omega_i \right) \ . \tag{2.7}$$

---

[7] There is an obvious ambiguity here because adding fields to the theory which have quadratic mass terms in W would not affect the infrared behavious.

If $\nu = 0$, then J is a constant, and the integration over ξ_1 in (2.6) yields a delta function $\delta W'(1, \xi_2 \cdots \xi_r)$, constraining the remaining fields to lie on the manifold $W = 0$.[8] Proceeding in like fashion with the other fields, we can cover field space with patches, in each of which we obtain a similar result. We note, though, that the change of variables we have used to simplify the path integral is not-to-one. In fact, upon inspection we see that ξ_i are invariant under the transformation

$$X_i \to e_i^{2\pi i \omega} X_i \; . \qquad (2.8)$$

The X_i are thus naturally interpreted as homogeneous coordinates on $W\mathbb{CP}^4_{d\omega_1 \cdots d\omega_r}$, where d is the smallest integer such that $d\omega_i \in \mathbb{Z} \; \forall i$. However, because of this invariance, the model we have shown to be equivalent to propagation on the manifold $W = 0$ in this projective space is not the theory (2.1) but rather the quotient of this by the transformation (2.8). Since the charge of X_i is ω_i, this is precisely the quotient by $g_0 = e^{2\pi i J_0}$ required to obtain a consistent (space-time supersymmetric) string vacuum [8] [20]. Many properties of the resulting model may be extracted from the superpotential alone, as discussed in [20]. An important role in this equivalence was played by the fact that the Jacobian for the transformation to homogeneous coordinates was constant. For $r = 5$ this is simply the condition that the central charge $c = 9$. It also turns out to be precisely the condition that the hypersurface $W = 0$ have vanishing first Chern class [12].

We note that our discussion seems to indicate that a unique Calabi–Yau manifold is associated to a chosen $N = 2$ superconformal theory of the form (2.1). The essence of mirror symmetry, however, is that *more* than one Calabi–Yau can correspond to

---

[8]We have ignored the fermionic components of the superfields here. Carrying them through the calculation yields a delta function constraining them to lie tangent to the manifold parametrized by the bosonic coordinates, as expected by supersymmetry.

the same conformal theory. For explicit examples of this we turn to the exactly solved models of [8]. These correspond to the case in which W is a polynomial of Fermat type

$$W = X_1^{l_1} + \cdots + X_r^{l_r} \ . \tag{2.9}$$

In this case the mirror manifold of M can be constructed quite simply [4]. The model respects a $\mathbb{Z}_{l_1} \times \cdots \times \mathbb{Z}_{l_r}$ discrete symmetry which acts by scaling the coordinate superfields. Through the identification of these with the homogeneous coordinates in the ambient projective space, this action may be lifted to an action on the manifold. Note that S does not act effectively. (a $\mathbb{Z}_d$ subgroup acts trivially.) We define $G \subset S$ as the maximal subgroup of "allowable" symmetries, i.e. the maximal subgroup of effective symmetries which preserve the integral charge projection (in conformal field theory language) or the holomorphic three-form (in geometric terms). The mirror manifold is then the quotient M/G. Details of the construction are reviewed in [1].

The moduli space of the physical problem of string propagation on M is (locally) a product of the space of inequivalent complex structures on M and a complexification of the space of Kähler structures. Deformations of the complex structure of M obtained by deforming the defining polynomial are very simply expressed in the Landau–Ginzburg formulation as deformations of the superpotential W. As is familiar, deformations of the defining polynomials do not, however, necessarily cover the entire space of complex structures. In the Landau–Ginzburg picture, the missing deformations appear in sectors twisted by g_0. In a similar fashion, all of the deformations of Kähler structure also appear in twisted sectors.

As emphasized in [22], there is a quantum symmetry associated with any orbifold of a conformal theory. In the simple abelian case this associates to each state in the Hilbert space the (exponential of) the number of the sector in which it appears. In particular, only untwisted states are invariant. The quotient of the orbifold by this symmetry yields the original theory. Thus, the Landau–Ginzburg theory may be viewed as a quotient of the string vacuum by a quantum symmetry which appears only at specific values of the modui – namely those at which th expectation values of all twisted fields vanish. The superconformal model at other values of the moduli will nto be obtainable by any deformation of the Landau–Ginzburg superpotential.

In particular, note that the explicit arguments of [4] hold only at one preferred point in the moduli space. The mirror property may be extended throughout a neighborhood of this point by deformation arguments (see [1] and [23].)

3. Nonlinear Field Transformations and Quantum Symmetries

3.1. Preliminaries

In this section we will introduce a method for studying the properties of quotients of Calabi–Yau manifolds which is not limited to the exactly soluble case reviewed in the last section. We will later use this technique, together with the known collection of mirror pair constructions, to extend the class of manifolds for which a mirror can be constructed. In particular, we will present a representation for quotients of an $N = 2$ landau–Ginzburg model by discrete symmetry groups whose action is determined by a linear transformation on the coordinate superfields. The representation we use is essentially an extension of the argument reviewed above relating Landau–Ginzburg Lagrangians to Calabi–Yau manifolds. As noted above, nonlinear changes of variables naturally induce an orbifolding operation. This then leads us to points in moduli space at which the associated conformal field theory respects certain quantum symmetries [22]. Since the discrete symmetry by which we quotient to obtain the Calabi–Yau model acts by scaling the coordinate superfields, the quantum symmetry may be understood quite simply as a related scaling symmetry whose action on the holomorphic superfields x_i is the inverse of its action on the antiholomorphic fields $\bar{x}_i$. This means, of course, that it cannot be represented as an action on the coordinates $X_i = x_i + \bar{x}_i$. However, as emphasized in [4], this *is* interpretable as a *geometric* or "classical" symmetry of the mirror manifold. That is, a quantum symmetry on a Calabi–Yau conformal field theory has a classical geometric realization on the mirror Calabi–Yau theory. We will make use of this idea in identifying mirror pairs below.

We can extend the above ideas to induce other quotients by changes of variables. As an example, consider [15] the level $2P$ minimal model with A series modular invariant, represented by the superpotential

$$P = X^{2P+2} + Y^2 \ , \tag{3.1}$$

where we have used our freedom to add an extra field to the model which will not alter the infrared behaviour. This model possesses a $\mathbb{Z}_2$ discrete symmetry given by $X \to -X$. The quotient of the model by this symmetry, as is well known, is the minimal model at level $2P$ with the D series modular invariant. We can induce this quotient by making the following change of variables

$$
\begin{aligned}
X &= U^{1/2} \\
Y &= U^{1/2}V \ .
\end{aligned}
\tag{3.2}
$$

This nonlinear field transformation is only one-to-one (away from $X = 0$) if one quotients by the $\mathbb{Z}_2$ discrete symmetry mentioned above. Note that as in the arguemt of [12] the fact that the Jacobian for this change of variables is a constant is crucial. The path integral over U, V will yield the quotient by the required $\mathbb{Z}_2$ symmetry of the original path integral. Rewriting the superpotential in terms of the new coordinates we indeed find

$$P = U^{P+1} + UV^2 \ , \tag{3.3}$$

which is the superpotential representing the D series modular invariant at level $2P$ [19] [13].

The Laudau–Ginzburg model does not allow exact calculation of all quantities in the theory. However, the superpotential does uniquely determine the infrared fixed point, so a change of variables with constant Jacobian should lead to an identical super-conformal theory. In the example just given, this is precisely what happens. The reason this deserves mention is that the change of variables we have performed is nonlinear,[9] and singular at $X = 0$. Our interest will be in quotients of models representing string propagation on Calabi–Yau manifolds, and in these cases we shall see that the naive interpretation of the result as a quotient will require modification. The arguments are most clearly demonstrated by an example, but are in fact valid generically as we shall discuss below. For this presentation, therefore, we will again shortly return to the simplest of three dimensional Calabi–Yau manifolds, the Fermat quintic hypersurface M. Before doing so we first consider a case which serves to highlight some essential points in a less subtle context. Namely, we consider the case of a one complex dimensional torus. This example will aso serve to make manifest the way the higher dimensional case involves new complexities.

3.2. The Case of a One-dimensional Torus

As a less trivial example let us consider the superconformal theory describing propagation on an elliptic curve. If we represent the torus as the surface in $\mathbb{CP}^2$ given by

$$x_1^3 + x_2^3 + x_3^3 = 0 \, , \qquad (3.4)$$

---

[9]We emphasize that although we speak of them as coordinates, X, Y, U and V are actually superfields – a fact which makes the use of fractional exponents a bit subtle. We proceed using the behaviour of the scalar compenents of these superfields, i.e. the coordinates, as our primary focus.

the mirror manifold is the quotient of this by a $\mathbb{Z}_3$ symmetry G whose generator acts as

$$x_j \rightarrow \alpha^{j-1} x_j \qquad (3.5)$$

with $\alpha^3 = 1$. This symmetry acts freely, and we find that the mirror manifold is itself a torus.[10] The parameter space of a torus comprises one complex structure modulus τ and a complexified Kähler modulus ρ (where $\mathrm{Re}(\rho)$ is the antisymmetric tensor field). The moduli space is the quotient of this by independent $SL(2, \mathbb{Z})$ transformations acting on τ and ρ. Mirror symmetry teaches us that the parameter space for Kähler deformations, and that for string propagation exchanging the two is a symmetry of the physics, a well known result [24] [25]. However we can still learn from this example.

Propagation on a torus with general complex structure may be represented [3] [12] [26] as (a $\mathbb{Z}_3$ quotient of) the infrared fixed point corresponding to the Landau–Ginzburg action with superpotential

$$W_a = X^3 + Y^3 + Z^3 + 6aXYZ \, , \qquad (3.6)$$

where a parametrizes the moduli space of complex structures for the torus; it is related to the standard modular parameter τ by [26]

$$j^{1/3}(\tau) = 48 \frac{a(1 - a^3)}{1 + 8a^3} \, , \qquad (3.7)$$

where j is the absolute modular invariant of $SL(2, \mathbb{Z})$. The point $a = 0$, which corresponds to the exactly solved model, is distinguished by a $\mathbb{Z}_3$ symmetry $X_1 \rightarrow \alpha X_1$, and corresponds to $\tau = e^{\frac{2\pi i}{3}}$. In fact, the Kähler structure of the torus given by (3.6) for any a is also distinguished by a $\mathbb{Z}_3$ (quantum) symmetry [22] and thus corresponds

---

[10]Complex dimensions one and two are special in that there is only one topological class of Calabi–Yau manifolds in each case and hence mirror manifolds are of the same topological type. In higher dimensions mirror manifolds are generically of different topological types.

to $\rho = e^{\frac{2\pi i}{3}}$. The symmetry (3.5), on the other hand, is present for any value of a. If we present the torus in a more familiar form as $\mathbb{C}/\Gamma$, where Γ is the lattice generated by 1 and τ, then there are three $\mathbb{Z}_3$ symmetries which are present at any value of τ, and quotienting by these leads to the following actions on τ

$$z \to z + 1/3 \Rightarrow \tau \to 3\tau$$
$$z \to z + \tau/3 \Rightarrow \tau \to \tau/3 \tag{3.8}$$
$$z \to z + (1+\tau)/3 \Rightarrow \tau \to (1+\tau)/3 \,,$$

where z is the coordinate on $\mathbb{C}/\Gamma$. However, in the Fermat case ($a = 0$), the mirror torus must describe the same conformal theory, hence it too has $\tau = \rho = e^{\frac{2\pi i}{3}}$. It turns out that only the third action fixes this value of τ (to within an $SL(2, \mathbb{Z})$ action). That is, the action (3.5) is given in this representation by the third equation in (3.8). In fact, since as mentioned, the symmetry persists for all complex structures, we can consider the quotient torus at any value of a. We note, however, that this quotient will no longer be the mirror of the torus being orbifolded. In fact, by our general arguments, varying the complex structure parameter a will be equivalent to varying the Kähler structure on the mirror torus while keeping its complex structure fixed. However, when interpreted directly as a modulus on the quotient, a variation of a will lead to a change in complex structure. Thus, W_a in (3.6) and W_a/G are only mirror manifolds at $a = 0$. By considering the quotient at various points in complex structure space with $a \neq 0$, we are probing the mirror images of various points in Kähler moduli space of the original torus, much as was done for the more complex case of a quintic in $\mathbb{CP}^4$ in [5].

Now, we can consider the quotient as induced by the change of variables

$$(X_1, X_2, X_3) = (Y_1, Y_2, Y_3^{1/3}, Y_3^{2/3}) \,, \tag{3.9}$$

as requiring this to be one-to-one forces the identifications in (3.5). Furthermore, we note that this induces an identification on the Y coordinates as well, under $Y_2 \to -Y_2$, $Y_3 \to -Y_3$. However, we are in fact interested in the two models *after* projecting to integral $U(1)$ charges; the identifications on Y are subsumed by this quotient. Mathematically, there latter identifications will be subsumed in weighted projective space identifications.

The advantage of using (3.9) is that it gives rise to an explicit realization of the mirror torus. We thus find that as descriptions of Calabi–Yau manifolds, the mirror torus of (3.6) is given by

$$W = Y_1^3 + Y_3 Y_2^3 + Y_3^2 + 6a Y_1 Y_2 Y_3 \,, \tag{3.10}$$

where the Y variables are now seen to be homogeneous coordinates on $W\mathbb{CP}^2_{2,1,3}$. By a linear change of variables this can be brought to the form

$$W'_a = Z_1^3 + Z_2^6 + Z_3^2 + \tilde{a} Z_1 Z_2^4 \,, \tag{3.11}$$

where

$$\tilde{a} = -3a \frac{x+1}{(\frac{2}{3}x^2 + x + \frac{1}{4})^{2/3}} \,, \tag{3.12}$$

and $x = 9a^3$. This is, of coruse, another torus; for this parametrization, the relation between $\tilde{a}$ and $\tilde{\tau}$ is given by [26]

$$j(\tilde{\tau}) = 12^3 \frac{4\tilde{a}^3}{4\tilde{a}^3 + 27} \,. \tag{3.13}$$

Putting this together with our knowledge that $\tilde{\tau} = (1 + \tau)/3$, we obtain a non-trivial identity for the modular function j, one which is well known in number theory [27].[11]

---

[11] We thank N. Elkies for help with this and several other mathematical questions.

The truth of this identity confirms that we have constructed an alternative representation of the family of mirror tori W_a/G by W_a'. There are two points of interest: First, let $M_a/\widetilde{G}$ and M_a' be the direct analogs of W_a/G and W_a' for Calabi–Yau's in higher dimensions. Unlike the case studied above, we will find that $M_a/\widetilde{G}$ and M_a' are *not* alternative representations of the same manifolds. Rather, they are located at *different* points in moduli space. This is due to the fact that whereas the singular points in the change of variables (3.9) do not lie on the trous, they generally will in higher dimensions thus giving rise to nontrivial birational transformations. The fact that $M_a/\widetilde{G}$ and M_a' are not equivalent will lead us to our new mirror manifold constructions as we shall see. Second, even in this simple example for which our procedure only yields an alternative representation for W_a/G, we can still extract interesting information. For example, (3.7) gives τ as a function of a where a is the complex structure parameter associated with the conformal field theory modulus $\Phi_C = XYZ$. Once might wonder what the relation is for ρ as a function of b where b is the Kähler parameter associated with the Kähler modulus Φ_R. We can immediately solve this by interpreting Ψ_R as the complex structure deformation on the mirror. Thus we have $\rho(b) = \tilde{\tau}(b)$ with $\tilde{\tau}(b)$ given by (3.13). This is the 'mirror map' on the torus.[12]

3.3. The Case of a Three-dimensional Quintic Hypersurface

The analysis we have described so far is essentially devoid of physics. The consistency of the results gives us further confidence, however, that the procedures we are employing make sense at the level of conformal field theory. In higher dimensions, though, an important difference is that the resultant spaces are singular. As we now

---

[12] We should say that once one realizes that $\tilde{\tau} = (1 + \tau)/3$, the mirror map is directly obtained without the variable transformations. Our point is that the procedure we give carries through in more complicated examples where one can not directly read off the complex structure of the orbifold in a useful manner.

describe, this gives rise to a phenomenon not encountered in the simpler case of the torus.

Let us now consider the case of complex dimension three, in particular the quintic hypersurface given by the vanishing locus of the Landau–Ginzburg superpotential

$$W = X_1^5 + X_2^5 + X_3^5 + X_4^5 + X_5^5 \tag{3.14}$$

interpreted as an equation in the homogeneous coordinates of $\mathbb{CP}^4$. In [4], we showed that the mirror manifold $\widetilde{M}$ is in fact $M/(\mathbb{Z}_5)^3$. This quotient may be induced by the change of variables [14] [5]

$$(X_1, X_2, X_3, X_4, X_5) = (Y_1 Y_3^{1/5}, Y_2^{4/5} Y_5^{1/5}, Y_3^{4/5} Y_4^{1/5}, Y_4^{4/5} Y_2^{1/5}, Y_5^{4/5}) \tag{3.15}$$

as again this transformation is only one-to-one (away from coordinate hyperplanes) after the $(\mathbb{Z}_5)^3$ orbifold is performed on the X's. In terms of these variables, the superpotential becomes the quasihomogeneous equation

$$\widetilde{W}'(Y) = Y_3 Y_1^5 + Y_5 Y_2^4 + Y_4 Y_3^4 + Y_2 Y_4^4 + Y_5^4 \,, \tag{3.16}$$

describing a hypersurface $\widetilde{M}'$ in $W\mathbb{CP}^4_{41,48,51,52,64}$ cut out by a polynomial of homogeneity degree 256. An alternative representation of the same quotient may be induced by the change of variables

$$(X_1, X_2, X_3, X_4, X_5) = (Y_1 Y_4^{1/5}, Y_2^{4/5} Y_5^{1/5}, Y_3, Y_4^{4/5} Y_2^{1/5}, Y_5^{4/5}) \,, \tag{3.17}$$

leading to the superpotential

$$\widetilde{W}''(Y) = Y_4 Y_1^5 + Y_5 Y_2^4 + Y_3^5 + Y_2 Y_4^4 + Y_5^4 \,, \tag{3.18}$$

of homogeneity degree 320, describing a hypersurface $\widetilde{M}''$ in $W\mathbb{CP}^4_{51,60,64,65,80}$. (Other changes of variable in this example will be discussed shortly.)

One can explicitly varify that both $\widetilde{M}'$ and $\widetilde{M}''$ have a one dimensional moduli space of complex structure deformations and 101 harmonic (1,1) forms labeling Kähler deformations. The naive reasoning described above would lead one to conclude the isomrophisms $\widetilde{M} \sim \widetilde{M}' \sim \widetilde{M}''$. It is immediately clear that this does not hold because the orbifold singularities $\widetilde{M}'$ and $\widetilde{M}''$ inherit from the ambient space do not have the orders given by the $(\mathbb{Z}_5)^3$ quotient. Indeed, we shall show that the three conformal theories represent different points in the moduli space of *Kähler* deformations of $\widetilde{M}$.[13]

The identification of this deformation poses a nontrivial challenge to the validity of the change of variables as an orbifolding procedure. It is evident that the hypersurface $\widetilde{M}'$ for example, is not he mirror of the Fermat minimal model quintic hypersurface M. This is because as a conformal field theory describing a Calabi–Yau manifold, the model (3.16) enjoys a $\mathbb{Z}_{256}$ quantum symmetry [22] generated by the projection on integral $U(1)$ charge. As mentioned above, such a symmetry on the mirror $\widetilde{M}$ should have an interpretation as a *geometrical* symmetry acting on M, but the manifold described by (3.14) lacks such an invariance. In line with the previous statement, one is led to conjecture that the superpotential (3.16) describes not the mirror manifold of (3.14), but a Kähler deformation of it. This is because a Kähler deformation of a manifold corresponds to a complex structure deformation of its mirror – and it is the latter kind of deformation which changes the geometrical discrete symmetries which a theory respects. The $\mathbb{Z}_{256}$ symmetry can be used to generate strong support for

---

[13]In the following we often speak of deforming the original pair of mirror manifolds to various points in their respective moduli spaces. We do not, however, present a precise description of these moduli spaces themselves. Recent work provides some indication that the structure of these moduli spaces can be rather subtle, and it is sometimes delicate to relate such forms to the conformal field theory moduli space. These questions are under active study; we stress these points here so that the reader may bear them in mind as our analysis leads us to consider certain unexpected points in various moduli spaces. In particular, whereas our complete understanding of toroidal moduli space allowed for an independent check on the results obtained from nonlinear coordinate transformations in the previous example, no such check can be performed at present in the higher dimensional examples.

this conjecture because the Landau–Ginzburg techniques of [28] allow us to compute explicitly the structure of the (antichiral, chiral) ring for this model, and in particualr to extract the spectrum of eigenvalues of the generator of this discrete symmetry on the ring elements (these are simply given by the sectors in which the various states appear). On the original quintic hypersurface we thur seek a point in the space of complex structure deformations at which we find a $\mathbb{Z}_{256}$ discrete symmetry with the *appropriate spectrum* when acting on the chiral ring. Such a point indeed exists, and in a basis diagonalizing the $\mathbb{Z}_{256}$ generator the superpotential is given by

$$W' = X_1^5 + X_1 X_2^4 + X_2 X_3^4 + X_3 X_4^4 + X_4 X_5^4 \, , \tag{3.19}$$

wher the symmetry acts on the fields as

$$(X_1, X_2, X_3, X_4, X_5) \to (X_1, \beta^{64} X_2, \beta^{112} X_3, \beta^{100} X_4, \beta^{231} X_5) \tag{3.20}$$

where β is a primitive 256th root of unity. This then is the manifold M' to which we propose $\tilde{M}'$ is related by mirror duality. In a similar fashion, we find that the quintic hypersurface M'' dual to $\widetilde{M''}$ is given by

$$W'' = X_1^5 + X_3 X_2^4 + X_1 X_3^4 + X_4^5 + X_2 X_5^4 \, , \tag{3.21}$$

where the symmetry acts on the fields as

$$(X_1, X_2, X_3, X_4, X_5) \to (X_1, \gamma^{20} X_2, \gamma^{240} X_3, \gamma^{64} X_4, \gamma^{315} X_5) \, , \tag{3.22}$$

with γ being a primitive 320th root of unity. These examples are prototypes of the new pairs of mirror manifolds proposed in this paper (further generalization appears in section 7).

Mathematically, as will be clear in section six, the three manifolds in each moduli space are all birationally but not holomorphically equivalent – their differences arising on the coordinate hyperplanes where our nonlinear change of variables is singular.

The existence of deformations of M with the correct symmetry properties[14] is highly nontrivial and provides strong support for our understanding of the change of variables as inducing a quotient to within a deformation. We can, however, make an even stronger case for such claims by appealing to the differential equations satisfied by the period integrals of these manifolds. We will do this in the next section. Once more, the arguments will center around the specific example discussed here. While we do not have a general proof of the validity of our arguments, we believe that they follow directly from the representation of the model as a Landau–Ginzburg theory and hold for any model possessing such a representation.

4. Differential Equations for Periods

In the previous section we have given convincing evidence that our nonlinear field transformations have given rise to the mirror manifolds of certain special yet unexpected points in the moduli space of the quintic hypersurface. Our analysis indicates that the field transformations have caused us to flow to different points in quintic complex structure space and hence different Kähler points on the mirror. If this reasoning is in fact correct, it leads to an interesting prediction which we shall presently explain in the context of the $\mathbb{Z}_{256}$ pair of the last section.

The moduli space containing $\widetilde{M}'$ has $h^{2,1} = 1$ and this single deformation direction can be represented as a deformation of the defining polynomial. Let $\widetilde{M}'_\alpha$ denote the vanishing locus of

$$\widetilde{W}'_\alpha = Y_3 Y_1^5 + Y_5 Y_2^4 + Y_4 Y_3^4 + Y_2 Y_4^4 + Y_5^4 + \alpha Y_1 Y_2 Y_3 Y_4 Y_5 \qquad (4.1)$$

---

[14] We stress that symmetry here refers not only to the existence of a point in moduli space which is invariant under the required symmetry, but also that the action of this symmetry on the cohomology elements be identical to that derived on the mirror.

where α is a complex parameter. Now, our previous reasoning indicates that $\widetilde{M}'_\alpha$ and $\widetilde{M}_\alpha$ differ only by a deformation of their respective Kähler structure, where $\widetilde{M}_\alpha$ is the $(\mathbb{Z}_5)^3$ quotient of the vanishing locus M_α of

$$W_\alpha = W + \alpha X_1 X_2 X_3 X_4 X_5 \ . \tag{4.2}$$

Let us define $F(\alpha)$ by

$$F(\alpha) = \int_A \Omega \tag{4.3}$$

where Ω is the holomorphic three form on $\widetilde{M}_\alpha$ and A is a nontrivial three cycle. Similarly define F' (and also F'') by replacing Ω by Ω' (or Ω''), the three form on $\widetilde{M}'_\alpha$ ($\widetilde{M}''_\alpha$), and by replacing A with a suitable three cycle on the relevent space. The functions F and F' satisfy hypergeometric differential equations known as the Picard-Fuchs equations. In the present context, these have been studied recently in [29]–[33]. The particular form of this equation for F was recently derived in [5]. Our claim is that F', although at first sight appearing to be a thoroughly distinct function of α, should satisfy *precisely the same differential equation as* F. The reason for this is immediate: the differential equation satisfied by a period is independent of the Kähler structure of the manifold being studied.[15]

Our next task, therefore, is to explicitly compute the differential equations satisfied by F and F'. To do so, we will make use of a procedure due to Griffiths [34].[16] We will first briefly describe this algorithm for generating the differential eqution. The analysis hinges on a result of Griffiths which states the following:

---

[15] From a less highbrow point of view, one might expect the form of the differential equation for the periods to be invariant under a change of coordinates. The point which makes this less clear is that the desired transformations are singular as are the resultant Calabi–Yau spaces. The latter point distinguishes these cases from the straightforward example of the one – complex dimensional torus studies earlier.

[16] We learned of this method from N.P. Warner and W. Lerche whom we thank for a number of detailed discussions on the theory and application of this approach.

Let W be the quasihomogeneous defining equation for an algebraic hypersurface in a weighted projective space of complex dimension $n+1$. Let $\xi_1, \ldots, \xi_{n+1}$ be projective coordinates, $\omega = \sum_{i=1}^{n+1}(-1)^i \xi_i d\xi_1 \wedge \cdots \wedge \hat{d\xi_i} \wedge \cdots \wedge d\xi_{n+1}$, $P(\xi)$ a quasihomogeneous polynomial and $l \leq n+1$. Then

$$\frac{\partial}{\partial \xi_i}\left(\frac{P}{W^l}\right)\omega \tag{4.4}$$

is exact. This result is useful due to the following representation of F [34]:

$$F = \int_{\Gamma(A)} \frac{\omega}{W} \, . \tag{4.5}$$

From this equation it is immediate that

$$\frac{d^p F}{d\alpha^p} = \int \frac{Q}{W^{p+1}}\omega \tag{4.6}$$

where Q is a quasihomogeneous function of the ξ_i. For sufficiently large p, Q will not be a primitive element of the chiral ring and hence can be expressed as

$$Q(\xi) = q^A \frac{\partial W}{\partial \xi_A} + r \tag{4.7}$$

where r is in the chiral ring. By homogeneity, though, clearly $r = 0$. We can rewrite (4.6) using (4.7) as

$$\frac{d^p F}{d\alpha^p} = -\frac{1}{p+1}\int \frac{\partial_A q^A}{W^p}\omega \tag{4.8}$$

where we have made use of the exactness claimed in (4.4). If the numerator in (4.8) is not a primitive element of the chiral ring, we apply the same analysis as before to reduce it further. Continuing in this manner we finally get something of the form

$$\frac{d^p F}{d\alpha^p} = \sum_{j=1}^{p}\int \frac{r_j(\xi)}{W^j}\omega \, . \tag{4.9}$$

Although often a tedious calculation, using (4.5) and (4.4) repeatedly, the right hand side of (4.9) can be reexpressed in terms of $\dfrac{d^k F}{d\alpha^k}$ for $k < p$. In this way, therefore, we arrive at a differential equation for F. The degree p of this differential equation is generally the dimension of the n-th cohomology group of the manifold.

Let us now apply this procedure to $\widetilde{M}_\alpha$; this is the case studied in [5] and will give us a nontrivial check on our manipulations. In this particular example we can actually perform our analysis on M_α so long as we restrict our manipulations to the subring of the chiral ring on M_α invariant under the action of G. Using the expression (4.5) for F we see that

$$\frac{d^j F}{d\alpha^j} = j! \int \frac{X_1^j X_2^j X_3^j X_4^j X_5^J}{W^{j+1}} \omega \; . \tag{4.10}$$

For $j = 4$ we see that the numerator in (4.10) is not a primitive element of the chiral ring and hence $p = 4$ (recall also that the dimension of $H^3(W_\alpha)$ is 4). Using the ring manipulations described above, we can thus reexpress (4.10) for $j = 4$ in terms of (4.10) for $j < 4$. Carrying out this procedure we find

$$(1 - \alpha^5)\frac{d^4 F}{d\alpha^4} - 10\alpha^4 \frac{d^3 F}{d\alpha^3} - 25\alpha^3 \frac{d^2 F}{d\alpha^2} - 15\alpha^2 \frac{dF}{d\alpha} - \alpha F = 0 \; . \tag{4.11}$$

Making the substitution $z = \alpha^5$ and $F = \alpha\omega_0$ (4.10) is in fact precisely the hypergeometric equation derived for the periods in [5].

One can now carry out the same analysis for $\widetilde{M}'_\alpha$ and for $\widetilde{M}''_\alpha$ (using the explicit form of their defining equations). As discussed, we predict that all three of these vastly different equations (recall that W is of degree 5 while these latter two equations are of degree 256 and 320 respectively) should give rise to the same differential equation (4.10). *They do.*

Our quantum symmetry argument of the last section combined with this analysis of periods thus provides solid evidence that the nonlinear field transformations under discussion have the advertised property of yielding explicit pairs of mirror manifolds at new points in moduli space. In particular, relative to the original explicit mirror pair, in the $\widetilde{M}$ moduli space we have moved to a different Kähler structure and in the M moduli space we have moved to a different complex structure.

From a purely formal point of view it is surprising that this flowing in moduli space is the sole manifestation of such singular field transformations. One might note, however, that the analysis of [4] already gave strong indication that such mirror pairs should *exist*;[17] from this vantage point the present work appears to confirm and give explicit representation to some such pairs. In fact, however, the present work allows us to go a good deal further and argue for the existence of mirror pairs beyond those explicitly or implicitly found in [4] as we shall discuss in the last section.

Organizing and Applying Nonlinear Transformations

In the previous two sections we have concentrated on nonlinear transformations which have taken us from a Fermat equation to particular non-Fermat forms. One might well wonder about the existence and suefulness of other changes of variables taking us to equations of different form. The appropriate language for systematizing such a discussion is singularity theory.[18] Recall that a crucial constraint on our analysis is that the image of any initially well defined superpotential have the property that it is a quasihomogeneous polynomial with an isolated singularity at the origin. Geometrically this ensures that the corresponding manifold is transverse. If one had a complete list of singularities in five variables which meet these conditions, one could systematically

---

[17] These implicit results were based on deformation arguments taking us away from the Fermat and Fermat mirror points in moduli space.

[18] We will see in the next section that singularity theory naturally arises in the mathematical formulation of this discussion as well.

consider the nonlinear transformations which take the Fermat quintic, say, to each of these singularities. Although such a complete list, unfortunately, does not presently exist, a partial list has been constructed [9]. The two examples studied in the previous sections map the Fermat singularity to two other viable singularity types (in the notation of table 1 of [9], these two are singularity types 13 and 1 plus 8 respectively). We would now like to focus on some of the other singularity types – especially those which do not change the homogeneity type of the original equation. That is, for the case being studied, we emphasize those singularity types which are homogeneous quintic forms in five variables.

Consider first the transformation

$$(X_1, X_2, X_3, X_4, X_5) = (Y_1^{4/5}Y_2^{1/5}, Y_2^{4/5}Y_3^{1/5}, Y_3^{4/5}Y_4^{1/5}, Y_4^{4/5}Y_5^{1/5}, Y_5^{4/5}Y_1^{1/5}) \ . \quad (5.1)$$

It is straightforward to see that this induces a $(\mathbb{Z}_5)^3$ orbifold on the X-coordinates and a $\mathbb{Z}_{205}$ orbifold on the Y coordinates. Notice that the image of the Fermat quintic in the Y variables is again a quintic, namely

$$M_{41} = Y_1^4 Y_2 + Y_2^4 Y_3 + Y_3^4 Y_4 + Y_4^4 Y_5 + Y_5^4 Y_1 \qquad (5.2)$$

(this is singularity type 19 in the notation of [9]). After performing the $\mathbb{Z}_5$ identifications to take us from the Landau–Ginzburg theory to the Calabi–Yau theory, the $\mathbb{Z}_{205}$ symmetry becomes a $\mathbb{Z}_{41}$ symmetry, thus accounting for the name used in (5.2). Explicitly, this $\mathbb{Z}_{41}$ acts according to

$$(Y_1, \ldots, Y_5) \rightarrow (\gamma Y_1, \gamma^{37} Y_2, \gamma^{16} Y_3, \gamma^{18} Y_4, \gamma^{10} Y_5) \qquad (5.3)$$

where γ is a primitive forty-first root of unity. By the reasoning expounded upon in the previous sections, we know therefore that $M_{41}/\mathbb{Z}_{41}$ lies in the moduli space of the mirror quintic $\widetilde{M}$. Following, again, our previous reasoning, this $\mathbb{Z}_{41}$ quotient must be the mirror manifold a point in quintic moduli space which has an enhanced $\mathbb{Z}_{41}$ geometric symmetry. This time, however, because we have stayed within the class of homoegeneous fifth order polynomials (unlike what we did in previous sections), we do not have to look far for this $\mathbb{Z}_{41}$ symmetric quintic: it is M_{41} itself. Thus, just as M and $M/(\mathbb{Z}_5)^3$ form a mirror pair, so do M_{41} and $M_{41}/\mathbb{Z}_{41}$. This is extremely suggestive. The maximal discrete symmetry groups by which we can orbifold M and M_{41} are $(\mathbb{Z}_5)^3$ and $\mathbb{Z}_{41}$ respectively. Thus, the construction of mirror pairs found in [4] and reviewed earlier appears to be a good deal more general than initially thought.

As another example, consider

$$(X_1, X_2, X_3, X_4, X_5) = (Y_1^{4/5}Y_2^{1/5}, Y_2^{4/5}Y_3^{1/5}, Y_3^{4/5}Y_4^{1/5}, Y_4^{4/5}Y_1^{1/5}, Y_5) . \tag{5.4}$$

This also induces a $(\mathbb{Z}_5)^3$ identification on the X-coordinates while yielding a Z_{51} identification on the Y-coordinates. The latter acts via

$$(Y_1, \ldots, Y_5) \to (\gamma Y_1, \gamma^{47} Y_2, \gamma^{16} Y_3, \gamma^{38} Y_4, Y_5) \tag{5.5}$$

where γ is a primitive fifty-first root of unity. Reasoning as before with M_{51} defined by

$$M_{51} = Y_1^4 Y_2 + Y_2^4 Y_3 + Y_3^4 Y_4 + Y_4^4 Y_1 + Y_5^5 \tag{5.6}$$

(in the notation of [9] this is singularity type 12 plus 1) we conclude that M_{51} and $M_{51}/\mathbb{Z}_{51}$ are a mirror pair. As before, $\mathbb{Z}_{51}$ is the maximal symmetry group by which we can orbifold M_{51}. Continuing on in this fashion we get numerous explicit pairs of mirror points in the quintic and mirror quintic moduli spaces. We list a number of these pairs in table 1.

We have split this table into two parts to emphasize an interesting point. At first sight the pairs in the top half and the pairs in the bottom half have little in common.

However there is one characteristic which they both share: up to global identifications, the singularity type of the defining equations of mirror pairs are identical. This leads us to speculate that singularity type is an invariant of mirror pairs.[19] For the cases being studied, we will give further support to this statement in the next section.

In table 2 we list one more example (this was also the second examples studied in [4]):

$$X_1^5 + X_2^{10} + X_3^{10} + X_4^{10} + X_5^2 = 0 \tag{5.7}$$

in $W\mathbb{CP}^4_{2,1,1,15}$. From the general result in [4] that mirrors of minimal model theories are obtained through orbifolding by their maximal scaling symmetry group, the mirror of this example is found (as shown in the first line of table 2) by a $\mathbb{Z}_{10} \times \mathbb{Z}_{10}$ orbifold. The other two lines are obtained via the methods described in the previous paragraphs.

6. A Mathematical Perspective on Nonlinear Changes of Variables

In this section we describe the natural mathematical setting for the analysis described in physical terms in the previous sections. We assume the reader is familiar with the essentials of toric geometry and with the formalism of [16].[20]

Let us briefly recall how mirror symmetry is described in the language of toric geometry [16]; we begin with some notation. Let

$$X : z_1^{d_1} + z_2^{d_2} + z_3^{d_3} + z_4^{d_4} + z_5^{d_5} = 0 \tag{6.1}$$

be a Calabi–Yau hypersurface of Fermat type in $W\mathbb{CP}^4_{n_1,\ldots,n_5}$ of degree d. The condition that X is Calabi–Yau leads to $d = \sum_{i=1}^{5} n_i$ with $d = d_i n_i$ and $d_i \geq 2$. Let $q_i = 1/d_i$;

---

[19] One must be careful with the meaning of such a statement, though, since a complex structure deformation on one manifold of a pair corresponds to a modal deformation of the singularity, while on the mirror only the Kähler structure is modified as the singularity is not affected. Thus, in the present context, we only aim to consider singularities which contain precisely five summands.

[20] As mentioned above, a more detailed presentation of the work described here appears in [7].

in physical terms q_i is the charge of the field z_i. Let T be the algebraic torus $\mathbb{C}^{*5}$ and

$$D = \left\{ \begin{pmatrix} t_1 \\ \vdots \\ t_5 \end{pmatrix} \in T \,|\, t_i^{d_i} = 1 \;\forall i \right\} . \tag{6.2}$$

Let

$$SD = \left\{ \begin{pmatrix} t_1 \\ \vdots \\ t_5 \end{pmatrix} \in D \,|\, t_1 t_2 t_3 t_4 t_5 = 1 \right\} ; \tag{6.3}$$

define $\exp_{\vec{q}} : \mathbb{R}^5 \to T$ by

$$\exp_{\vec{q}} \begin{pmatrix} x_1 \\ \vdots \\ x_5 \end{pmatrix} = \begin{pmatrix} e^{2\pi i q_1 x_1} \\ \vdots \\ e^{2\pi i q_5 x_5} \end{pmatrix} \tag{6.4}$$

and let Q denote the group generated by

$$\exp_{\vec{q}} \begin{pmatrix} 1 \\ 1 \\ 1 \\ 1 \\ 1 \end{pmatrix} . \tag{6.5}$$

In studying orbifolds of X by subgroups $Q \subset G \subset SD$ (quotient by these preserve the Calabi–Yau condition) toric geometry leads us to consider a pair of dual lattices and their associated vector spaces, either of which may be used to reconstruct the orbifold. These are

$$N_G = \mathrm{Hom}(\mathbb{C}^*, T/G) , \tag{6.6}$$

the group of one-parameter subgroups of T/G and

$$M_G = \mathrm{Hom}(T/G, \mathbb{C}^*) , \tag{6.7}$$

the group of characters of T/G. In both of these definitions Hom denotes algebraic group homomorphisms. We note that N_G and M_G are dual under the inner product

$$\langle x, y \rangle_{\vec{q}} = \sum q_i x_i y_i . \tag{6.8}$$

Also, we can alternatively describe the two lattices as

$$N_G = \exp_{\bar{q}}^{-1}(G)$$

$$M_G = \left\{ \begin{pmatrix} k_1 \\ \vdots \\ k_5 \end{pmatrix} \in \mathbb{Z}^5 \mid z_1^{k_1} z_2^{k_2} z_3^{k_3} z_4^{k_4} z_5^{k_5} \text{ is } G\text{-invariant} \right\} . \qquad (6.9)$$

In the favorable circumstances we are studying, information regarding the Kähler structure of the orbifold can be extracted from N_G and information regarding its complex structure from M_G. Mirror symmetry in this language therefore takes the following form [16]. Let $\widetilde{G}$ denote the subgroup $Q \subset \widetilde{G} \subset SD$ such that

$$N_{\widetilde{G}} = M_G$$
$$M_{\widetilde{G}} = N_G . \qquad (6.10)$$

Then the minimal toroidal resolutions $\widehat{X/G}$ and $\widehat{X/\widetilde{G}}$ are a mirror pair of Calabi–Yau manifolds. It is important to emphasize that the definition of mirror manifolds states that if M and $\widetilde{M}$ form a mirror then they correspond to the same conformal field theory. In [16] it was *not* shown that $\widehat{X/G}$ and $\widehat{X/\widetilde{G}}$ satisfy this more precise definition of mirror pairs. Rather, a canonical isomorphism between $H^{1,1}(\widehat{X/G})$ and $H^{2,1}(\widehat{X/\widetilde{G}})$ as well as between $H^{2,1}(\widehat{X/G})$ and $H^{1,1}(\widehat{X/\widetilde{G}})$ was established. A mathematical proof of the deeper aspects of mirror symmetry – namely the equality of quantum n-point functions on the manifolds comprising a mirror pair – has not yet been found. These properties have been established only using the methods of conformal field theory. Nonetheless we expect that (6.10) is the signature of mirror symmetry as all known examples can be so described.

The purpose of this section is to show that the construction of mirror pairs studied in the previous sections fits naturally into this description of mirror symmetry. We will show that the nonlinear transformation used in the construction of the new mirror pairs amounts to a change of bases of the lattices N and M. Geometrically, this corresponds to particualr birational transformations performed on the original mirror pair.

Notice that $G \subset SD$ implies that $z_i^{d_i}$ and $z_1 z_2 z_3 z_4 z_5$ are G-invariant monomials. Therefore, M_G contains the elements

$$b_i = \begin{pmatrix} 0 \\ 0 \\ d_i \\ 0 \\ 0 \end{pmatrix} \qquad i = 1,\dots,5$$

$$t = \begin{pmatrix} 1 \\ 1 \\ 1 \\ 1 \\ 1 \end{pmatrix}. \tag{6.11}$$

When we perform a nonlinear change of variables we are expressing the b_i as a linear combination with integral coefficients of a $\mathbb{Q}$-base $\{g_1,\dots,g_5\}$ of $M_G \underset{\mathbb{Z}}{\otimes} \mathbb{Q}$ such that t again has the form $t = \sum_i g_i$. This latter requirement is equivalent to the requirement that the change of variable have constant Jacobian.

From the point of view of $N_G \underset{\mathbb{Z}}{\otimes} \mathbb{Q}$ we are introducing a new basis $\{g^1,\dots,g^5\}$ dual to g_i under the inner product $\langle\ ,\ \rangle_{\bar{q}}$. The essential point is that the lattice N_G is originally generated by the standard coordinate basis vectors $\{e^1,\dots,e^5\}$ (with $e^i_{(j)} = \delta_{ij}$) together with additional vectors $\{f^1,\dots,f^r\}$ (which account for G in (6.6)). We now generate the same lattice using the $\{g^i\}$ in place of the $\{e^i\}$ and additional vectors $\{h^1,\dots,h^s\}$ as required to generate N_G. Just as the $\{f^i\}$ represented a quotient by G, the $\{h^i\}$ correspond to a quotient by a new group G'. In general, G and G' are distinct; this is the mathematical origin of the results in the previous sections in which we generate distinct points in the moduli space of the mirror by making use of a variety of different group actions.

In a moment we will illustrate these points in an example – we will explicitly show that all of the entries in table 1 and 2 can be described as arising from different choices of the basis $\{g^i\}$ in the lattice $N_{(\mathbb{Z}_5)^3}$. Before we do this, we have yet to discuss half of the story. Namely the mirror manifolds to spaces we are generating by this change of basis (in our example, the left-hand column in table 1 and 2). To generate the mirros in this framework we follow the basic paradigm embodied in (6.10). That is, we take the N-type (M-type) lattice discussed above as the M-type (N-type) lattice of the mirror. Furthermore, an important point to emphasize is the following. The elements $\{b_i\}$ of the lattice M_G are dual to the original standard basis $\{e^i\}$ of N_G. The $\{b_i\}$ can be used to reconstruct the defining equation of the hypersurface being studied through its image in $\mathrm{Hom}(T/G, \mathbb{C}^*)$. Simply stated, b_i corresponds to $z_i^{d_i}$ and the basis $\{b_i\}$ thus gives rise to the Fermat hypersurface determined by $f = \sum z_i^{d_i}$. Now, the $\{g_i\}$ are the new basis vectors for N_G. On the mirror, then, they form the new basis for M. Hence, the basis vectors $\{g^i\}$ can be used to reconstruct the defining equation of the mirror just as the $\{b_i\}$ were used above. This leads us to the defining equations for the mirrors. In particular, we will see that this procedure leads to the left-hand column in tables 1 and 2. Thus this geometrical procedure agrees with the physical approach based upon quantum/classical symmetries. In fact, it is not at all hard to show directly that the hypersurface constraints obtained by this lattice procedure necessarily possess the classical symmetries predicted by the physical reasoning of section three. In summary, then, the candidate mirror pairs obtained by the nonlinear changes of variables fit into precisely the same mathematical framework [16] that describes the known (proven) mirror pairs of [4].

Let us now follow this analysis through in an example; in particular, let us consider the pair $(M_{41}, M_{41}/\mathbb{Z}_{41})$. In this case

$$X : f(z_1, \ldots, z_5) = z_1^5 + z_2^5 + z_3^5 + z_4^5 + z_5^5 = 0$$
$$G = SD = (\mathbb{Z}_5)^3 \tag{6.12}$$

and we consider the nonlinear change of variables

$$(z_1^5, z_2^5, z_3^5, z_4^5, z_5^5) = (x_1^4 x_2, x_2^4 x_3, x_3^4 x_4, x_4^4 x_5, x_5^4 x_1) . \tag{6.13}$$

We clearly have $b_i = 5e_i$ (where e_i are the standard coordinate basis vectors). The $\mathbb{Q}$-basis $\{g_i\}$ of $M_G \underset{\mathbb{Z}}{\otimes} \mathbb{Q}$ corresponding to this change of variables is determined by

$$(b_1, \ldots, b_5) = (g_1, \ldots, g_5) \begin{pmatrix} 4 & 0 & 0 & 0 & 1 \\ 1 & 4 & 0 & 0 & 0 \\ 0 & 1 & 4 & 0 & 0 \\ 0 & 0 & 1 & 4 & 0 \\ 0 & 0 & 0 & 1 & 4 \end{pmatrix} \tag{6.14}$$

where the b_i and the g_i are column vectors. Solving this yields

$$(g_1, \ldots, g_5) = \frac{1}{205} \begin{pmatrix} 256 & 1 & -4 & 16 & -64 \\ -64 & 256 & 1 & -4 & 16 \\ 16 & -64 & 256 & 1 & -4 \\ -4 & 16 & -64 & 256 & 1 \\ 1 & -4 & 16 & -64 & 256 \end{pmatrix} . \tag{6.15}$$

Notice that we have $t = \sum_i g_i$ as required. The dual basis $\{g^i\}$ for N_G is given by $\langle g_i, g^j \rangle_{\bar{q}} = \delta_i^j$ with $q_i = 1/5 \; \forall i$. Thus

$$(g^1, \ldots, g^5) = \begin{pmatrix} 4 & 1 & 0 & 0 & 0 \\ 0 & 4 & 1 & 0 & 0 \\ 0 & 0 & 4 & 1 & 0 \\ 0 & 0 & 0 & 4 & 1 \\ 1 & 0 & 0 & 0 & 4 \end{pmatrix} . \tag{6.16}$$

From this we immediately determine the following three results:

(a) The weighted projective space in which the resulting hypersurface is embedded. Since $t = \sum_j q_j' g^j$ with $q_j' = 1/5 \; \forall j$ we find that the weights of all the variables are equal and hence the embedding space is ordinary projective space $\mathbb{CP}^4$.

(b) The symmetry group by which we quotient. To see this, note that $N_G/(\sum_j \mathbb{Z} g^j)$ is generated by the cosets of t and the vector $b_1 = 5e_1$. Now, in the $\{g^i\}$ basis, we have

$$b_1 = \frac{1}{205}(256g^1 + g^2 - 4g^3 + 16g^4 - 64g^5) \tag{6.17}$$

and $t = 1/5 \sum_j g^j$. As in our earlier discussion, these vectors determine an orbifolding operation. Explicitly, they generate

$$(x_1, \ldots, x_5) \to (\alpha^{256} x_1, \alpha x_2, \alpha^{-4} x_3, \alpha^{16} x_4, \alpha^{-64} x_5)$$
$$(x_1, \ldots, x_5) \to (\beta x_1, \beta x_2, \beta x_3, \beta x_4, \beta x_5) \tag{6.18}$$

with $\alpha^{205} = \beta^5 = 1$. Recalling the projective space identifications on the x_i this reduces to a $\mathbb{Z}_{41}$ action generated by

$$(x_1, \ldots, x_5) \to (\gamma x_1, \gamma^{37} x_2, \gamma^{16} x_3, \gamma^{18} x_4, \gamma^{10} x_5) \tag{6.19}$$

where $\gamma^{41} = 1$, in agreement with the result found previously (5.3).

(c) The mirror manifold. To find this we take the basis $\{g^i\}$ of $N_G \underset{\mathbb{Z}}{\otimes} \mathbb{R}$ and consider it as a basis of $M \underset{\mathbb{Z}}{\otimes} \mathbb{R}$ for the mirror manifold. We then read off the defining equation from (6.16), and find

$$\tilde{f}(z_1, \ldots, z_5) = z_1^4 z_5 + z_2^4 z_1 + z_3^4 z_2 + z_4^4 z_3 + z_5^4 z_4 \; . \tag{6.20}$$

We thus find that $f = 0$ in $\mathbb{CP}^4$ and its quotient by the above $\mathbb{Z}_{41}$, action are mirror manifolds in the limited sense of (6.10). As discussed above, this property appears to be the mathematical "smoking gun" of mirror symmetry and this analysis thus provides additional support for the validity of the procedure for generating mirror manifolds which we have proposed. All of the other examples studied previously similary fit into this formalism; our discussion immediately generalizes to (moduli spaces of) the Calabi–Yau mirror manifolds of [4].

Furthermore, it is immediately clear that the anlaysis presented here ensures that the respective singularity types of the proposed mirror pairs will always be the same. This is the mathematical justification of our earlier observation regarding this fact. In particular, we note that the last two entries in table 1, which were presented in section 3, arise directly in this approach. Finally, we note that mirror symmetry in the framework studied here is manifested in the fact the (6.14) and (6.16) are tranposes of each other. Recently the authors of [35] have conjectured that just such an operation might produce mirror manifolds. The work in this section reaches the same conclusion thus supplying at least a partial justification for such conjectures.[21]

7. Extending Mirror Symmetry

The lynchpin securing all of the mirror identifications described above is the mirror symmetry mechanism found in [4]. That is, all of our analysis is tied in one way or another to minimal models. For instance, in the quintic example everything relies on our knowledge of mirror symmetry on the Fermat quintic (first line of table 1). Nonetheless, the methods described here use the result of [4] combined with unusual field transformations to generate what appear to be new explicit mirror pairs. In fact we see quite generally that if we are given a Calabi–Yau theory Q_1 which admits a

---

[21] We say partial justification because, as emphasized previously, we do not have a complete proof that our analysis yields true (in the sense of conformal field theory) mirror manifolds and also because there appear to be some examples in [35] which would not be obtained using our anlysis.

change of variables to Fermat form (up to global identifications) then we can construct the mirror of Q_1. More precisely, if we begin with the model Q_1, then this change of variables yields the result

$$Q_1/G_1 \sim Q_2/G_2 \, , \tag{7.1}$$

where the quotient groups represent the induced identifications on both sets of coordinates and Q_2 is of Fermat type, i.e. has a point in its moduli space corresponding to a minimal model construction. The $\sim$ in (7.1) implies equivalence up to deformations by twisted fields, but for our purpose of establishing the existence of the mirror manifold, the precise point in its moduli space is inessential. The arguments of [4] allow us to find the mirror manifold of the orbifold appearing on the right hand side of (7.1), expressed as a quotient of Q_2 by the appropriate 'dual' subgroup to G_2. However, we are interested in constructing the mirror of Q_1, not of its quotient. It is here that the work of the previous sections gives the solution. The left hand side of (7.1) respects a quantum G_1 symmetry [22]; its quotient by this is Q_1. As we observed, though, on the mirror manifold of (7.1), this will be realized as a *geometrical* symmetry at some point in the moduli space. At this point, the quotient by this symmetry will mirror the quotient by the quantum G_1, yielding the mirror Q_1'. To summarize, in [4] we showed that so long as a manifold lies in a moduli space which contains a minimal model point, we are assured that the manifold admits a mirror pair. Now we see that this condition can be relaxed: so long as some *orbifold* of our initial theory lies in a moduli space which contains a point described as some other *orbifold* of a minimal model theory, we are assured that the mirror to the original theory exists. This, of course, is a far weaker condition.

In this paper we have seen explicit examples of this approach. For more general cases one purely technical complication which shall arise is that some complex structure fields may come from twisted sectors and hence will not admit a monomial representation. This will make it technically more complicated to explicitly locate points in moduli space with the requisite enhanced discrete symmetries; our general arguments, however, ensure that they do exist.

An important question which immediately follows is: how general is the space of Calabi–Yau theories (with Landau–Ginzburg representation) which admit a (nonlinear) transformation to a Fermat form (up to global field identifications)? Does it exhaust, for example, the space of hypersurfaces in weighted four space? This is an interesting and important question to resolve.

8. Conclusions

In [4] the existence of mirror manifolds was established via direct construction. This construction involves generating the mirror of a highly symmetric Calabi–Yau manifold (e.g. Fermat form in a weighted porjective space) through orbifolding by the maximally allowed group of discrete scaling symmetries. By deformation arguments [4] this then establishes the existence of mirror pairs of manifolds in at least a local neighborhood of the initial Calabi–Yau space.

In this paper we have built upon the work of [4] and presented a method by which strong evidence for new explicit mirror pairs (away from the special Fermat points) has been given. Furthermore, we do not even require that there exist a Fermat point in the moduli space of either member of such a pair. Rather, we require a much weaker condition described in the last section: an orbifold of our initial theory should lie in a moduli space containing a point described by another orbifold of a minimal model theory. If this condition is met (and we stress that we only require the existence of *some* orbifolding which meets the condition) then the arguments of this paper apply and we have strong evidence that the mirror of our initial theory exists. For the explicit cases we have analysed it appears that the crux of the construction used in [4] plays a role in more general situations. Namely, we have found non-minimal model mirror pairs related through orbifolding by maximally allowed discrete symmetry groups – just as we initially found for the minimal model cases.

We have seen that the language of singularity theory appears to provide a powerful organizing principle for the mirror manifolds constructed in this manner. Up to global orbifold identifications, the singularity type of a manifold and its mirror are identical. The $\mathbb{Z}_{256}$ an $\mathbb{Z}_{320}$ symmetric quintic hypersurfaces are prototypes of such results. We have seen that this result follows directly from the mathematical formulation based on toric geometry. We have also seen that the procedure we have employed has a natural and attractive mathematical for mulation. In this framework, the approach we are following is encoded in certain special basis changes in the lattices/vector spaces suppled by toric geometry. We have found that our new candidate mirror pairs, from this viewpoint, stand in precisely the same relation [16] as all of the previously proven cases of mirror symmetry [4]. This lends strong support to our claims.

This work reemphasizes two important points that present obstacles to a complete generalization of mirror symmetry. First, on the physics side, we do not have sufficiently explicit understanding of non-minimal-model Landau–Ginzburg theories. This prevents us from proceeding directly at the level of conformal field theory, for example as in [4]. Second, on the mathematics side, there is only a partial proof [16] of mirror symmetry (avoiding conformal field theory methods) in the known cases. Namely, there is no mathematical understanding, at present, of the physical result of the equality of n-point functions on the manifolds of a mirror pair. This prevents us from proceeding directly at the level of mathematics along the lines of section 5.

Progress on either the physics or mathematics fronts could likely yield dramatic advances in our understanding of the theory of mirror manifolds.

It is a pleasure to thank P. Candelas, I.M. Singer, S.-T. Yau for organizing a productive and enjoyable workshop, and K. Intrilligator, C. Vafa and S.-T. Yau for a number of important discussions. S-S.R. thanks the MSI and the Physics Department of Cornell University for hospitality and support. B.R.G. is supported by the National Science Foundation and M.R.P. is supported in part by DOE grant DE-AC-02-76ERO3075.

References

[1] B.R. Greene and M.R. Plesser, elsewhere in these proceedings.

[2] L.J. Dixon, in *Superstrings, Unified Theories, and Cosmology* 1987, (G. Furlan et. al., eds.), World Scientific, 1988, p. 67.

[3] W. Lerche, C. Vafa, and N.P. Warner, Nucl. Phys. **B324** (1989) 427.

[4] B.R. Greene and M.R. Plesser, Nucl. Phys. **B338** (1990) 15.

[5] P. Candelas, X.C. de la Ossa, P.S. Green, and L. Parkes, Nucl. Phys. **B359** (1991) 21; Phys. Lett. **258B** (1991) 118.

[6] P.S. Aspinwall, C.A. Lütken, and G.G. Ross, Phys. Lett. **241B** (1990) 373.

[7] B.R. Greene, M.R. Plesser, and S.-S. Roan, in preparation.

[8] D. Gepner, Phys. Lett. **B199** (1987) 380; Nucl. Phys. **B296** (1987) 380.

[9] P. Candelas, M. Lynker, and R. Schimmrigk, Nucl. Phys. **B341** (1990) 383.

[10] B. Greene, Comm. Math. Phys. **130** (1990) 335.

[11] M. Kreuzer, R. Schimmrigk, and H. Skarke, *Abelian Landau–Ginzburg Orbifolds and Mirror Symmetry*, Preprint NSF-ITP-91-102 TUW-91-10.

[12] B.R. Greene, C. Vafa and N.P. Warner, **B324** (1989), 371.

[13] E. Martinec, in *V.G. Knizhnik Memorial Volume* (L. Brink et. al., eds.); Phys. Lett. **B217** 431.

[14] B.R. Greene, unpublished.

[15] R. Schimmrigk and M. Lynker, Phys. Lett. **B249** (1990) 237.

[16] S.-S. Roan, Int. Jour. of Mathematics 2 (1991) 439; see also the contribution of this author in these proceedings.

[17] S. Cecotti, Nucl. Phys. **B355** (1991) 755.

[18] B.R. Greene, S.-S. Roan, and S-T. Yau, Cornell Preprint CLNS-91-1045.

[19] C. Vafa and N.P. Warner, Phys. Lett. **B218** (1989) 51.

[20] C. Vafa, Mod. Phys. Lett. **A4** (1989) 1169.

[21] P. Green and T. Hübsch, Comm. Math. Phys. **113** (1987) 505.

[22] C. Vafa, Mod. Phys. Lett. **A4** (1989) 1615.

[23] B.R. Greene and M.R. Plesser, in *Proceedings of the Second International* PASCOS *conference*, Northeastern University, 1991.

[24] R. Dijkgraaf, E. Verlinde and H. Verlinde, Comm. Math. Phys. **115** (1988) 649.

[25] A. Shapere and F. Wilczek, Nucl. Phys. **B230** (1989) 669.

[26] W. Lerche, D. Lüst, and N.P. Warner, Phys. Lett. **B231** (1989) 417.

[27] J.-F. Mestre, in *Proceedings of the International Conference on Class Numbers and Fundamental Units of Algebraic Number Fields, Katat, Japan,* (1986).

[28] K. Intrillgator and C. Vafa, Nucl. Phys. **B339** (1990) 95.

[29] B. Blok and A. Varehenko, *Topological Conformal Field Theories and the Flat Coordinates*, IAS preprint IASSNS-HEP-91/5.

[30] A.C. Cadavid and S. Ferrara, Phys. Lett. **B267** (1991) 193.

[31] S. Ferrara, Mod. Phys. Lett. **A6** (1991) 2175.

[32] W. Lerche, D.-J. Smit, and N. Warner, *Differential Equations and Flat Coordinates in Two-dimensional Topological Matter Theories*, Preprint UCB-PTH-91-39.

[33] D.R. Morrision, *Picard-Fuchs equations and mirror maps for hypersurfaces*, Duke Preprint DUK-M-91-14.

[34] P.A. Griffiths, Ann. of Math. **90** (1969), 460.

[35] P. Berglund and T. Hübsch, *A Generalized Construction of Mirror Manifolds*, Preprint CERN-TH-6341/91 UTTG-32-91.

M	$\widetilde{M}$	G Generators
$z_1^5 + z_2^5 + z_3^5 + z_4^5 + z_5^5 = 0$	$M/(\mathbb{Z}_5^3)$	$[0,0,0,1,4]$ $[0,0,1,0,4]$ $[0,1,0,0,4]$
$z_1^4 z_2 + z_2^4 z_3 + z_3^4 z_4 + z_4^4 z_5 + z_5^4 z_1 = 0$	$M/(\mathbb{Z}_{41})$	$[1,37,16,18,10]$
$z_1^4 z_2 + z_2^4 z_3 + z_3^4 z_4 + z_4^4 z_1 + z_5^5 = 0$	$M/(\mathbb{Z}_{51})$	$[1,37,16,38,0]$
$z_1^4 z_2 + z_2^4 z_3 + z_3^4 z_1 + z_4^5 + z_5^5 = 0$	$M/(\mathbb{Z}_5 \times \mathbb{Z}_{13})$	$[0,0,0,1,4]$ $[1,9,3,0,0]$
$z_1^4 z_2 + z_2^4 z_1 + z_3^5 + z_4^5 + z_5^5 = 0$	$M/(\mathbb{Z}_5^2 \times \mathbb{Z}_3)$	$[0,0,1,0,4]$ $[0,0,0,1,4]$ $[1,2,0,0,0]$
$z_1^4 z_2 + z_2^4 z_1 + z_3^4 z_4 + z_4^4 z_5 + z_5^4 z_3 = 0$	$M/(\mathbb{Z}_3 \times \mathbb{Z}_{13})$	$[1,2,0,0,0]$ $[0,0,1,9,3]$
$z_1^5 + z_1 z_2^4 + z_2 z_3^4 + z_3 z_4^4 + z_4 z_5^4 = 0$	$y_3 y_1^5 + y_5 y_2^4 + y_4 y_3^4 + y_2 y_4^4 + y_5^4 = 0$ in $\mathrm{W}\mathbb{CP}^4_{41,48,51,52,64}$	
$z_1^5 + z_3 z_2^4 + z_1 z_3^4 + z_4^5 + z_2 z_5^4 = 0$	$y_4 y_1^5 + y_5 y_2^4 + y_3^5 + y_2 y_4^4 + y_5^4 = 0$ in $\mathrm{W}\mathbb{CP}^4_{51,60,64,65,80}$	

Table 1

Quintic Hypersurfaces M in $\mathbb{CP}^4$ and their Mirrors $\widetilde{M}$

The mirror $\widetilde{M}$ of M is indicated either explicitly or as a quotient of M.
The group actions are denoted by the powers of an appropriate primitive
root of unity by which they multiply the coordinates of $\mathbb{CP}^4$.

M	$\widetilde{M}$	G Generators
$z_1^5 + z_2^{10} + z_3^{10} + z_4^{10} + z_5^2 = 0$	$M/(\mathbb{Z}_{10}^2)$	$[0,0,1,9,0]$ $[0,1,9,0,0]$
$z_1^5 + z_2^9 z_3 + z_3^9 z_4 + z_4^9 z_2 + z_5^2 = 0$	$M/(\mathbb{Z}_{73})$	$[0,1,64,8,0]$
$z_1^5 + z_2^9 z_3 + z_3^9 z_4 + z_4^{10} + z_5^2 = 0$	$y_1^5 + y_2^{10} y_3 + y_3^9 y_4 + y_4^9 + y_5^2 = 0$ in $\mathrm{W}\mathbb{CP}^4_{162,73,80,90,405}$	

Table 2

Hypersurfaces M in $\mathrm{W}\mathbb{CP}^4_{2,1,1,1,5}$ and their Mirrors $\widetilde{M}$

The mirror $\widetilde{M}$ of M is indicated either explicitly or as a quotient of M.
The group actions are denoted by the powers of an appropriate primitive
root of unity by which they multiply the coordinates of $\mathrm{W}\mathbb{CP}^4_{2,1,1,1,5}$.

AMS/IP Studies in Advanced Mathematics
Volume **9**, 1998

Deformations of Calabi-Yau Kleinfolds

Ziv Ran*

Math Department

University of California

Riverside, CA 92521

Recall that a *Calabi-Yau manifold* is a compact Kähler manifold X whose canonical bundle K_X is trivial (we do not assume any extra conditions such as on the Hodge numbers of X). A well-known theorem, due to Bogomolov, Tian and Todorov, asserts that such X always has *unobstructed deformations*, i.e. its local moduli (Kuranishi) space is smooth. By a *Calabi-Yau Kleinfold* we shall mean a compact analytic variety X_0 with $\omega_X = \mathcal{O}_X$ which admits a resolution of singularities by a compact Kähler manifold, and whose singularities are *Kleinian*, i.e. isolated simple hypersurface singularities (type A-D-E), and satisfy an additional local condition of "goodness" (see *Definition* below, this condition is automatically satisfied at least for all 3-fold Kleinian singularities and A_1 singularities in all dimensions, and may well hold for all Kleinian singularities). By a theorem of Grauert, X_0 admits a local moduli space or semiuniversal deformation. Our purpose here is to prove the following.

Theorem 1. *Any Calabi-Yau Kleinfold of dimension ≥ 3 has unobstructed deformations.*

For an excellent background discussion and motivation (especially in dimension 3), see [F]. In the special case of Kleinfolds with only *ordinary* double point singularities, unobstructedness was proven independently by Kawamata [K], the author (unpublished) and Tian [T]. While Tian's proof is analytic, the former two proofs are essentially algebraic in character and are based on the author's "T^1-lifting criterion" [R1] (used previously to give a simple proof of the smooth case), together with the theory of degeneration of Hodge structures as developed by Steenbrink *et al.* Here we use an analogous approach, except

---

* Supported in part by NSF

© 1998 American Mathematical Society
and International Press

that, rather than the T^1-lifting criterion, we use its "dual" as developed in [R2] (which might be called the "T^2-injecting criterion"); To do so, we establish the fact, of independent interest, that Kleinfolds admit a good (pure) Hodge theory below the middle dimension, analogous to Steenbrink's [S] Hodge theory for V-manifolds (cf. Proposition 4 below). It is the shift from T^1 to T^2 that allows us to avoid the troublesome middle dimension.

Definition. *Let (X_p) be a isolated singularity with resolutions $\tilde{X}$, $E \subset \tilde{X}$ the exceptional divisor and $\hat{X}$ the formal completion of $\tilde{X}$ along E. (X, p) is said to be* **good** *if for all (i, j), $0 < i \neq j$, $i + j < \dim X$, the map induced by exterior derivative.*

$$H^i(\Omega^j_{\hat{X}}) \xrightarrow{d} H^i(\Omega^{j+1}_{\hat{X}})$$

is injective.

It is easy to check that this condition is independent of the resolution (equivalently, that a *smooth* point is good, in fact, $H^i(\Omega^j_{\hat{X}}) = 0$ for all $0 < i \neq j$) and is trivially satisfied for all 3-fold rational singularities. It is an elementary calculation that an A_1-singularity (any dimension) is good, the only nonzero $H^i(\Omega^j_{\hat{X}})$ with $0 < i \neq j$, $i + j < n = \dim X$, being those with $(i, j) = (n/2 - 1, n/2)$, $(n/2 - 1, n/2 + 1)$. We believe all Kleinian singularities are good, but have not yet checked this.

Let X_0 be a Calabi-Yau Kleinfold. As is well-known, infinitesimal deformations of X_0 are measured by the group.

$$T^1(X_0) = \mathrm{Ext}^1(\Omega^1_{X_0}, \mathcal{O}_{X_0});$$

moreover, as X_0 has only isolated hypersurface singularities, its local T^2 vanishes, hence obstructions to deforming X_0 lie in the group.

$$T^2(X_0) = \mathrm{Ext}^2(\Omega^1_{X_0}, \mathcal{O}_{X_0}).$$

Now put $S_j = \mathrm{Spec}\,(\mathbb{C}[\epsilon]/(\epsilon^{j+1}))$ and let X_j/S_j be a flat deformation of X_0. It is then a very special case of Theorem 1.1, (ii) of [R2] that Theorem 1 is a consequence of the following

Claim 2. *The group* $\mathrm{Ext}^2(\Omega^1_{X_j/S_j}, \mathcal{O}_{X_j})$ *is a free* S_j-*module and its formation commutes with base-change.*

Now by Serre duality, $\mathrm{Ext}^2(\Omega^1_{X_j/S_j}, \mathcal{O}_{X_j}) = \mathrm{Ext}^2(\Omega^1_{X_j/S_j}, \omega_{X_j/S_j})$ is dual to $H^{n-2}(\Omega^1_{X_j/S_j})$, $n = \dim X_0$ (see proof of Lemma 6 below). So Claim 2 would follow if we can prove

Claim 3. $H^{n-2}(\Omega^1_{X_j/S_j})$ *is a free* S_j-*module and its formation commutes with base-change.*

Let $j : (X_0)\mathrm{reg} \to X_0$ be the inclusion. Taking a clue from Steenbrink [S] we define a modified de Rham complex for X_0 as

$$\tilde{\Omega}^\bullet_{X_0} = j_* \Omega^\bullet_{(X_0)\mathrm{reg}}.$$

The following Proposition then shows that this leads to a good (pure) Hodge theory for X_0, at least below the middle dimension.

Proposition 4. *Let* X *be a compact Kleinfold of dimension* $n \geq 2, \pi : \tilde{X} \to X$ *a resolution of singularities with* $\tilde{X}$ *Kähler. Then*

(i) $\tilde{\Omega}^\bullet_X = \pi_* \Omega^\bullet_{\tilde{X}}$;

(ii) $\tilde{\Omega}^\bullet_X$ *is a resolution of the constant sheaf* $\mathbb{C}_X$;

(iii) *The Hodge-DeRham spectral sequence*

$$E_1^{p,q} = H^q(X, \tilde{\Omega}^p_X) \Rightarrow H^{p+q}(X, \mathbb{C})$$

degenerates at E_1 *through degree* $n - 1$ *(i.e. all differentials* d_r^{pq}, $p + q \leq n - 1$, *vanish).*

Assuming the Proposition for now, let us proceed to prove Claim 3. We similarly define

$$\tilde{\Omega}^\bullet_{X_j/S_j} = j_*(\Omega^\bullet_{X_j/S_j}|_{(X_0)\mathrm{reg}})$$

(considering $\tilde{\Omega}^\bullet_{X_j/S_j}$ as a complex of sheaves on X_0).

Lemma 5. $\tilde{\Omega}^\bullet_{X_j/S_j}$ *is a resolution of the constant sheaf* $\mathbb{C}\,[\epsilon]/(\epsilon^{j+1})$ *on* X_0.

Proof. For $j = 0$ this is in Proposition 4. The general case follows from this by considering the adic filtration on S_j which induces a filtration on $\tilde{\Omega}^\bullet_{X_j/S_j}$ whose associated gradeds are just $\tilde{\Omega}^\bullet_{X_0/S_0}$ (note that the differentials in $\tilde{\Omega}^\bullet_{X_j/S_j}$ are S_j-linear).

Lemma 6. $H^q(\tilde{\Omega}^p_{X_j/S_j})$ *is a free S_j-module and its formation commutes with base-change provided either $p + q \leq n - 1$ or $(p, q) = (n, 0)$.*

Proof. For $p + q \leq n - 1$ this follows from Proposition 4 and Lemma 5 by an easy argument of Deligne (essentially just semi-continuity); cf [D], Theorem 5.5. As the unique n-form on X_0 is obviously closed, the case $(p, q) = (n, 0)$ then follows.

Finally, to prove Claim 3 it suffices to prove

Lemma 7. We have $\Omega^1_{X_j/S_j} = \tilde{\Omega}^1_{X_j/S_j}$ for $n \geq 3$.

Proof. We work locally at a singular point $p \in X_o$. We may assume X_o is given locally by an equation $f(x) = 0$ in $\mathbb{C}^{n+1}$; by the well-known structure of the versal deformation of a Kleinian singularity [Tj], we may assume the deformation X_j/S_j in given by equations

$$f(x) = \epsilon^k \qquad k \leq j + 1$$

$$\epsilon^{j+1} = 0$$

in (x, ϵ)-space. We have an exact sequence

$$0 \to \mathcal{O}_{X_j} \xrightarrow{\nabla f} (n+1)\mathcal{O}_{X_j} \to \Omega^1_{X_j/S_j} \to 0.$$

As $n \geq 3$, we see immediately by depth considerations that $H^1_p(\Omega^1_{X_j/S_j}) = 0$, proving thee lemma.

Proof of Propositon 4. (i) We work locally at a singular point $p \in X$ and use induction on n. For $n = 2$ the result has been proven by Steenbrink [S], since a Kleinian singularity is then a V-manifold singularity. In the general case we may cut the germ of X at p with a general hyperplane through p to conclude that for any p-form ω holomorphic on $X - p$, $p \leq n - 1$, the pullback $\pi^*\omega$ has no pole at a generic point of any component of the exceptional divisor, hence is holomorphic on $\tilde{X}$ by Hartogs' theorem. That the same is true for $p = n$ is standard (just the rationality of the singularity).

(ii) Again the assertion is local at a singularity $p \in X$ and is in essence due to Naruki (cf. [L], Lemma 9.9, p. 177). The point is that the Kleinian singularity p admits a *good* (positive-weight) $\mathbb{C}^*$-action, hence an Euler derivation ζ. As ζ lifts to the canonical resolution, it acts on $\tilde{\Omega}^\bullet_{X,p}$ and the proof in *loc. cit.* applies.

(iii) This is the crux of the matter. We will prove in detail that the differentials in E_1 of total degree $\leq n - 1$ vanish, the case of E_r, $r \geq 2$, being similar. Consider the Leray spectral sequence

$$E_2^{pq} = H^p(R^q\pi_*\Omega^j_{\tilde{X}}) \Rightarrow H^i(\Omega^j_{\tilde{X}}).$$

and the analogous one for the sheaves $\hat{\Omega}^\bullet_{\tilde{X}}$ of closed forms. As π is an isomorphism off the finite singular locks of X_1 we have a diagram

$$
\begin{array}{ccccc}
H^0(R^{i-1}\pi_*\Omega^j) & \longrightarrow & H^0(R^{i-1}\pi_*\hat{\Omega}^{j+1}) & \longrightarrow & H^0(R^{i-1}\pi_*\Omega^{j+1}) \\
\downarrow & & \downarrow & & \downarrow \\
H^i(\tilde{\Omega}^j_X) & \longrightarrow & H^i(\tilde{\hat{\Omega}}^{j+1}_X) & \longrightarrow & H^i(\tilde{\Omega}^{j+1}_X) \\
\pi^*_{ij}\downarrow & & \downarrow & & \downarrow \\
H^i(\Omega^j_{\tilde{X}}) & \xrightarrow{\alpha} & H^i(\hat{\Omega}^{j+1}_{\tilde{X}}) & \longrightarrow & H^i(\Omega^{j+1}_{\tilde{X}})
\end{array}
\qquad (*)
$$

with exact columns, where $\tilde{\hat{\Omega}}^\bullet := \pi_*\hat{\Omega}^\bullet$. Kählerity of $\tilde{X}$ yields the vanishing of α, while goodness of the singularities of X yields the vanishing of β, provided $i + j < n$, $j \neq i - 2$. From this an easy diagram chase around $(*)$ yields the vanishing of d_1^{ij} ($=$ composite of middle-row arrows) for all $i + j < n$, $j \neq 1 - 2$.

We proceed now to prove the vanishing of $d_1^{i\ i-2}$, $2 \leq i \leq \frac{n+1}{2}$, using induction on i, beginning with $i = 2$. Note that the part of $H^2(\mathcal{O}_X)$ which contributes to $H^2(\mathbb{C}_X)$ must obviously be contained in ker $d_1^{2\ o}$, i.e.

$$(\ker\ d_1^{20} \supseteq\ im\ (H^2(\mathbb{C}_X) \xrightarrow{\alpha} H^2(\mathcal{O}_X)).$$

Consider the diagram

$$
\begin{array}{ccc}
H^2(\mathbb{C}_X) & \xrightarrow{\alpha} & H^2(\mathcal{O}_X) \\
\pi^*_2\downarrow & & Z| \\
H^2(\mathbb{C}_{\tilde{X}}) & \xrightarrow{\tilde{\alpha}} & H^2(\mathcal{O}_{\tilde{X}})
\end{array}
$$

Note that $H^2(\mathbb{C}_{\tilde{X}})$ is generated by $H^2(\mathbb{C}_X)$ plus classes supported on the exceptional divisor E. As E is a union of quadrics all its cohomology is algebraic, hence yields (p,p)-classes on $\tilde{X}$. In particular, $im(\tilde{\alpha}_o\pi^*_2) = im(\tilde{\alpha}) = H^2(\mathcal{O}_{\tilde{X}})$, hence α is surjective and $d_1^{20} = 0$.

The general induction step on i is similar. We will illustrate it by doing the next case $i = 3$, i.e. the vanishing of $d_1^{3,1} :\ H^3(\tilde{\Omega}^1_X) \to H^3(\tilde{\Omega}^2_X)$. Identifying $H^\bullet(\mathbb{C}_X) = H^\bullet(\tilde{\Omega}^\bullet_X)$ and using the exact sequence of complexes

$$0 \to F^1\tilde{\Omega}^\bullet_X \to \tilde{\Omega}^\bullet_X \to \mathcal{O}_X \to 0\ ,$$

F^1 being the stupid filtration, we get a diagram

$$
\begin{array}{ccccccc}
H^3(\mathcal{O}_X) & \longrightarrow & \tilde{F}_X^{14} & \longrightarrow & H^4(\mathbb{C}_X) & \longrightarrow & H^4(\mathcal{O}_X) \\
& & \downarrow & & \pi_4^* \downarrow & & Z| \\
0 & \longrightarrow & F_{\tilde{X}}^{14} & \longrightarrow & H^4(\mathbb{C}_{\tilde{X}}) & \longrightarrow & H^4(\mathcal{O}_{\tilde{X}}) \longrightarrow 0 \;,
\end{array}
$$

where $\tilde{F}_X^{14} = IH^4(F^1\tilde{\Omega}_X^\bullet)$ and similarly for $F_{\tilde{X}}^{14}$. As above, the top left horizontal arrow vanishes while the top right horizontal arrow is surjective, and moreover $H^4(\mathbb{C}_{\tilde{X}})$ is generated by $im\ \pi_4^*$ plus some $(2,2)$-classes. Hence $F_{\tilde{X}}^{14}$ is generated by the image of $\tilde{F}_X^{14}$ plus some $(2,2)$ classes, hence in the diagram

$$
\begin{array}{ccc}
\tilde{F}_X^{14} & \longrightarrow & H^3(\tilde{\Omega}_X^1) \\
\downarrow & & Z| \\
F_{\tilde{X}}^{14} & \longrightarrow & H^3(\Omega_{\tilde{X}}^1) \;,
\end{array}
$$

surjectivity of the top horizontal arrow follows from that of the bottom, hence $d_1^{31} = 0$.

Acknowledgement

We are grateful to Professor D. R. Morrison for comments on an earlier version of this paper. Thanks are also due to Professor R. Friedman for initially mentioning the problem to us (in the case of ordinary double points).

References

[D] Deligne, P. "Théoréme de Lefschetz et critères de dégénérescence de suites spectrales". Publ. Math. IHES **35** (1968), 197-226.

[F] Friedman, R.: "On threefolds with trivial canonical bundle" (preprint).

[K] Kawamata, Y.: "Unobstructed deformations — a remark on a paper by Z. Ran" (preprint).

[L] Looijenga, E. J. N.: Isolatcd singular points on complete itersections. Cambridge University Press 1984.

[R1] Ran, Z.: "Deformations of manifolds with torsion or negative canonical bundle", J. Algebraic Geometry **1** (1992), 279-291.

[R2] ____: "Hodge theory and deformations of maps" , Compositio Math (to appear).

[S] Steenbrink, J.H.M.: "Mixed Hodge structure on the vanishing cohomology". In: "Real and complex singularities", Oslo 1976, 513–536.

[T] Tian, G.: (to appear).

[Tj] Tjurina, G.: "Resolution of singularities of plane deformations of double rational points." Funct. Anal. Appl. **4** (1970), 68–73.

AMS/IP Studies in Advanced Mathematics
Volume **9**, 1998

Smoothing 3-folds with trivial canonical bundle and ordinary double points

Gang Tian*

Department of Mathematics

State University of New York

Stony Brook NY 11794

*This work is partially supported by a grant from NSF

© 1998 American Mathematical Society
and International Press

0. Introduction.

In the last decade, there has been a great success in the classification of threefolds by their Kodaira dimension, thanks to the program carried out by Mori and others. However, the structure of threefolds with Kodaira dimension zero remains a mystery. Among all such threefolds, those whose canonical bundle is actually trivial are an especially interesting subject. If X is a smooth threefold with trivial canonical bundle and contains disjoint rational curves $C_1, \ldots, C_l$ of type (-1, -1), then one can contract those rational curves to obtain a singular threefold X_0 with l ordinary double points. Here, by type (-1, -1), we mean that the normal bundle of each $C_1, \ldots, C_l$ is a direct sum of two tautological line bundles on the rational curve. A natural question is whether or not there exists a smoothing of X_0, in other words, there exists a complex fourfold $\mathcal{X}$, together with a proper flat map $\pi : \mathcal{X} \mapsto \Delta$, where Δ is the unit disk in C, such that $\pi^{-1}(0) = X_0$ and $\pi^{-1}(t) = X_t$ is smooth for $t \neq 0$. It is showed in [Fr] that there is an infinitesimal smoothing of X_0 if and only if the fundamental classes $[C_i]$ in $H^2(X; \Omega_X^2)$ satisfy a relation $\sum_i \lambda_i [C_i] = 0$ such that, for every i, $\lambda_i \neq 0$. This infinitesimal smoothing can be realized by a real smoothing in case the obstruction group $\mathbf{T}^2_{X_0}$ happens to be zero. This was used by Clemens to find new compact complex manifolds (cf. the end of §4). The obstruction group can be identified with the quotient of $H^2(X; \Omega_X^2)$ by those fundamental classes $[C_i]$ ($i = 1, \ldots, l$). Therefore, this is often not zero, in fact, if X_0 is projective, this is never zero.

The purpose of this paper is to show that the infinitesimal smoothing can always be realized by a real smoothing

Theorem 0.1 Let X_0 be a singular threefold with l ordinary double points as only singular points $p_1, \ldots, p_l$. Let X be the small resolution of X_0 by replacing p_i by a smooth rational curve C_i. Assume that X is Kähler and has trivial canonical line bundle. Furthermore, we assume that the fundamental classes $[C_i]$ in $H^2(X, \Omega_X^2)$ satisfy a relation $\sum_i \lambda_i [C_i] = 0$ such that, for every i, $\lambda_i \neq 0$. Then X_0 admits a smoothing.

We should point out that the Kählerian of X can be replaced by the weaker condition that

X is only cohomologically Kähler, namely, the $\partial\overline{\partial}$-lemma holds for X. This is because, as the readers will see, we only use the property of a Kähler manifold, the $\partial\overline{\partial}$-lemma, in the proof.

The same proof also yields the following unobstructedness theorem of deformation of n-dimensional complex varieties with trivial dualizing sheaf and ordinary double points. In fact, the proof for higher dimensions is easier.

Theorem 0.2 Let X_0 be a singular n-fold ($n \geq 3$) with trivial dualizing sheaf and ordinary double points as only singular points. Let X be the resolution of X_0 obtained by replacing each singular point be a smooth quadratic hypersurface in CP^{n+1}. Assume that X is coholomogically Kähler. Then the flat deformation of X_0 is unobstructed.

In fact. Theorem 0.2 can be generalized to the case the singular variety X_0 may have more complicated singularities of special type, such as, double points of rational type.

The organization of this paper is as follows. In section 1, we reduce the problem of deformation to the formal deformation of the regular part of X_0 using a theorem of Schlessinger and a theorem of Grauert, Froster and Knorr. In section 2, we briefly discuss the deformation of ordinary double points and the associated Kodaira-Spencer map. Theorem 0.2 is proved in section 3, while Theorem 0.1 is proved in section 4 as a corollary of Theorem 0.2 and a result of R. Friedman. In section 5, we briefly study the behavior of the Peterson-Weil metric in the smoothing of singular 3-folds with double points. In particular, we proved that the Peterson-Weil metric is incomplete and has linearly logarithmic growth at the singular variety X_0(cf. Corollary 5.1).

After the author just finished the preliminary version of this paper, he learned that Professor Kawamata also proved Theorem 0.2 as an application of Deligne-Kawamata-Ran T^1-lifting principle on the formal deformation. Finally, the author would like to thank Professor Friedman for bringing to me the problem of smoothing singular varieties and some very helpful conversations.

1. Reduction to formal deformation

In this section, we shall assume that X_0 is the singular n-fold with only isolated singular points. We shall reduce the deformation of X_0 to solving a sequence of $\bar{\partial}$-equations on the regular part of X_0 by using some general theorems of Schlessinger, Grauert, Forster and Knoor, and following the arguments of Kodaira-Spencer in [KS].

Define the global and local extension sheaves on X_0:

$$\mathbf{T}^i_{X_0} = \mathrm{Ext}^i(\Omega^1_{X_0}, \mathcal{O}_{X_0}) \text{ and } T^i_{X_0} = \mathcal{E}xt^i(\Omega^1_{X_0}, \mathcal{O}_{X_0}),$$

where $\Omega^1_{X_0}$ is the sheaf of Kähler differentials on X_0. These are the groups and sheaves arising from the cotangent complex in the theory of Lichtenbaum-Schlessinger.

The infinitesimal deformations of X_0 are classified by $\mathbf{T}^1_{X_0}$. Additionally, since the singularities of X_0 are isolated complete intersections, by using the spectral sequence associated to the hypercohomologies $\mathbf{T}^i_{X_0}$, one can show that there is an exact sequence

$$0 \longrightarrow H^1(X_0, T^0_{X_0}) \longrightarrow \mathbf{T}^1_{X_0} \longrightarrow H^0(X_0, T^1_{X_0}) \longrightarrow$$
$$\longrightarrow H^2(X_0, T^0_{X_0}) \longrightarrow \mathbf{T}^2_{X_0} \tag{1.1}$$

The following proposition is due to M. Schlessinger [Sc].

Proposition 1.1 Let U be the open subset in X_0 consisting of all regular points, we assume the dimension of X_0 is not less than 3. Then

(1) $\mathbf{T}^1_{X_0} = \mathrm{H}^1(U, T^0_U) = \mathrm{H}^1(U, \Theta_U)$

(2) There is a long exact sequence

$$0 \longrightarrow H^1(X_0, T^0_{X_0}) \longrightarrow \mathbf{T}^1_{X_0} \longrightarrow \oplus^l_{i=1} Cp_i \longrightarrow$$
$$\longrightarrow H^2(X_0, T^0_{X_0}) \tag{1.2}$$

where p_i for $i = 1, \cdots, l$ are the singular points of X_0.

Proof: There is an exact sequence

$$J/J^2 \longrightarrow \Omega^1_{C^4} \longrightarrow \Omega^1_{X_0} \longrightarrow 0, \tag{1.3}$$

where J is the ideal sheaf of X_0. Let N_0 be the image of J/J^2 in $\Omega^1_{C^4}$. Since the variety X_0 has only isolated singular points, the dual of N_0 is isomorphic to the dual of J/J^2. It follows that we have an exact sequence

$$
\begin{aligned}
0 \longrightarrow T^0_{X_0} \longrightarrow\ & \mathrm{Ext}^0(\Omega^1_{C^4}, \mathcal{O}_{X_0}) \longrightarrow \mathrm{Ext}^0(J/J^2, \mathcal{O}_{X_0}) \longrightarrow \\
\longrightarrow\ & T^1_{X_0} \longrightarrow \mathrm{Ext}^1(\Omega^1_{C^4}, \mathcal{O}_{X_0}) \longrightarrow \mathrm{Ext}^1(J/J^2, \mathcal{O}_{X_0})
\end{aligned}
\tag{1.4}
$$

Since X_0 is a variety of dimension ≥ 3 with only isolated ordinary double points, the extension sheaf $\mathrm{Ext}^i(F, \mathcal{O}_{X_0})$ are isomorphic the $\mathrm{Ext}^i(F, \mathcal{O}_U)$ for $i = 0,\ 1$ and F being $\Omega^1_{C^4}$ or J/J^2. On the other hand, $\mathrm{Ext}^1(\Omega^1_U, \mathcal{O}_U)$ is just $H^1(U, \Theta_U)$. Then (1) follows. For (2), we remark that all extension sheaves $\mathcal{E}xt^i(\Omega^1_{X_0}, \mathcal{O}_{X_0})$ vanish for $i > 1$ and $\mathcal{E}xt^1(\Omega^1_{X_0}, \mathcal{O}_{X_0})$ is a skyscraper sheaf with stalk C supported at the singular points p_i $(1 \leq i \leq l)$ of X_0. Then (2) follows from the exact sequence (1.1).

This argument in the proof of (1) may be extended to prove the following.

Proposition 1.2   The formal deformation theories of X_0 and of U are equivalent.

Next, let us recall some of Kodaira-Spencer deformation theory. Let $\{V_i\}_{i \in I}$ be a finite covering of U by stein open subsets. Let $\{h_{ij}\}_{i,j \in I}$ be the transition functions of U associated with this covering. Then the deformation of U can be described in terms of transition functions as follows. Denote by Δ_ϵ, the ball with radius ϵ in the complex plane C^1, and by z_i^α $(\alpha = 1, 2, \dots, n)$ the coordinate functions in V_i, where $i \in I$. A new complex manifold $\mathcal{X}$ can be constructed by gluing $V_i \times \Delta_\epsilon$ by identifying (z_j, t) in $V_j \times \Delta_\epsilon$ and (z_k, t) in $V_k \times \Delta_\epsilon$ if $z_j^\alpha = f_{jk}^\alpha(z_k, t)$, where $f_{jk}^\alpha(z_k, t)$ is a holomorphic function defined on $(V_j \cap V_k) \times \Delta_\epsilon$. These functions should satisfy

$$
f_{jk}^\alpha(z_k, 0) = h_{jk}^\alpha(z_k) \quad on \quad (V_j \cap V_k) \times \Delta_\epsilon
\tag{1.5}
$$

$$
f_{ik}^\alpha(z_k, t) = f_{ij}^\alpha(f_{jk}(z_k, t), t) \quad on \quad (V_i \cap V_j \cap V_k) \times \Delta_\epsilon
\tag{1.6}
$$

This new manifold $\mathcal{X}$ is a deformation of U. Therefore, in order to find the deformation of U, it suffices to solve (1.5) and (1.6) for transition functions f_{jk}. A formal solution of (1.6) is a set of the formal power series

$$
f_{jk}^\alpha(z_k, t) = \sum_{\nu=0}^{\infty} f_{jk|\nu}^\alpha(z_k) t^\nu,
\tag{1.7}
$$

which satisfies the following equations

$$f^\alpha_{jk|0}(z_k) = h^\alpha_{jk}(x_k) \tag{1.8}$$

$$f^\alpha_{ik|\nu}(z_k) = f^\alpha_{ij|\nu} z_j + \sum_{\mu=0}^{\nu-1}\sum_{k=1}^{\nu-\mu}\sum_{k_1+\cdots+k_n=k}\sum_{\nu_1+\cdots+\nu_n=\nu}\frac{\partial^k f^\alpha_{ij|\mu}}{\partial^{k_1}z_j^1\cdots\partial^{k_n}z_j^n}(z_j)$$
$$\cdot\,(f^1_{jk|\nu_1}\cdots f^1_{jk|\nu_{k_1}}\cdots f^n_{jk|\nu_n})(z_k) \tag{1.9}_\nu$$

A formal deformation $\mathcal{X}_{\text{for}}$ of U is simply a set of formal transition functions given by formal series in (1.7) satisfying (1.8) and (1.9)$_\nu$ for all $\nu \geq 1$. Associated with this formal deformation is an 1-cocycle θ in $\mathrm{H}^1(U, \Theta_U)$ as follows: define holomorphic vector fields θ_{jk} on $V_j \cap V_k$ by

$$\theta_{jk} = \sum_{\alpha=1}^{n} f^\alpha_{jk}(z_k)\frac{\partial}{\partial z_j^\alpha} \tag{1.10}$$

Now (1.9)$_1$ implies

$$\theta_{ik} = \theta_{ij} + \theta_{jk} \quad \text{on} \quad V_i \cap V_j \cap V_k$$

Then $\theta = \{\theta_{jk}\}_{j,k\in I}$. This 1-cocycle is the infinitesimal deformation of U associated to $\mathcal{X}_{\text{for}}$. Let ψ_1 be the Dolbeault representation of θ.

Proposition 1.3. Let ψ_1 be an infinitesimal deformation in $\mathrm{H}^1(U, \Theta_U)$. Then there is a flat deformation of X_0 in the direction of $\theta = [\psi_1]$ if and only if the following equations are solvable for Θ_U-valued $(0,1)$-forms $\psi_2,\ldots,\psi_\nu,\ldots$ in U.

$$\bar{\partial}\psi_\nu + \frac{1}{2}\sum_{\mu=1}^{\nu-1}[\psi_\mu, \psi_{\nu-\mu}] = 0 \tag{1.11}_\nu$$

where U is the regular part of X_0 as above and $[\cdot,\cdot]$ is the Lie bracket on Θ_U-valued $(0,1)$-forms.

Proof: The necessity is rather easier. We only sketch a proof of the sufficiency here. By a famous theorem of Grauert [Gr] and Froster and Knorr [FK], it suffices to show that the flat deformation of X_0 is formally unobstructed. By Proposition 1.2, the formal deformation theories of X_0 and of U are equivalent. Therefore, we only need to prove that the formal deformation of U along the infinitesimal direction ψ_1 in $\mathrm{H}^1(U, \Theta_U)$ follows from the solvability of $(1.11)_\nu$ for $\nu \geq 2$. Let $\psi_\nu(\nu \geq 2)$ be the solutions of $(1.11)_\nu$. Since all V_j are stein and ψ_1 is $\overline{\partial}$-closed, there are smooth functions $\xi_{j|1}^\alpha$ on V_j such that

$$\overline{\partial}\xi_{j|1}^\alpha = \psi_1^\alpha \quad \text{on} \quad V_j \tag{1.12}$$

Then

$$f_{jk|1} = \xi_{k|1} - \xi_{j|1} \quad \text{on} \quad V_j \cap V_k \tag{1.13}$$

Let us construct $\xi_{j|\nu}$ on V_j inductively. They should satisfy:

$$\overline{\partial}\xi_{j|\nu}^\alpha = \sum_{\mu=1}^{\nu-1}\sum_{\lambda=1}^{n} \psi_\mu^\lambda \partial_\lambda \xi_{j|\nu-\mu}^\alpha \text{ on } V_j \tag{1.14$_\nu$}$$

Assume that we have already found $\xi_{j|1}, \cdots, \xi_{j|\tau-1}$ on V_j.

Lemma 1.1. We have

$$\overline{\partial}\left(\sum_{\mu=1}^{\tau-1}\sum_{\lambda=1}^{n} \psi_\mu^\lambda \partial_\lambda \xi_{j|\tau-\mu}^\alpha\right) = 0 \text{ on } V_j \tag{1.15}$$

<u>**Proof:**</u> Using $(1.11)_\nu$ and $(1.14)_\nu$ for ν between 2 and $\tau - 1$, we obtain

$$\overline{\partial}\left(\sum_{\mu=1}^{\tau-1}\sum_{\lambda=1}^{n} \psi_\mu^\lambda \partial_\lambda \xi_{j|\tau-\mu}^\alpha\right)$$

$$= \sum_{\mu=1}^{\tau-1}\sum_{\lambda=1}^{n}\left(-\frac{1}{2}\sum_{\beta=1}^{\mu-1}[\psi_\beta, \psi_{\mu-\beta}]^\lambda \partial_\lambda \xi_{j|\tau-\mu}^\alpha\right.$$

$$\left. +\psi_\mu^\lambda \wedge \sum_{\gamma=1}^{\tau-\mu-1}\sum_{\beta=1}^{n} \partial_\lambda(\psi_\gamma^\beta \partial_\beta \xi_{j|\tau-\mu-\gamma})\right) \tag{1.16}$$

$$= -\frac{1}{2}\sum_{\mu=1}^{\tau-1}\sum_{\lambda=1}^{n}\sum_{\beta=1}^{\mu-1}[\psi_\beta, \psi_{\mu-\beta}]^\lambda \partial_\lambda \xi_{j|\tau-\mu}^\alpha$$

$$+ \sum_{\mu=1}^{\tau-1}\sum_{\lambda,\gamma=1}^{n}\sum_{\beta=1}^{\mu-1} \psi_\beta^\lambda \wedge \partial_\lambda \psi_{\mu-\beta}^\gamma \partial_\gamma \xi_{j|\tau-u}$$

$$= 0 \quad \text{(by the definition of } [\psi_\beta, \psi_{\mu-\beta}])$$

Therefore, by the steinness of V_j, we can solve $(1.14)_\tau$ for $\xi_{j|\tau}$. The induction is completed. In fact, if we define a formal power series by

$$\xi_j = \sum_{\nu=0}^{\infty} \xi_{j|\nu} t^\nu \tag{1.17}$$

406   GANG TIAN

where $\xi_{j|0} = z_j$. Then ξ_j is the formal solution of the equation

$$\overline{\partial}\xi_j = \sum_{\lambda=1}^{n} \psi^\lambda \partial_\lambda \xi_j \tag{1.18}$$

where ψ is formally the T_U-valued $(0,1)$-form defined by

$$\psi = \sum_{\nu=1}^{\infty} \psi_\nu t^\nu$$

Next, we define successively the functions $f^0_{jk|\nu}$ on $V_j \cap V_k$ by the formulas:

$$\xi_{j|\nu}(z_k) = \sum_{\beta=1}^{n} \frac{\partial z_j}{\partial z_k^\beta} \xi^\beta_{k|\nu}(z_k) + f_{jk|\nu}(z_k)$$

$$+ \sum_{\mu=0}^{\nu-1} \sum_{l=\max\{1,2-\mu\}}^{\nu-\mu} \sum_{\beta_1+\cdots+\beta_n=1} \frac{1}{\beta_1! \cdots \beta_n!} \frac{\partial^l f_{jk|\mu}}{(\partial z_k^1)^{\beta_1} \cdots (\partial z_k^n)^{\beta_n}}(z_k) \tag{1.19}$$

$$\sum_{\substack{\lambda_1+\cdots+\lambda_1=\nu-\mu \\ \lambda_1,\ldots,\lambda_1>0}} \xi^1_{k|\lambda_1} \cdots \xi^1_{k|\lambda_{\beta_1}} \cdots \xi^n_{k|\lambda_l}(z_k)$$

In fact, the transition functions $f_{jk|\nu}$ are just the formal compositions $\xi_j \circ \xi_k^{-1}$. Therefore, they are holomorphic on the intersection $V_j \cap V_k$ and satisfy $(1.11)_\nu$, that is, there is a formal deformation of U along the direction θ The proposition is proved

2. Local deformation of an ordinary double point

Let z_i $(i = 1, 2, \ldots, n+1)$ be the coordinates of C^{n+1}. Define $\mathcal{A}$ to be the affine variety:

$$\{(z_1, z_2, \ldots, z_{n+1}, t) \in C^{n+1} \times C \mid \sum_{i=1}^{n+1} z_i^2 = t\}$$

There is a natural map $\pi : \mathcal{A} \mapsto C$ mapping (z,t) to t. Then $A_t = \pi^{-1}(t)$ is smooth for $t \neq 0$ and A_0 has only one singular point at the origin. Therefore, the variety $\mathcal{A}$ is a smoothing of the ordinary double point of A_0 at the origin. In this section, we derive some properties of the Kodaira-Spencer map associated with this smoothing.

First, let us define a family of maps ζ_t from $A_0 \setminus \{o\}$ into A_t:

$$\zeta = (\zeta^1, \zeta^2, \ldots, \zeta^{n+1}), \quad \zeta_t(\cdot) = \zeta(\cdot, t),$$
$$\zeta_t^i(z) = z_i + \frac{t\overline{z_i}}{2|z|^2} \qquad i = 1, \ldots, n+1$$

By Kodaira-Spencer deformation theory, there is, at least formally, a unique TA_o-valued $(0,1)$-form $\varphi(z,t)$ on the nonsingular part $U \times C$ of $A_0 \times C$ such that any locally holomorphic function ζ satisfies,

$$\overline{\partial}\zeta = \sum_{j=1}^{n} \varphi^j(z,t)\partial_j\zeta$$

whenever this is defined. Here, the $(0,1)$-forms φ^j $(j = 1, 1, \ldots, n)$ are the components of φ. Put $U_i = \{z_i \neq 0\}$. This φ can be explicitly given on each $U_i \cap A_0$. For example, let us express this on $U_{n+1} \cap A_0$. Replacing ζ by ζ^i $(i = 1, 2, \ldots, n)$. we have

$$\overline{\partial}\zeta^i = \sum_{j=1}^{n} \varphi^j(z,t)\partial_j\zeta^i$$

that is, in local coordinates $(z_1, z_2, \ldots, z_n)$

$$\delta_{ij} - \frac{\overline{z}_i z_j}{|z|^2} + \frac{\overline{z}_i \overline{z}_j z_{n+1}}{\overline{z}_{n+1}|z|^2}$$
$$= \sum_{l=1}^{n} \varphi^l_{\overline{j}}\left(\delta_{il} - \frac{t\overline{z}_i \overline{z}_l}{|z|^2} + \frac{t\overline{z}_i z_l \overline{z}_{n+1}}{z_{n+1}|z|^2}\right)$$

Write $\varphi^i_{\overline{j}} = \sum_{\alpha=1}^{n} \varphi^i_{\alpha\overline{j}}$, then

$$\varphi^i_{1\overline{j}} = \delta_{ij} - \frac{\overline{z}_i z_j}{|z|^2} + \frac{\overline{z}_i \overline{z}_j z_{n+1}}{\overline{z}_{n+1}|z|^2}$$

and for all $\alpha \geq 2$,

$$\varphi^i_{\alpha\overline{j}} = \sum_{l=1}^{n} \varphi^l_{\alpha-1\overline{j}}\left(\frac{\overline{z}_i z_l}{|z|^2} - \frac{\overline{z}_i z_l \overline{z}_{n+1}}{z_{n+1}|z|^2}\right)$$

A straightforward computation shows $\varphi^i_{2\overline{j}} = 0$ on A_0 consequently, all $\varphi^i_{\alpha\overline{j}}$ vanish for $\alpha \geq 2$. It follows

$$\varphi(z,t) = \sum_{i,j=1}^{n} \varphi_j^i \frac{\partial}{\partial z_i} d\bar{z}_j$$

$$= t \sum_{i,j=1}^{n} \frac{1}{2|z|^2} \left(\delta_{ij} - \frac{\bar{z}_i z_j}{|z|^2} + \frac{\bar{z}_i \bar{z}_j z_{n+1}}{\bar{z}_{n+1}|z|^2} \right) \frac{\partial}{\partial z_i} d\bar{z}_j \tag{2.1}$$

$$= t\bar{\partial} \left(\sum_{i=1}^{n} \frac{\bar{z}_i}{2|z|^2} \frac{\partial}{\partial z_i} \right)$$

Furthermore, the Lie bracket $[\varphi,\varphi]$ is identically zero on the nonsingular part of A_0. In general, we have

$$\varphi(z,t) = t\bar{\partial} \left(\sum_{i \neq j} \frac{\bar{z}_i}{2|z|^2} \frac{\partial}{\partial z_i} \right) \quad \text{on } U_j \setminus \{o\} \tag{2.2}$$

There is a nonvanishing holomorphic n-form Ω on the regular part $U = A_0 \setminus \{o\}$. In the open subset U_{n+1}, this is given by

$$\Omega = \frac{dz_1 \wedge dz_2 \wedge \cdots \wedge dz_n}{2z_{n+1}}$$

Let $\varphi \lrcorner \Omega$ be the interior product of the n-form and the deformation form φ, in local coordinates $(z_1, z_2, \ldots, z_n)$.

$$\varphi \lrcorner \Omega$$

$$= \frac{t}{4z_{n+1}|z|^2} \sum_{i,j=1}^{n} (-1)^{i-1} \left(\delta_{ij} - \frac{\bar{z}_i z_j}{|z|^2} + \frac{\bar{z}_i \bar{z}_j z_{n+1}}{\bar{z}_{n+1}|z|^2} \right) dz_1 \wedge \cdots \widehat{dz_i} \cdots \wedge dz_n \wedge d\bar{z}_j$$

$$= t\bar{\partial} \left(\sum_{i=1}^{n} (-1)^{i-1} \frac{\bar{z}_i}{4z_{n+1}|z|^2} dz_1 \wedge \cdots \wedge \widehat{dz_i} \wedge \cdots \wedge dz_n \right) \tag{2.3}$$

Lemma 2.1   The $(n-1,1)$-form $\varphi \lrcorner \Omega$ are both ∂- and $\bar{\partial}$-closed on $A_0 \setminus \{o\}$.

Proof   The $\bar{\partial}$-closedness follows directly from (2.3). An easy computation shows

$$\partial \left(\sum_{i=1}^{n} (-1)^{i-l} \frac{\bar{z}_i}{4z_{n+1}|z|^2} dz_1 \wedge \cdots \wedge \widehat{dz_i} \wedge \cdots \wedge dz_n \right)$$

$$= \sum_{i=1}^{n} \frac{\partial}{\partial z_i} \left(\frac{\bar{z}_i}{4z_{n+1}|z|^2} \right) dz_1 \wedge dz_2 \wedge \cdots \wedge dz_n \tag{2.4}$$

$$= \frac{1}{z_{n+1}^2} \Omega$$

Therefore, using (2,3), we have

$$\partial(\varphi \lrcorner \Omega) = t\overline{\partial}(\frac{1}{z_{n+1}^2}\Omega) = 0 \tag{2.5}$$

The lemma is proved.

Lemma 2.2 The $(n-2,2)$-form $\wedge^2\varphi \lrcorner \Omega$ are both ∂- and $\overline{\partial}$-closed on $A_0 \setminus \{o\}$.

Proof The $\overline{\partial}$-closedness follows from (2.1). The ∂-closedness follows from $[\varphi, \varphi] = 0$ and the identity (cf. Lemma 3.5 in section 3):

$$[\varphi, \varphi] \lrcorner \Omega = \partial(\wedge^2\varphi \lrcorner \Omega),$$

in case $\partial(\varphi \lrcorner \Omega) = 0$.

Remark: By the definition, one has

$$\wedge^2\varphi \lrcorner \Omega = \varphi \lrcorner (\varphi \lrcorner \Omega).$$

Let $\tilde{A}_0$ be the resolution of the singular variety A_0 by blowing up C^{n+1} at the origin and $\pi : \tilde{A}_0 \mapsto A_0$ be the natural projection. Then the exceptional divisor $\pi^{-1}(o)$ is a quadratic hypersurface in CP^n.

Lemma 2.3: The cohomology group $H^{n-2}(\tilde{A}_0, \Omega^1_{\tilde{A}_0})$ vanishes if $n \geq 4$ and is naturally isomorphic to $H^{n-2}(\pi^{-1}(o), \Omega^1_{\pi^{-1}(o)})$ if $n = 3$.

Proof: First we remark that $H^{n-2}(\tilde{A}_0, \Omega^1_{\tilde{A}_0})$ is same as $H^{n-2}(\pi^{-1}(o), \Omega^1_{\tilde{A}_0}|_{\pi^{-1}(o)})$. Put V to be $\pi^{-1}(o)$. Using the negativity of the normal bundle $N_{V|\tilde{A}_0}$, one can easily show that $H^{n-2}(V, \Omega^1_{\tilde{A}_0}|_V)$ is equal to $H^{n-2}(V, \Omega^1_V)$. Then the lemma follows from a computation of the cohomology group of the quadratic hypersurface in CP^n. In fact, one can show that $H^{n-1}(V, R)$ is zero by computing the euler number of V and using Lefschetz Theorem.

We define two (n-1, 2)-currents ρ_1 and ρ_2 as follows: for any C^∞ smooth (1, n-2)-form ω in $\tilde{A}_0$,

$$\rho_1(\omega) = \lim_{\epsilon \to +0} \int_{|z|=\epsilon} (\varphi \lrcorner \Omega) \wedge \omega \tag{2.7}$$

$$\rho_2(\omega) = \lim_{\epsilon \to +0} \int_{|z|=\epsilon} (\wedge^2\varphi \lrcorner \Omega) \wedge \omega \tag{2.8}$$

Formally, the currents ρ_1 and ρ_2 are $\overline{\partial}(\varphi \lrcorner \Omega)$ and $\partial(\wedge^2\varphi \lrcorner \Omega)$, respectively.

Proposition 2.1. For $n \geq 4$. both $\rho_1(\omega)$ and $\rho_2(\omega)$ are zero for any d-closed smooth $(1, n-2)$-form ω on $\tilde{A}_0$.

Proof: It follows from Lemma 2.1, 2.2, 2.3.

In case $n = 3$, the exceptional divisor $\pi^{-1}(o)$ is a rational surface $CP^1 \times CP^1$. Blowing down $\tilde{A}_0$ along its two rulings, we obtain two small smooth resolutions $\tilde{A}_{01}$ and $\tilde{A}_{02}$ of A_0. Let π_1 and π_2 be the two projections. Then both $\pi_i^{-1}(o)$ (i=1, 2) are smooth rational curves.

Proposition 2.2 In case $n = 3$, we have: for any closed smooth $(1,1)$-form ω on $\tilde{A}_{0i}$ (i=1, 2), we have

$$\rho_2(\omega) = 0 \tag{2.9}$$

$$\rho_1(\pi^*(\pi_{i*}\omega)) = c \int_{\pi_i^{-1}(o)} \omega \tag{2.10}$$

where c is some universal constant $\neq 0$, whose exact value is not important to us.

Proof: We may assume that $i = 1$. First we remark that if ω' is another $(1,1)$-form which is cohomologeous to ω, i.e,

$$\omega' = \omega + d_\varsigma$$

then $\rho_j(\omega') = \rho_j(\omega)$ for $j = 1, 2$. Next, let us describe the small resolution $\tilde{A}_{01}$. Introduce a new coordinate system (w_1, w_2, w_3, w_4) by the transformation

$$w_1 = z_1 + \sqrt{-1}z_2, \quad w_2 = z_1 - \sqrt{-1}z_2$$
$$w_3 = z_3 + \sqrt{-1}z_4, \quad w_4 = z_3 - \sqrt{-1}z_4 \tag{2.11}$$

Then the affine variety A_0 is defined by the quadratic polynomial $w_1 w_2 + w_3 w_4$. Without losing the generality, we may assume that $\tilde{A}_{01}$ is the small resolution covered by the following two coordinate charts $\{V_1; (w_1, w_4, \varsigma)\}$ and $\{V_2; (w_2, w_3, \eta)\}$. They are both isomorphic to C^3. The transition function F on $V_1 \cap V_2$ is

$$w_3 = w_1\varsigma; \quad w_2 = -w_4\varsigma; \quad \eta = \varsigma^{-1} \tag{2.12}$$

In particular, $\tilde{A}_{01}$ is the total space of the direct sum of two tautological line bundles on $CP^1 = \pi_1^{-1}(o)$, therefore, ω is cohomological to ω_δ, where $\delta > 0$ and

$$\alpha = \int_{\pi_1^{-1}(0)} \omega$$
$$\omega_\delta = \frac{\sqrt{-1}}{2\pi} \partial\bar{\partial} \log(\delta + |\xi - 1|^2)$$

(2.13)

By the remark at the beginning of this proof, for $j = 1, 2$, we have

$$\rho_j(\omega) = \lim_{\epsilon \to +0} \lim_{\delta \to +0} \int_{|z|=\epsilon} \alpha\,\omega_\delta$$
$$= \alpha \lim_{\epsilon \to +0} \int_{|z|=\epsilon, \xi=1} \wedge^j \varphi \dashv \Omega$$

(2.14)

In the plane $S = V_1 \cap \{\xi = 1\}$, we choose z_1 and z_2 to be the coordinates and have

$$z_4 = -\sqrt{-1}z_1; \quad z_3 = \sqrt{-1}z_2; \quad |z|^2 = 2(|z_1|^2 + |z_2|^2)$$
$$\varphi \dashv \Omega = \frac{\bar{z}_2}{|z_1|^2|z|^4}(|z_1|^2 dz_1 \wedge d\bar{z}_1 \wedge dz_2 - \mathrm{Re}\,(z_1\bar{z}_2)dz_1 \wedge dz_2 \wedge d\bar{z}_2)$$

(2.15)

$$\wedge^2 \varphi \dashv \Omega = \frac{1}{2|z|^4}\left\{\sum_{1 \leq i < j \leq 3} \frac{\partial}{\partial z_i} \wedge \frac{\partial}{\partial z_j}(d\bar{z}_i \wedge d\bar{z}_j - (\bar{z}_j d\bar{z}_i - \bar{z}_i d\bar{z}_j) \wedge \frac{\bar{\partial}|z|^2}{|z|^2})\right\} \dashv \Omega$$
$$= \frac{1}{4z_{n+1}|z|^4}\left\{dz_3 \wedge (d\bar{z}_1 \wedge d\bar{z}_2 - (\bar{z}_2 d\bar{z}_1 - \bar{z}_1 d\bar{z}_2) \wedge \frac{\bar{\partial}|z|^2}{|z|^2}\right.$$
$$+ dz_2 \wedge (\bar{z}_3 d\bar{z}_1 - \bar{z}_1 d\bar{z}_3) \wedge \frac{\bar{\partial}|z|^2}{|z|^2}$$
$$\left. - dz_1 \wedge (\bar{z}_3 d\bar{z}_2 - \bar{z}_2 d\bar{z}_3) \wedge \frac{\bar{\partial}|z|^2}{|z|^2}\right\}$$
$$= \frac{1}{4z_1|z|^6}\{|z_1|^2 d\bar{z}_1 \wedge dz_2 \wedge d\bar{z}_2 + z_1\bar{z}_2 dz_1 \wedge d\bar{z}_1 \wedge d\bar{z}_2\}$$

(2.16)

Then (2.9) and (2.10) follow from (2.15) and (2.16) and an easy computation using the Residue Theorem.

3. The proof of Theorem 0.2

By Proposition 1.3, in order to prove Theorem 0.2, it suffices to solve the sequence of the equations in $(1.11)_\nu$ for $\nu > 1$. Suppose that there is an infinitesimal deformation ψ_1 in $H^1(U, \Theta_U)$, where U is the regular part of X_0 as in the statement of Theorem 0.2. This ψ restricts to a cohomology class in $H^0(T^1_{X_0})$. Let $\pi : X \mapsto X_0$ be the cohomologically Kähler resolution as we assumed. Let Ω be the nonvanishing holomorphic n-form on X_0. At each singular point p_i $(i = 1, \ldots, l)$, there are holomorphic functions $z^i_1, z^i_2, \ldots, z^i_{n+1}$ defined in the neighborhood V_i in X_0 such that they define a holomorphic embedding of V_i into C^{n+1} with properties:

$$V_i = \{(z^i_1)^2 + (z^i_2)^2 + \cdots + (z^i_{n+1})^2 = 0 \text{ in } C^{n+1}; |z^i_\beta| < 1 \text{ for } 1 \le \beta \le n+1\}$$

$$\Omega|_{V_i} = (-1)^{\beta-1} \frac{dz^i_1 \wedge \cdots \wedge \widehat{dz^i_\beta} \wedge \cdots \wedge dz^i_{n+1}}{2z^i_\beta}, \quad \beta = 1, \ldots, n+1 \tag{3.1}$$

The local cohomology group $H^0(T^1_{X_0})$ is generated by Θ_U-valued $(0,1)$-forms φ_i defined in $V_i \setminus \{o\}$ (cf. section 1 and 2), where is the open subset $\{z^i_{n+1} \ne 0\}$,

$$\varphi_i = \overline{\partial} \left(\sum_{j=1}^n \frac{\overline{z}^i_j}{2|z|^2} \frac{\partial}{\partial z^i_j} \right)$$

By adding a coboundary term to ψ_1, we may assume

$$\psi_1|_{V_i} = \lambda_i \varphi_i, \quad \lambda_i \text{ is in } C \tag{3.2}$$

The following lemma can be proved by elementary computations.

Lemma 3.1 Any holomorphic n-form on $\tilde{V}_i$ $(1 \le i \le l)$ is of the form ∂u for some holomorphic $(n-1,0)$-form u, where $\tilde{V}_i = \pi^{-1}(V_i)$ is in X.

Remark: This is actually a special case of a general theorem on the degeneration of the spectral sequence associated to the double complex $C^q(\Omega^p)(\tilde{V}_i)$, where $\tilde{V}_i$ is the pull-back of V_i by π. We refer the readers to [GH], p448-449.

Lemma 3.2 ($\partial\overline{\partial}$-Lemma, [GH], p149). If η is any $\overline{\partial}$-closed (resp. ∂-closed) (p,q)-form on a compact Kähler manifold, and η is ∂-exact (resp ∂-exact), then

$$\eta = \partial\overline{\partial}\gamma$$

for some (p-1,q-1)-form γ.

Define open neighborhoods W_δ of the singular set of X_0 by

$$W_\delta = \{x \in \bigcup_{i=l}^{l} V_i \mid |z_i|(x) < \frac{\delta}{2} \text{ if } x \in V_i\}$$

where $\delta \leq 2$. Then $W_2 = \bigcup_{i=1}^{l} V_i$. Put $\tilde{W}_\delta = \pi^{-1}(W_\delta)$. Without losing the generality, we may assume that W_2 consists of disjoint union of open neighborhoods of p_i $(i = 1, 2, \ldots, l)$.

Lemma 3.3 The cohomology groups $H^{n-1}(\tilde{W}_\delta, \mathcal{O}_{\tilde{W}_\delta})$ and $H^1(\tilde{W}_\delta, \Omega_{\tilde{W}_\delta}^{n-1})$ vanish.

It follows from an elementary computation. We omit the proof.

Lemma 3.4 By a modification, we may assume that $\psi_1 \dashv \Omega$ is also ∂-closed and still coincides with $\varphi_i \dashv \Omega$ in the connected component of W_1 containing p_i $(i = 1, 2, \cdots, l)$.

Proof: By (3.2) and Lemma 2.1, $\pi^*(\partial(\psi_l \dashv \Omega))$ is a d-closed (n,1)-form on X. For any d-closed (0,n-1)-form ω on X, we have

$$\int_X \pi^*(\partial(\psi_1 \dashv \Omega)) \wedge \omega$$
$$= \lim_{\epsilon \to +0} \int_{\partial W_\epsilon} (\psi_1 \dashv \Omega) \wedge \pi_* \omega \tag{3.3}$$

where ∂W_ϵ denotes the boundary of W_ϵ. Now by Lemma 3.3, $\omega = \overline{\partial} u$ on $\tilde{W}_2$ for some (0,n-2)-form u. Then by Lemma 2.1 and Proposition 2.1 and 2.2, we have

$$\int_X \pi^*(\partial(\psi_1 \dashv \Omega)) \wedge \omega$$
$$= \lim_{\epsilon \to +0} \int_{\partial W_\epsilon} (\psi_1 \dashv \Omega) \wedge d\pi_* u \tag{3.4}$$
$$= 0$$

Therefore, $\pi^*(\partial(\psi_1 \dashv \Omega))$ is d-exact on X. Since X is cohomologically Kähler, $\pi^*(\partial(\psi_1 \dashv \Omega))$ is actually $\overline{\partial}$-closed and ∂-exact. By Lemma 3.2, there is a smooth (n-1,0)-form v_1 on X satisfying:

$$\pi^*(\partial(\psi_1 \dashv \Omega)) = \partial\overline{\partial} v_1 \tag{3.5}$$

Since $(\psi_1 \dashv \Omega)$ coincides with $\varphi_i \dashv \Omega$ in V_i, by Lemma 2.1, the (n,0)-form ∂v_1 is $\overline{\partial}$-closed in $\tilde{W}_2$. By Lemma 3.1, there is a holomorphic (n-1,0)-form u_1 on $\tilde{W}_2$ satisfying:

$$\partial v_1 = \partial u_1 \text{ on } \tilde{W}_2 \tag{3.6}$$

By Lemma 3.3, $H^{n-1}(\tilde{W}_2, \mathcal{O}_{\tilde{W}_2})$ is zero, so $v_1 - u_1 = \partial w_1$ for some (n-2,0)-form w_1 on $\tilde{W}_2$. Let η be a smooth cut-off function with is identically one in $W_{3/2}$ and vanishes outside W_2. Now define $v = v_1 - \partial(\eta_1 w_1)$. Then the (n-1,1)-form $(\psi_1 \dashv \Omega) - \overline{\partial} v$ is what we want.

The next lemma provides a crucial identity in solving the equations $(1.11)_\nu$. This lemma was first used in [Ti] to deform Kähler manifolds with vanishing first Chern class, including those Calabi-Yau manifolds.

Lemma 3.5 ([Ti], also [Fr]). Let Y be any n-dimensional complex manifold with a nonvanishing holomorphic n-form Ω_Y. Then for any two Θ_Y-valued $(0,1)$-forms u_1 and u_2, we have identity:

$$[v_1, v_2] \dashv \Omega_Y = \partial((v_1 \wedge v_2) \dashv \Omega_Y)$$
$$+ \partial(v_1 \dashv \Omega_Y) \wedge \#\partial(v_2 \dashv \Omega_Y) - \#\partial(v_1 \dashv \Omega_Y) \wedge \partial(v_2 \dashv \Omega_Y) \tag{3.7}$$

where $\#$ is the obvious map identifying (n,q)-forms with $(0,q)$-forms using the nonvanishing holomorphic n-form Ω_Y, i.e., $\#\ (v \wedge \Omega_Y) = v$.

We are ready to solve the equations $(1.11)_\nu$. We will do it by induction on ν. Suppose that we already found $\psi_1, \ldots, \psi_{\nu-1}$ satisfying:

(1) They satisfy the equations $(1.11)_1, \ldots, (1.11)_{\nu-1}$;

(2) The support of ψ_λ is outside $W_{\frac{1}{2^{\lambda-1}}}$ for λ between 1 and $\nu - 1$.

(3) they also satisfy: $\partial(\psi_\lambda \dashv \Omega) = 0$ for λ from 1 to $\nu - 1$.

The equation $(1.11)_\nu$ is solvable if the well-defined right-handed side in $(1.11)_\nu$ can be pulled back to be ∂-exact on the resolution X. If ν is equal to 2, then by Lemma 3.5, the right-handed side in $(1.11)_2$ can be written as $\partial((\wedge^2\psi_1) \dashv \Omega)$. This is ∂-exact by using Proposition 2.2 and Lemma 2.2. If ν is greater than 2, by the property (3) and Lemma 3.5, the right-handed side in $(1.11)_\nu$ can be written as $\partial\varphi_\nu$, where

$$\varphi_\nu = \sum_{\lambda=1}^{\nu-1}(\psi_\lambda \wedge \psi_{\nu-\lambda}) \dashv \Omega \tag{3.8}$$

If $\nu > 2$, the (n-2,2)-form φ_ν is identically zero in $W_{\frac{1}{2^{\nu-2}}}$ by using the property (2) in the assumption of the induction, so it pulls back a smooth form on X and the right-handed side of $(1.11)_\nu$ is ∂-exact. Therefore, by the $\partial\overline{\partial}$-Lemma, there is a solution $\tilde{\psi}_\nu$ of $(1.13)_\nu$ satisfying

(1) and (3) stated above. Since $\overline{\partial}(\tilde{\psi}_\nu \dashv \Omega)$ is zero in $W_{\frac{1}{2^{\nu-2}}}$, by Lemma 3.3, there is a smooth $(n\text{-}1,0)$-form u in $W_{\frac{1}{2^{\nu-2}}}$ such that

$$\tilde{\psi}_\nu \dashv \Omega = \overline{\partial}u \quad \text{in} \quad W_{\frac{1}{2^{\nu-2}}}$$

It follows $\overline{\partial}(\partial u) = 0$ in $W_{\frac{1}{2^{\nu-2}}}$. Now ∂u is a $(n,0)$-form, so by Lemma 3.1, there is a holomorphic $(n\text{-}1,0)$-form v with $\partial u = \partial v$. Consequently, since $H^{n-1}(\tilde{W}_\delta, \mathcal{O}_{\tilde{W}_\delta})$ vanishes, $u - v = \partial w$ for some $(n\text{-}2,0)$-form U. Let η_ν be a cut-off function satisfying: $\eta_\nu = 1$ in $W_{\frac{1}{2^{\nu-1}}}$ and $\eta_\nu = 0$ in $W_{\frac{1}{2^{\nu-2}}}$. Then $\psi_\nu = \tilde{\psi}_\nu - \overline{\partial}\partial(\eta_\nu w)$ is the solution of $(1.11)_\nu$ with the properties (1), (2) and (3).

The induction is completed. So Theorem 0.2 is proved.

Remark: In fact, one should be able to generalize the proof of Theorem 0.2 to deforming singular varieties with trivial dualizing sheaf and more complicated singularities.

4. The proof of Theorem 0.1

In this section, we apply Theorem 0.2 to prove Theorem 0.1. Now the dimension of X_0 is 3 and X is a small resolution of X_0. First we recall a result of R. Friedman. Let U be the regular part of X_0 as before, then $H^2(X_0, T^0_{X_0})$ can be identified with $H^2(U, T^0_U)$ in the natural way. On the other hand, this U can be regarded as a Zariski open subset in the small resolution X. The complement of U in X consists of l smooth rational curves. It follows that $H^2(X, \Theta_X) = H^2(U, T^0_U)$, where Θ_X is the tangent sheaf of X. Therefore, $H^2(X_0, T^0_{X_0}) = H^2(X, \Theta_X)$.

Proposition 4.1 ([Fr]. Proposition 8.7). With hypotheses as above, suppose in addition that the canonical line bundle of X is trivial. Then $H^2(X_0, T^0_{X_0})$ is equal to $H^2(X, \Omega^2_X)$ and the natural map from $H^0(T^1_{X_0}) = \oplus^l_{i=1} C p_i$ to $H^2(X, \Omega^2_X)$ can be identified with the map ρ which sends the basis element $[pi]$ to the fundamental class of C_i in $H^2(X, \Omega^2_X)$. Consequently, there is a first order deformation of X_0 which smooths the singular points p_i $(i = 1, \ldots, l)$ to first order if fundamental classes $[C_i]$ in $H^2(X, \Omega^2_X)$ satisfy a relation $\sum_i \lambda_i[C_i] = 0$ with $\lambda_i \neq 0$ for every i.

Remark: The obstruction group $T^2_{X_0}$ to smoothing is isomorphic to the cokernel of ρ (cf. [Fr]).

Let us sketch an elementary proof of this proposition. An element in $H^0(T^1_{X_0})$ can be represented by a $(2,1)$-form $\sum_{i=1}^l \lambda_i \varphi_i$ defined on $W_2 \backslash \mathrm{Sing}(X_0)$, where W is the disjoint union of stein neighborhoods of p_i $(i = 1, \dots, l)$ defined as in section 3, and for each i, φ_i is defined as follows (cf. section 2): we can identify W at the singular point p_i with an open neighborhood of the origin in the affine variety defined by the polynomial $z_1^2 + z_2^2 + z_3^2 + z_4^2 = 0$ in C^4. Then in the open subset $\{z_j \neq 0\}$, where $j = 1, \dots, 4$,

$$\varphi_i = \overline{\partial}\left(\sum_{k \neq j} \frac{\overline{z}_k}{|z|^2} \frac{\partial}{\partial z_k} \right) \tag{4.1}$$

Let η be a cut-off function such that $\eta(|z|)$ is identically 1 in W_1 and 0 outside W_2. define a $(2,2)$-form ψ on X by

$$\psi = \lambda_i \overline{\partial}(\eta \varphi_i \dashv \Omega)$$

on the component of $\tilde{W}_2$ containing $\pi^{-1}(p_i)$. Note that it is well defined since by Lemma 2.1, $\varphi_i \dashv \Omega$ vanishes near p_i. This $(2,2)$-form supports in the exceptional curve C_i, i.e.,

$$\overline{\partial}\psi(f) = c \sum_{i=1}^l \lambda_i \int_{C_i} f = 0 \tag{4.2}$$

where c is some constant and f is any d-closed smooth $(1,1)$-form on X. This implies that there is a smooth $(2,1)$-form u satisfying $\psi = \overline{\partial}u$ on X. Using Lemma 2.1, $\overline{\partial}u$ vanishes in $\tilde{W}_1$, that is, u is in $H^1(\tilde{W}_1, \Omega^2_{\tilde{W}_1})$. However, this last group is trivial by an easy computation. This enables us to modify u such that it vanishes in $\tilde{W}_{1/2}$. Therefore, we obtain a $\overline{\partial}$-closed $(2,1)$-form which is $(\eta \varphi_i \dashv \Omega) - u$ in the neighborhood of p_i for each i. This $(2,1)$-form coincides with $\varphi_i \dashv \Omega$ near p_i for each i. On the other hand, the homomorphism from $H^1(U, \Theta_U)$ into $H^0(T^1_{X_0})$ is simply the restriction map. So the proposition is proved.

Now Theorem 0.1 follows from Theorem 0.2 and Proposition 4.1.

Theorem 0.1 can be used to construct new complex manifolds from a given one. Namely, if the 3-dimensional Kähler manifold X has disjoint smooth rational curves of type $(-1, -1)$, then we can contract those rational curves to obtain a singular threefold X_0 with a few ordinary double points as singularities. By Theorem 0.1, under certain conditions on the configuration of the rational curves, this X_0 can be smoothed to new complex manifolds denoted by X_t, where $t \neq 0$. Such an idea of constructing new complex manifolds was due to Clemens. Topologically, X_t is obtained by cutting out some copies of $D^4 \times S^2$ from X and filling in copies of $S^3 \times D^3$. For reader's convenience, we include here a proposition relating the geometry and topology of X_t to those of X.

Proposition 4.2 Let X and X_0 be given as in Theorem 0.1, and X_t be the smoothing of X_0. Then

(1) The Betti numbers of X_t are: $b_2(X_t) = \mathrm{rk}(\mathrm{coker}\,(\rho))$, $b_3(X_t) = b_3(X) + 2\,\mathrm{rk}(\ker\,(\rho))$, where ρ is the map defined in Theorem 0.1;

(2) if X_0 is projective and $h^{0,2}(X) = 0$, then X_t is also projective;

(3) if $C_1(X) = 0$ and $h^{0,1}(X) = 0$, then $C_1(X_t) = 0$, i.e., X_t is Calabi-Yau.

5. Asympototic behavior of Peterson-Weil metric

Let X_0 be the singular threefold with l ordinary double points as only singularities $p_1, \ldots, p_l$ as in Theorem 0.1. By Theorem 0.1, one can construct a local universal family $\pi : \mathcal{X} \mapsto B$, where B is a neighborhood of the origin in $H^1(U, \Theta_U)$, satisfying (1) $\mathcal{X}$ is smooth; (2) $\pi^{-1}(o) = X_0$. By Proposition 1.1 and 4.1, there is an exact sequence:

$$0 \longrightarrow H^1(X_0, T^0_{X_0}) \longrightarrow \mathbf{T}^1_{X_0} \longrightarrow \ker(\rho) \longrightarrow 0$$

where ρ was defined in Proposition 4.1. The subgroup $H^1(X_0, T^0_{X_0})$ classifies the deformations of X_0 preserving all singular points. Let $\{\sum_{i=1}^{l} \alpha_{ji} p_i\}_{1 \leq j \leq k}$ be a basis of $\ker(\rho)$, where α_{ji} are complex numbers and k is the rank of the kernel of ρ. Therefore, there is a coordinate system $(t_1, \ldots, t_k, s_1, \ldots, s_m)$, where m is the rank of $H^1(X_0, T^0_{X_0})$, such that $B \cap H^1(X_0, T^0_{X_0})$ has coordinates $s_1, \ldots, s_m$. By Theorem 0.1, we further have that for any (t, s) in B, $X_{(t,s)} = \pi^{-1}(t, s)$ is the deformation of X_0 smoothing those singular points p_i with $\sum_{j=1}^{k} t_j \alpha_{ji} \neq 0$ and preserving the others. In particular, the Zariski open subset

$$B_0 = \{(t,s) \mid \sum_{j=1}^{k} t_j \alpha_{ji} \neq 0 \text{ for all } i\}$$

is the set of all local deformations in B smoothing X_0.

There is a natural Kähler metric on B_0, that is, the Peterson-Weil metric G_{PW}. In case of Calabi-Yau manifolds, this metric can be defined as follows (cf. [Ti]): the tangent space $T_{(t,s)}B$ at (t,s) in B_0 can be naturally identified with $H^2(X_{(t,s)}, \Omega^1_{X_{(t,s)}})$, then for any d-closed $(2,1)$-forms v and w in $H^2(X_{(t,s)}, \Omega^1_{X_{(t,s)}})$,

$$G_{PW}(\mu, w) = \frac{\sqrt{-1}}{2\pi} \int_{X_{(t,s)}} v \wedge \overline{w} \tag{5.1}$$

Theorem 5.1 By shrinking B and scaling the coordinates $(t_1, \ldots, t_k, s_1, \ldots, s_m)$ on B if necessary, we have the formula for the Kähler form of G_{PW}

$$\omega_{PW} = -\frac{\sqrt{-1}}{2\pi}\partial\overline{\partial}\left(\log\left(-\sum_{i=1}^{l}(\frac{\sqrt{-1}}{2\pi})^3 \int_{\substack{|w_1|,|w_2|,|w_3|<1 \\ |\sum_{j=1}^{3} w_j - \sum_{j=1}^{4} t_j \alpha_{j,i}|<|}} \frac{dw_1 \wedge dw_2 \wedge dw_3 \wedge d\overline{w}_1 \wedge d\overline{w}_2 \wedge d\overline{w}_3}{|w_1||w_2||w_3||\sum_{j=1}^{k} t_j\alpha_{ji} - w_1 - w_2 - w_3|} + f\right)\right) \tag{5.2}$$

where f is a positive C^∞-smooth function on B.

Recall that a smoothing X_0 is a smooth Kähler manifold Y of dimension 4 with a holomorphic map $\pi : Y \mapsto \Delta$, where Δ is the unit disc in the complex plane, such that the fibers are smooth except the central one and the central one is biholomorphic to X_0.

Corollary 5.1 Let $\pi : Y \mapsto \Delta$ be a smoothing of X_0 as above. Then the restriction of the Peterson-Weil metric to the punctured disc Δ^* is asympototically given by

$$\omega = \frac{c\sqrt{-1}}{2\pi}(-\log|z| + O(1))dz \wedge d\overline{z} \tag{5.3}$$

where c is a positive constant and $O(1)$ denotes a bounded function.

This follows from an evaluation of the integrals in (5.2). We omit the proof.

In the rest of this section, we sketch a proof of Theorem 5.1. First let us construct the Kähler potential of G_{PW} near the origin of B by using one result from [Ti]. Let Ω_0 be a nonvanishing holomorphic 3-form on X_0. This n-form can be extended to be a holomorphic n-from on the local universal deformation space $\mathcal{X}$. Denote this form by Ω. Then by shrinking B if necessary, we may assume that the restriction of Ω to each fiber is nonzero at smooth points.

Lemma 5.1([Ti], Theorem 2). The Kähler form of the Peterson-Weil metric is given by

$$\omega_{PW} = -\frac{\sqrt{-1}}{2\pi}\partial\overline{\partial}\left(\log\left(-(\frac{\sqrt{-1}}{2\pi})^3\int_{X_{(t,s)}}\Omega\wedge\overline{\Omega}\right)\right) \tag{5.4}$$

At each singular point p_i of X_0, one can choose a local coordinate chart V_i of $\mathcal{X}$ with coordinate functions $z_1^i,\cdots,z_{m+k+3}^i$ satisfying:

(1) $V_i = \{|z_j^i| < \epsilon \mid j = 1,2,\ldots,m+k+3\}$, where ϵ is a uniform small constant > 0;

(2) for each (t,s) in B,

$$X_{(t,s)}\cap V_i = \{z\in V_i \mid \sum_{j=1}^{4}(z_j^i)^2 = \sum_{j=1}^{k}t_j\alpha_{ji}\} \tag{5.5}$$

Moreover, we may assume

$$\Omega|_{X_{(t,s)}\cap V_i} = \frac{dz_1^i\wedge dz_2^i\wedge z_3^i}{2z_4^i}\ \text{ whenever } z_4^i\neq 0 \tag{5.6}$$

The exponential of the Kähler potential is

$$\int_{X_{(t,s)}}\Omega\wedge\overline{\Omega}$$

$$=\sum_{i=1}^{l}\left(\int_{V_i\cap X_{(t,s)}}\Omega\wedge\overline{\Omega}\right) + \int_{(\cup_{i=1}^{l}V_i)\cap X_{(t,s)}}\Omega\wedge\overline{\Omega}$$

$$=\sum_{i=1}^{l}\int_{\substack{|z_j^i| < \epsilon \\ j=1,2,3,4}}\frac{dz_1^i\wedge dz_2^i\wedge dz_3^i\wedge d\overline{z}_1^i\wedge d\overline{z}_2^i\wedge d\overline{z}_3^i}{4|z_4^i|} + \int_{(\cup_{i=1}^{l}V_i)\cap X_{(t,s)}}\Omega\wedge\overline{\Omega} \tag{5.7}$$

$$=\sum_{i=1}^{l}\int_{\substack{|z_1^i|<\epsilon,|z_2^i|<\epsilon,|z_3^i|<\epsilon \\ |\sum_{j=1}^{k}t_j\alpha_{ji}-\sum_{j=1}^{3}(z_j^i)^2|<\epsilon}}\frac{dz_j^i\wedge dz_2^i\wedge dz_3^i\wedge d\overline{z}_1^i\wedge d\overline{z}_2^i\wedge d\overline{z}_3^i}{4|\sum_{j-1}^{k}t_j\alpha_{ji}-(z_1^i)^2-(z_2^i)^2-(z_3^i)^2|} + (\frac{\pi}{2\sqrt{-1}})^3 f$$

where f is defined in the obvious way. Since the n-form Ω is smooth in the complement of $\cup_{i=1}^{l}V_i$ in $\mathcal{X}$, the function f is clearly positive and C^∞-smooth in B.

Now Theorem 5.1 follows from (5.4), (5.7) and the change of variable.

Reference:

[Fk] 0. Froster, K. Knorr, *Konstruktion verseller Familien kompakter komplexer Räume*, Lec. in Math., vol. 705, Springer-Verlag.

[Fr] R. Friedman, *On threefolds with trivial canonical bundle*, Preprint.

[GH] P. Griffiths, J. Harris, *Principle of Algebraic Geometry*, John Wiley and Sons, New York. 1978.

[Gr] H. Grauert, *Der Satz von Kuranishi für kompskte komplexe Räume*, Inv. Math., vol. 25 (1974) 107-125.

[KS] Kodaira, D. Spencer, *On deformations of complex analytic structures*, I-I, III Annals of Math., vol. 67 (1958) 328-466, vol. 71 (1960) 43-76.

[Sc] M. Schlessinger, *Rigidity of quotient singularities*, Inv. Math., vol. 14 (1971) 17-26.

[Ti] G. Tian, *Smoothness of the universal deformation space of compact Calabi-Yau manifolds and its Peterson-Weil metric*, in "Mathematical Aspects of String Theory" (S.T. Yau, ed.), World Scientific, Singapore, 1987, pp629-626.

AMS/IP Studies in Advanced Mathematics
Volume **9**, 1998

Modified Calabi-Yau Manifolds With Torsion

Martin Roček[1]

School of Natural Sciences

Institute for Advanced Studies

Olden Lane, Princeton NJ 08540[2]

Abstract

I discuss a natural modification of Calabi-Yau geometries that leads to complex local product spaces with a pair of torsionful connections, and present the analog of the Monge-Ampère equation. I also briefly review the superspace formalism that leads to these results.

---

[1] Work supported in part by NSF grant No. PHY 89-08495, Permanent Address: Institute for Theoretical Physics, State University of New York at Stony Brook, Stony Brook, NY 11794-3840. Email: rocek@dirac.physics.sunysb.edu

[2] Email: rocek@iassns.bitnet

© 1998 American Mathematical Society
and International Press

1. Introduction and Results

From a string theoretic viewpoint, Calabi-Yau manifolds, *i.e.*, Ricci-flat Kähler manifolds, are a special class of approximate solutions to the classical string equations of motion. In this brief contribution, I will try to explain what I mean by "special class", and point out another special class on an essentially *equal* footing. These manifolds are bihermitian manifolds admitting a natural connection with torsion, and may be candidates for the mirror partners of Calabi-Yau spaces that lack deformations of the complex structure. The results here are taken largely from three papers [1, 2, 3].

Before describing the derivation of these results, I briefly summarize them. The new manifolds have two commuting integrable complex structures $[J^+, J^-] = 0$ and a metric g that is hermitian with respect to both. The corresponding symplectic forms $\omega^\pm$ obey $\partial^\pm \bar\partial^\pm \omega^\pm = 0$, where $\partial^\pm$, $\bar\partial^\pm$ are the ∂, $\bar\partial$ operators formed with respect to $J^\pm$. Furthermore, the manifolds admit two connections with torsion: $\Gamma^\pm = \Gamma^0 \pm T$. Here Γ^0 is the Levi-Civita connection of g, and the torsion tensor T with all indices lowered is totally antisymmetric. When T is regarded as a 3-form, it is closed. These connections preserve $J^\pm$, respectively: $\nabla^\pm J^\pm = 0$. The conditions that $J^\pm$ are integrable and are preserved by $\Gamma^\pm$ imply that the 3-form T can be written as $T = \pm \frac{1}{2} \hat{J}^\pm d\omega^\pm$, where $\hat{J}$ is the obvious tensor product action of J on the antisymmetric tensor product of the cotangent bundle $\wedge^3 T^*$ of the manifold. Many identities satisfied by the torsion T can be found in [1] (but note that T in [1] is $-T$ here).

It follows that the manifolds admit a *local product structure* $P = -J^+ J^-$: A local product structure is a tensor that squares to the identity, $P^2 = 1$, and is integrable. A compatible metric is only a product at a point, and will be a product globally only if P is covariantly constant; an example of a local product manifold is any bundle, but generic local product manifolds need not be bundles.

As shown in [1], the conditions on $J^\pm, g$, T, and $\Gamma^\pm$ imply that the metric can be expressed (locally) in terms of a single real Kähler-like potential K as follows:

$$ds^2 = K_{i\bar{j}} du^i d\bar{u}^j - K_{a\bar{b}} dv^a d\bar{v}^b \tag{1}$$

where u, v are complex coordinates defined with respect to the local product structure

$$J^{\pm} du = i du, \quad J^{\pm} dv = \pm i dv \quad \Longrightarrow \quad P du = du, \quad P dv = -dv \tag{2}$$

and $K_i = \frac{\partial}{\partial u^i} K$, $K_a = \frac{\partial}{\partial v^a} K$, etc. The potential K is defined modulo any function $f(u, v) +$ $g(u, \bar{v}) + c.c.$ The submanifolds defined by constant u or v are Kähler. The torsion is also expressed in terms of the potential K:

$$T_{i\bar{j}a} = -\frac{1}{2} K_{i\bar{j}a}, \quad T_{i\bar{j}\bar{a}} = \frac{1}{2} K_{i\bar{j}\bar{a}}, \quad T_{a\bar{b}\bar{i}} = -\frac{1}{2} K_{a\bar{b}\bar{i}}, \quad T_{a\bar{b}i} = \frac{1}{2} K_{a\bar{b}i}, \tag{3}$$

all other non-vanishing components defined by antisymmetry.

The final condition that the metric satisfies is [4]

$$R^{+}_{\mu\nu} = \nabla^{+}_{\mu} \partial_{\nu} \Phi \tag{4}$$

where Φ is some scalar (at least locally, i.e., $d\Phi$ is globally defined), $R^{+}_{\mu\nu}$ is the Ricci tensor for the connection Γ^{+} with torsion, and $\{\mu\} = \{i, \bar{i}, a, \bar{a}\}$. This can be rewritten explicitly in terms of the Levi-Civita connection Γ^{0} and the torsioin as

$$R^{0}_{\mu\nu} + T^{\sigma}_{\mu\rho} T^{\rho}_{\nu\sigma} = \nabla^{0}_{\mu} \partial_{\nu} \Phi, \qquad \nabla_{\rho} T^{\rho}_{\mu\nu} = T^{\rho}_{\mu\nu} \partial_{\rho} \Phi,$$

and hence the symmetric part of eq. (4) is even in the torsion T, whereas the antisymmetric part is odd in T; thus (4) is preserved by $\Gamma^{+} \rightarrow \Gamma^{-}$. For the particular geometry described her, $d\Phi$ can be written covariantly as the contraction $d\Phi = -\frac{1}{2}\iota((\omega^{\pm})^{-1} d\omega^{\pm})$, where $(\omega^{\pm})^{-1}$ is the inverse of the symplectic form associated to $J^{\pm}$, $d\Phi$ can also be expressed in terms of the contraction of the 3-form $T : d\Phi = \pm J^{\pm}\iota((\omega^{\pm})^{-1} T)$. Explictly, this is $\partial_{\tau}\Phi = \mp g^{\sigma\nu} J^{\pm\mu}{}_{\sigma} J^{\pm\rho}{}_{\tau} T_{\mu\nu\rho}$.

In the adapted coordinates defined by (1), (2), eq. (4) can be integrated to [2,3]

$$det(K_{i\bar{j}}) - det(-K_{a\bar{b}}) = 0 \tag{5}$$

with

$$\Phi = -\ln det(K_{i\bar{j}}). \tag{6}$$

Eq. (5) is analogous to the Monge-Ampère equation in the usual Calabi-Yau case.

At least one nontrivial compact example exists [5]: the Hopf manifold ($S_3 \times S_1$) with the potential

$$K = -\int^{\frac{v\bar{v}}{u\bar{u}}} dx \frac{\ln(1+x)}{x} + \ln u \ln \bar{u} \tag{7}$$

leading to the metric $ds^2 = (dud\bar{u}+dvd\bar{v})/(u\bar{u}+v\bar{v})$. In this case, (5) reduces to (flat) Laplace's equation. For the case of complex 3-folds, (5) reduces to

$$det(K_{i\bar{j}}) + K_{v\bar{v}} = 0 \quad \text{for } i = 1,2 \quad \text{or} \quad det(K_{a\bar{b}}) - K_{u\bar{u}} = 0 \quad \text{for } a = 1,2. \tag{8}$$

It would be extremely interesting to know how many manifolds of this sort exist; a rather trivial compact example can be constructed from the Hopf manifold $\times$ the complex torus (note that this example is not completely trivial, as the local product structure P is distinct from the obvious (global) product structure).

2. Yet Another Quick Introduction to Superspace

Having described the main results, I will now try to explain where they come from. This will take me on an excursion through various ideas familiar to string theorists, but often mysterious to mathematicians. One of many attempts at bridging the barrier is described in [6], and a quick glance there is suggested.

One approach to classical string theory begins with the generalized harmonic map problem, *i.e.*, maps from $\mathbb{R}^{1,1}$ (the "worldsheet") into some target manifold that extremize some energy functional (called the action by physicists). One then attempts to construct a conformal field theory based on that action. Proceeding perturbatively, one finds conditions for this. In particular, for an action

$$S = \frac{1}{2} \int d^2z(g_{\mu\nu} + b_{\mu\nu})\partial_+ X^\mu \partial_- X^\nu, \quad \partial_\pm = \frac{\partial}{\partial z^\pm}, \tag{9}$$

the first nontrivial condition is eq. (4) (see [4] and references therein). Here $b_{\mu\nu}$ is a (local) potential for the torsion: $T_{\mu\nu\rho} = \frac{3}{2}b_{[\mu\nu,\rho]}$, and $z^\pm$ are null coordinates on the world sheet.

Supersymmetric strings are described similarly, but have anticommuting (*i.e.*, fermionic) quantities in addition to the coordinates X on the target space. In one formulation, (the NSR string), these fermions are most easily included by adding anticommuting "coordinates" to the two usual coordinates of the worldsheet. This augmented minkowski plane is *superspace*; different superspaces arise depending on the number and types of fermionic coordinates. Remarkably, as was first noted in [7], and further studied to [8] and [1], different superspaces (*i.e.*, properties of the worldsheet) are related to different kinds of target spaces. *From this point of view, Calabi-Yau manifolds arise from one representation of supersymmetry, while the modified geometries described above arise from another; both appear to be equally natural.*

The usual starting point for discussions of supersymmetry is superalgebra. This is an extension of the usual Poincaré algebra (here in two dimensions) by a number of anticommuting spinor charges. For the examples that I want to discuss, I consider what is called $N = 2$ or $(2,2)$ supersymmetry. That is, in addition to the translation generators $P_\pm$ and the rotation generator L, I add a complex left-handed anticommuting spinor charge Q_+ and a complex right-handed spinor charge Q_-. The charges obey the algebra;

$$\{Q_\pm, \bar{Q}_\pm\} = \pm P_\pm, \quad \{Q_\pm, \bar{Q}_\mp\} = \{Q, Q\} = \{\bar{Q}, \bar{Q}\} = 0. \tag{10}$$

Just as the Poncaré algebra can be represented by translations and rotations of coordinates in the plane, the super-Poincaré algebra can be represented by formal "motions" of the super-plane. This is described by, in addition to the usual coordinates $z^\pm$, a complex left-handed anticommuting spinor coordinate θ^+ and a complex right-handed spinor coordinate θ^-. The hermitian conjugates of these are $(\theta^\pm)^\dagger = \bar{\theta}^\pm$. These spinor coordinates can be used to define derivatives $D_\pm, \bar{D}_\pm$

$$D_\pm = \frac{\partial}{\partial\theta^\pm} \pm \frac{i}{2}\bar{\theta}^\pm\partial_\pm, \quad \bar{D}_\pm = \frac{\partial}{\partial\bar{\theta}^\pm} \pm \frac{i}{2}\theta^\pm\partial_\pm, \tag{11}$$

which (anti)commute with the generators of the superalgebra and obey an isomorphic algebra:

$$\{D_\pm, \bar{D}_\pm\} = \pm i\partial_\pm, \quad \{D_\pm, \bar{D}_\mp\} = \{D, D\} = \{\bar{D}, \bar{D}\} = 0. \tag{12}$$

Just as the simplest representation of the Poincaré algebra is a function of the coordinates into the real numbers, *i.e.*, a scalar field, the simplest irreducible representations of $N = 2$

supersymmetry are "superfields", or formal function of the coordinates of superspace. Here I am interested in two representations given by constrained complex scalar superfields [1,6]: chiral scalar superfields Φ and twisted chiral superfields Λ. They obey

$$\bar{D}_\pm \Phi = D_\pm \bar{\Phi} = 0, \quad \bar{D}_+ \Lambda = D_- \Lambda = D_+ \bar{\Lambda} = \bar{D}_- \bar{\Lambda} = 0. \tag{13}$$

These superfields may be formally expanded in the odd coordinates; we shall see that the θ-independent components of the superfields can be identified with the complex coordinates of the first section:

$$\Phi|_{\theta=\bar{\theta}=0} = u, \quad \Lambda|_{\theta=\bar{\theta}=0} = v. \tag{14}$$

Expanding to first order, we find anticommuting spinors:

$$D_\pm \Phi|_{\theta=\bar{\theta}=0} = \psi_\pm, \quad D_+ \Lambda|_{\theta=\bar{\theta}=0} = \chi_+, \quad \bar{D}_- \Lambda|_{\theta=\bar{\theta}=0} = \chi_-, \quad \text{etc.} \tag{15}$$

Supersymmetry transformations (10) act formal rotations of the even components (14) and the old components (15). The real part of the generators Q give one left-handed and one right-handed supersymmetry with a standard form:

$$Re(Q_\pm)\begin{pmatrix} u \\ v \end{pmatrix} = \begin{pmatrix} \psi_\pm \\ \chi_\pm \end{pmatrix}. \tag{16}$$

The second supersymmetry involves precisely the complex structures $J^\pm$; this is the first sign of how worldsheet properties link to target space properties:

$$Im(Q_\pm)\begin{pmatrix} u \\ v \end{pmatrix} = -iJ^\pm \begin{pmatrix} \psi_\pm \\ \chi_\pm \end{pmatrix}. \tag{17}$$

The formal superspace action, whose purely bosonic part reduces to (9), is written as

$$S = \int d^2z\, D_+ D_- \bar{D}_+ \bar{D}_- K(\Phi, \bar{\Phi}, \Lambda, \bar{\Lambda}). \tag{18}$$

One can add to K any function $f(\Phi,\Lambda) + g(\Phi,\bar{\Lambda}) + c.c$; then because of the constraints (13) and the algebra (12), the action does not change ($D_+ D_- \bar{D}_+ \bar{D}_- (f + g + c.c.)$ is a total derivative on the world sheet). Imposing the requirement that to leading order this action gives rise to a conformal field theory, one finds precisely the condition (5). If K has no nontrivial dependence on the twisted chiral superfields Λ^a, then the resulting geometry is Kähler; on the other hand, if K depends on both chiral *and* twisted chiral superfields, then one finds the modified geometry described in the introduction. In this sense, both cases are on equal footing.

To conclude, I will sketch the computation of the component action from (18). This calculation is well known, and is included to give a flavor of manipulations in superspace, as well as to derive eqs. (1) and (3). The basic idea is to simply perform the spinor differentiations in (18), and use the constraints (13) and the algebra (12) whenever possible. One could proceed directly, but it is illuminating to first go from $N = 2$ superspace to $N = 1$ superspace. Introducing *real* spinor derivatives

$$\mathcal{D}_\pm = D_\pm + \bar{D}_\pm, \quad \mathbb{D}_\pm = i(\bar{D}_\pm - D_\pm) \quad \Rightarrow \quad \{\mathcal{D}, \mathbb{D}\} = 0, \quad \mathcal{D}_\pm^2 = \mathbb{D}_\pm^2 = \pm i\partial_\pm \tag{19}$$

the constraints (13) become

$$\mathbb{D}_\pm \Phi = -i\mathcal{D}_\pm \Phi, \quad \mathbb{D}_\pm \Lambda = \mp i\mathcal{D}_\pm \Lambda, \tag{20}$$

and, up to total derivatives on the worldsheet, the action (18) becomes

$$S = -\frac{1}{4} \int d^2z\, \mathcal{D}_+ \mathcal{D}_- \mathbb{D}_+ \mathbb{D}_- K. \tag{21}$$

Performing the $\mathbb{D}$ differentiations and using (20), one finds

$$\begin{aligned}
S = &-\frac{i}{4} \int d^2z\, \mathcal{D}_+ \mathcal{D}_- \mathbb{D}_+ [-K_i \mathcal{D}_- \Phi^i + K_{\bar{i}} \mathcal{D}_- \bar{\Phi}^i + K_a \mathcal{D}_- \Lambda^a - K_{\bar{a}} \mathcal{D}_- \bar{\Lambda}^a] \\
= &-\frac{1}{4} \int d^2z\, \mathcal{D}_+ \mathcal{D}_- [-K_i \mathcal{D}_+ \mathcal{D}_- \Phi^i - K_{\bar{i}} \mathcal{D}_+ \mathcal{D}_- \bar{\Phi}^i + K_a \mathcal{D}_+ \mathcal{D}_- \Lambda^a + K_{\bar{a}} \mathcal{D}_+ \mathcal{D}_- \bar{\Lambda}^a \\
&- K_{ij} \mathcal{D}_+ \Phi^i \mathcal{D}_- \Phi^j + K_{i\bar{j}} (\mathcal{D}_+ \Phi^i \mathcal{D}_- \bar{\Phi}^j + \mathcal{D}_+ \bar{\Phi}^j \mathcal{D}_- \Phi^i) \\
&- K_{\bar{i}\bar{j}} \mathcal{D}_+ \bar{\Phi}^i \mathcal{D}_- \bar{\Phi}^j + K_{ia} (\mathcal{D}_+ \Phi^i \mathcal{D}_- \Lambda^a - \mathcal{D}_+ \Lambda^a \mathcal{D}_- \Phi^i) \\
&+ K_{\bar{i}\bar{a}} (\mathcal{D}_+ \bar{\Phi}^i \mathcal{D}_- \bar{\Lambda}^a - \mathcal{D}_+ \bar{\Lambda}^a \mathcal{D}_- \bar{\Phi}^i) \\
&+ K_{i\bar{a}} (\mathcal{D}_+ \bar{\Lambda}^a \mathcal{D}_- \Phi^i - \mathcal{D}_+ \Phi^i \mathcal{D}_- \bar{\Lambda}^a) \\
&+ K_{\bar{i}a} (\mathcal{D}_+ \Lambda^a \mathcal{D}_- \bar{\Phi}^i - \mathcal{D}_+ \bar{\Phi}^i \mathcal{D}_- \Lambda^a) + K_{ab} \mathcal{D}_+ \Lambda^a \mathcal{D}_- \Lambda^b \\
&- K_{a\bar{b}} (\mathcal{D}_+ \Lambda^a \mathcal{D}_- \bar{\Lambda}^b + \mathcal{D}_+ \bar{\Lambda}^b \mathcal{D}_- \Lambda^a) + K_{\bar{a}\bar{b}} \mathcal{D}_+ \bar{\Lambda}^a \mathcal{D}_- \bar{\Lambda}^b]
\end{aligned} \tag{22}$$

Using (19), one can integrate $\mathcal{D}$ by parts and simplify this to

$$S = -\frac{1}{2} \int d^2 z \mathcal{D}_+ \mathcal{D}_- [(g_{\mu\nu} + b_{\mu\nu})\mathcal{D}_+ X^\mu \mathcal{D}_- X^\nu], \tag{23}$$

where $\{X\} = \{\Phi, \bar{\Phi}, \Lambda, \bar{\Lambda}\}|_{Im\theta=0}$ and

$$g_{i\bar{j}} = K_{i\bar{j}}, \quad g_{a\bar{b}} = -K_{a\bar{b}}, \quad b_{i\bar{a}} = -K_{i\bar{a}}, \quad b_{a\bar{i}} = K_{a\bar{i}}, \tag{24}$$

all other components related by symmetries of g and b or vanishing. Note the formal resemblence to (9); indeed, g and b in (23) are the metric and torsion potential, respectively, leading to (1) and (3).

One may now proceed and find the component action by performing the last two spinor differentiations, using (19), and integrating by parts on the world sheet. One finds

$$S = \frac{1}{2} \int d^2 z [(g_{\mu\nu} + b_{\mu\nu})\partial_+ X^\mu \partial_- X^\nu + i g_{\mu\nu}(\psi_+^\mu \nabla_- \psi_+^\nu - \psi_-^\mu \nabla_+ \psi_-^\nu) \\ -\frac{1}{2} R_{\mu\nu\rho\sigma}^- \psi_+^\mu \psi_+^\nu \psi_-^\rho \psi_-^\sigma - g_{\mu\nu} F^\mu F^\nu], \tag{25}$$

where $\psi = \mathcal{D}X|_{Re\theta=0}$ are fermions, $F^\mu = i(\mathcal{D}_+\mathcal{D}_- X|_{Re\theta=0} + \Gamma_{\nu\rho}^{-\mu}\psi_+^\nu \psi_-^\rho)$ is a real commuting auxiliary variable that can be dropped,

$$\nabla_- \psi_+^\nu = \partial_- \psi_+^\nu + \Gamma_{\mu\rho}^{+\nu}\partial_- X^\mu \psi_+^\rho, \quad \nabla_+ \psi_-^\nu = \partial_+ \psi_-^\nu + \Gamma_{\mu\rho}^{-\nu}\partial_+ X^\mu \psi_-^\rho, \tag{26}$$

and R^- is the curvature of Γ^-.

Acknowledgements

It is a pleasure to thank S.T. Yau, G. Tian, and C.M. Hull for discussions, and E. Witten and A. Shapere for numerous discussions and comments on the manuscript.

References

[1] S.J. Gates, C.M. Hull, M. Roček, Nucl. Phys. **B248** (1984) 157.

[2] T.H. Buscher, Phys. Lett. **159B** (1985) 127.

[3] C.M. Hull, *Superstring Compactifications with Torsion and Spacetime Supersymmetry*, Turin Superunif. (1985) 347.

[4] C.G. Callan, D. Friedan, E.J. Martinec, M.J. Perry, Nucl. Phys. **B262** (1985) 593.

[5] M. Roček, K. Schoutens, A. Sevrin, *Off-shell WZW Models in Extended Superspace*, Phys. Lett. B (to appear).

[6] N.J. Hitchin, A. Karlhede, U. Lindström, M. Roček, Commun. Math. Phys. **108** (1987), 535 (In particular, see chapters 4 and 5).

[7] B. Zumino, Phys. Lett. **87B** (1979), 203.

[8] L. Alvarez-Gaumé and D.Z. Freedman, Commun. Math. Phys. **80** (1981), 443.

AMS/IP Studies in Advanced Mathematics
Volume **9**, 1998

Calabi-Yau Threefolds and Complex Multiplication

Ciprian Borcea

Department of Mathematics and Physics

Rider College

2083 Lawrenceville Road

Lawrenceville, New Jersey 08648

Abstract

A Calabi-Yau threefold X is of complex multiplication type when the Hodge group associated to the rational Hodge structure on $H^3(X)$ is commutative. This property, expressed by the period point of X, is also reflected in the Weil and Griffiths intermediate Jacobians. It adumbrates the "enhanced symmetry" of models where the mirror scenario involves "geometrical modding".

© 1998 American Mathematical Society
and International Press

Introduction

The notion of complex multiplication originated with elliptic curves that have an "enhanced" endomorphism ring. When the curve is presented as the quotient E of the complex plane by a lattice generated by 1 and τ, the endomorphism ring is seen to be larger than the integers when *complex multiplicators* exist, that is: non real scalars sending the lattice to itself.

This is the case if and only if τ belongs to a totally imaginary quadratic extension of the rational field Q, and the extension is then isomorphic with $Hom(E, E) \otimes_Z Q$.

We recall first the generalization of this notion to Abelian varieties of arbitrary dimension and its expression in terms of Hodge groups of rational Hodge sructures [6, 4, 2, 3, 9].

In section two, Hodge structures of weight 3 are considered, and in particular those with $h^{3,0} = 1$. The main result on complex multiplication types is Theorem 2.3. It resembles the corresponding statement for $K3$ surfaces [1; see also 9].

In the final section we look at some examples.

1. Abelian varieties and Hodge structures of CM type

Let F be a totally imaginary number field (i.e. no embedding of F in C is contained in the reals) of degree $[F : Q] = 2n$ and let $\{\varphi_1, \cdots, \varphi_n\}$ a set of representatives for the conjugacy pairs of embeddings $F \hookrightarrow C$.

An Abelian variety A of dimension n over C is called of *complex multipication* (CM) type $(F, \{\varphi_i\})$ if an inclusion $F \hookrightarrow Hom(A, A) \otimes_Z Q$ is given such that the resulting representation of F on the tangent space to A at the origin is equivalent to the direct sum $\varphi_1 \oplus \cdots \oplus \varphi_n$.

This may happen [6] if and only if:

F contains two subfields K and K^σ such that:

(CM1) K^σ is totally real and K is a totally imaginary quadratic extension of K^σ. (The notation K^σ refers to the fixed ponts of the natural involution σ on K corresponding, for any embedding $K \hookrightarrow C$, to the complex conjugation.)

(CM2) There are no two isomorphisms smong the φ_i's which are complex conjugate on K.

Furthermore, under these circumstances, the Abelian variety A is *isogenous* to a product $B \times \cdots \times B = B^h$ of a *simple* Abelian variety B with itself, while $Hom(B, B) \otimes_Z Q$ is naturally a subfield of F that takes the role of K above. If $\{\psi_1, \cdots, \psi_m\}, m = dim_C B$ is the set of distinct embeddings $K \hookrightarrow C$ induced by φ_i's (each ψ_j comes up $h = n/m$ times), B is of CM Type $(K, \{\psi_j\})$.

CM types that correspond to *simple* Abelian varieties are called *primitive*.

Mumford noticed [4] that it is convenient to extend the above notion of CM type and allow, up to isogeny, arbitrary products of primitive CM types. One has, namely, the following characterization:

Let A be an Abelian variety of dimension n. The conditions below are equivalent:

CM(i) A is isogenous to a product of primitive CM types;

CM(ii) $Hom(A, A) \otimes_Z Q$ contains a commutative semi-simple Q-algebra $\mathcal{R}$ such that $[\mathcal{R} : Q] \geq 2n$, in which case $[\mathcal{R} : Q]$ actually equals $2n$;

CM(iii)   The *Hodge group* of the rational Hodge structure on $H^1(A)$ is commutative (in fact, a torus algebraic group i.e. some power of the multiplicative group $G_m \approx C^*$).

Remark.   The Abelian variety A is determined, up to isomorphism, by the *integral* Hodge structure on $H^1(A)$, and up to isogeny, by the *rational* Hodge structure on $H^1(A)$.

Condition CM(iii) may be formulated for arbitrary rational Hodge structures, thus allowing a notion of complex multiplication for other varieties. We recall the basic definitions [2, 3, 9].

Let $S = R_{C/R} G_m$ be a real algebraic goup obtained from the multiplicative group by restriction of scalars from C to R. Thus, the real points $S(R)$ correspond to C^*.

S may be defined in $GL(2, C)$ by the equations : $a_{12} = -a_{21}, a_{11} = a_{22}$.

Let H_R be a finite dimensional *real* vector space. A *Hodge structure* on H_R can be defined, equivalently, by any of the following data:

$HS(i)$ A *Hodge decomposition* of $H_C = H_R \otimes_R C$

$$H_C = \bigoplus_{p-q} H^{p,q}, \text{ such that } H^{p,q} = \overline{H^{q,p}}, \ (p,q) \in Z \times Z$$

$HS(ii)$ A morphism of *real* algebraic groups

$$h : S \to GL(H_R)$$

The equivalence obtains by letting

$$H^{p,q} = \{x \in H_C | h(x) \cdot x = z^{-p}\overline{z}^Q \cdot x, \text{ for all } z \in C^* = S(R)\}.$$

The inclusion $R^m \hookrightarrow C^*$ corresponds to an inclusion of real algebraic groups $G_m \hookrightarrow S$, and the restriction of h to G_m is called the *weight* $\omega : G_m \to GL(H_R)$ of h. If $\omega(\lambda)$ is multiplication by λ^{-n}, that is, if $H^{p,q} = 0$ for $p + q \neq n$, the Hodge structure is said to be (homogeneous) of weight n.

In our considerations we'll have $H^{p,q} = 0$ for p or q less than zero.

The *level* of a Hodge structure h is defined as

$$max\{|p - q| : H^{p,q} \neq 0\}.$$

For $A = Z$ or Q, an *integral*, respectively *rational* Hodge structure, is an A-module of finite type H_A, endowed with a Hodge structure on $H_R = H_A \otimes_A R$.

For a *rational* Hodge structure of weight $n : h : S \to GL(H_Q)$, the *Hodge group* $Hg = Hg(h)$ is defined to be the smallest algebraic subgroup of $GL(H_Q)$, *defined over* Q, such that the real points $H_g(R) \supset h(U^1)$, where $U^1 = \{z \in C^* | z\overline{z} = 1\} \subset S(R)$.

A *polarization* of a rational Hodge structure of weight n is a bilinear map

$$<,>: H_Q \times H_Q \to Q$$

which is symmetric for n even and skew-symmetric for n odd, satisfying the *Hodge-Riemann bilinear relations:*

(HR1) $< H^{p,q}, H^{r,s} >= 0$, unless $r = n - p$ and $s = n - q$;

(HR2) $i^{p-q} < x, \bar{x} >> 0$, for any nonzero $x \in H^{p,q}$.

A polarizable rational Hodge structure h of weight n is called of CM type when its Hodge group is commutative. $Hg(h)$ is then a torus algebraic group.

Let h be a rational polarized Hodge structure of weight n.

The *Hodge filtration* $F^\bullet$ is a decreasing filtration on H_C defined by

HS(iii) $F^p = \bigoplus_{r \geq p} H^{r,q}$

The Hodge decomposition HS(i) is then recovered by virtue of:

$$H^{p,q} = F^p \cap \overline{F^q}$$

The *period domain* (classifying space) D to which h belongs is the space of all flags like $F^\bullet$ that satisfy the Hodge-Riemann bilinear relations (HR1) and (HR2). D is an *open* set in its *compact dual* $\check{D}$, which consists of flags satisfying (HR1).

$\check{D}$ is a homogeneous space of the (connected) linear algebraic group

$$G_C = \{\gamma \in GL(H_C)| < \gamma x, \gamma y >=< x, y >\}$$

D is a homogeneous space of G_R: the real points in G_C.

The rational points G_Q are dense in G_R.

Denote by $G_C(h)$ the isotropy group of $h \in \check{D}$, and by $G_R(h)$ the isotropy goup of $h \subset D$: $G_R(h)$ is a compact subgroup of G_R.

Proposition 1.1. *For a rational polarized Hodge structure of weight n, $h \in D \subset \check{D}$, the following conditions are equivalent:*

(i) h is of CM type;

(ii) The Hodge group of h is contained in the isotropy group of $h \in \check{D}$;

$$Hg(h) \subset G_C(h)$$

(iii) h *is an isolated fixed point in* D *of a rational transformation in* $Hg(h)$;

(iv) *Locally,* h *is the only common fixed point of some set of transformations in* G_Q.

Proof. (i) $\Rightarrow$ (ii): Since $Hg(h)$ is commutative, it commutes with $h(U^1)$, hence preserves the Hodge decomposition and (ii) obtains.

(ii) $\Rightarrow$ (iii): Consider the representation of $G_C(h)$ on the tangent space $TD(h) = T\check{D}(h)$ at h. Note that h is an isolated fixed point for most elements in $h(U^1)$, hence there are elements in $Hg(h)$ that do not have 1 as eigenvalue on $TD(h)$. By Zariski-density of rational points, h will be an isolated fixed point for some *rational* transformation in $Hg(h)$ as well.

(iii) $\Rightarrow$ (iv) is obvious.

(iv) $\Rightarrow$ (ii): Suppose $\sum$ is a set of transformations in G_Q with h as the only common fixed point in some neighbourhood of h in D. the elements of $\sum$ commute with $h(U^1)$, hence with $Hg(h)$. For $\gamma \in Hg(h)$ sufficiently close to the identity, γh is close to h and fixed by $\sum$, therefore $\gamma h = h$ and (ii) follows.

(ii) $\Rightarrow$ (i): A rational transformation in $Hg(h)$ will commute with $Hg(h)$ since it commutes with $h(U^1)$. But rational points are Zariski-dense in $Hg(h)$ and we have (i).

Proposition 1.2. *Let* h_1 *and* h_2 *be rational polarized Hodge structures of weight* n_1, *respectively* n_2.

Then $h_3 = h_1 \otimes h_2$ *is of CM type if and only if both* h_1 *and* h_2 *are of CM type.*

Proof. Since $Hg(h_3) \subset Hg(h_1) \otimes Hg(h_2)$, we only have to prove that h_3 of CM type implies h_1 and h_2 of CM type.

By the previous proposition, h_3 is an isolated fixed point of some rational transformation $\gamma \in Hg(h_3)$. We have $\gamma = \gamma_1 \otimes \gamma_2$, and h_i has to be an isolated fixed point of the rational transformation $\gamma_i \in Hg(h_i)$, for $i = 1, 2$. Consequently, h_1 and h_2 are of CM type.

2. Hodge structures of weight 3

Let X be a projective threefold and put $H_Z = H^3(X, Z)$.

The Hodge decomposition then reads:

$$H_C = H^{3,0} \oplus H^{2,1} \oplus H^{1,2} \oplus H^{0,3} \tag{1}$$

Assuming the first Betti number $b_1(X) = 0$ (as will be the case for Calabi-Yau threefolds with *finite fundamental group*), one has a natural (unimodular) polarization:

$$< x, y >= - \int_X x \wedge y \tag{2}$$

The *Hodge filtration* $H_C = F^0 \supset F^1 \supset F^2 \supset F^3 \supset \{0\}$ reads:

$$F^1 = H^{3,0} \oplus H^{2,1} \oplus H^{1,2}$$
$$F^2 = H^{3,0} \oplus H^{2,1} \tag{3}$$
$$F^3 = H^{3,0}$$

and the first Hodge-Riemann bilinear relation (HR1) gives:

$$F^1 = (F^3)^{\perp} \text{and } F^2 = (F^2)^{\perp} \tag{4}$$

where orthogonality is taken with respect to (2).

On the *real* analytic torus $J = H_R/H_Z$, of $dim_R J = b_3(X)$ there are two relevant complex structures, namely:

the *Weil intermediate Jacobian* : $J(w)$, corresponding to the weight 1 decomposition:

$$H_C = \mathcal{H}_w^{1,0} \oplus \mathcal{H}_w^{0,1} \tag{5}$$

with $\mathcal{H}_w^{1,0} = H^{0,3} \oplus H^{2,1}$ and $\mathcal{H}_w^{0,1} = H^{3,0} \oplus H^{1,2}$;

and the *Griffiths intermediate Jacobian*: $J(g)$, corresponding to the weight 1 decompositon:

$$H_C = \mathcal{H}_g^{1,0} \oplus \mathcal{H}_g^{0,1} \tag{6}$$

with $\mathcal{H}_g^{1,0} = F^2 = H^{3,0} \oplus H^{2,1}$ and $\mathcal{H}_g^{1,0} = \overline{F^2} = H^{0,3} \oplus H^{1,2}$.

When (1), (5) and (6) are given in terms of morphisms of S (see section 1, HS(ii)), say: h, w respectively g, we have:

$$w(s) \text{ commutes with } g(s), \text{ for any } s \in S \tag{7}$$

and:

$$h(u)w(u) = g(u)^2, \text{ for any } u \in U^1 \subset S(R) \tag{8}$$

Clearly, the data $\{w, g\}$ is equivalent to h.

The Weil intermediate Jacobian $J(w)$ is a pricipally polarized Abelian variety (by (HR2)), but, in general, does not vary holomorphically in a variation of Hodge structure.

The Griffiths intermediate Jacobian $J(g)$ does vary hollomorphically, since the Hodge filtration does so, but, in general, need not be an Abelian variety (2) induces on $J(g)$ a $(1,1)$ form of Hermition type $(h^{2,1}, h^{3,0})$; it is negative on $H^{3,0} \oplus H^{0,3} \cap H_R$.

In order to investigate CM types, it will be useful to consider the following *splitting over* Q of a given Hodge structure $h \in D$ with $h^{3,0} = 1$:

Define $T(h)$ to be the smallest rational Hodge substructure of h that contains the *line* $H^{3,0}$. $T(h)$ is polarized by the restriction of the polarization on h, its orthogonal $T(h)^\perp$ is a rational Hodge substructure of h of *level* 1 and

$$h = T(h) \oplus T(h)^\perp \text{ over } Q \tag{9}$$

Remark. When h comes from geometry, as considered in (1), Grothendieck's version of the Hodge conjecture [7,5] says that $T(h)^\perp$ should be the "algebraic" part of h, i.e. $\bigcup_W ker(H^3(X, C) \to H^3(X - W, C))$, with the union taken over all codimension 1 subvarieties $W \subset X$ or, on the intermediate Jacobian $J(g)$, $T(h)^\perp$ should correspond with the image of the Abel-Jacobi map defined on (rational equivalence classed of) codimension 2 algebraic cycles of X, algebraically equivalent to zero.

$T(h)$ might therefore be thought of as the "transcendental" part of the weight 3 rational Hodge structure h.

Obviously, h is of CM type if and only if both $T(h)$ and $T(h)^\perp$ are of CM type.

Now, $T(h)$ is *simple* (i.e. has no non-trivial rational substructure), while $T(h)^\perp$, which is of *level* 1, is (once converted to a weight 1 rational Hodge structure) the isogeny type of an isogeny factor present on both $J(g)$ and $J(w)$; in particular, if one of the intermediate Jacobians is of CM type, so is $T(h)^\perp$.

To deal with $T(h)$ is tantamount to the investigation of h *under the assumption of simplicity*.

Let therefore $h \in D$ be *simple* as a rational Hodge structure and consider the Q-algebra:

$$End_{Hg(h)}(H_Q) = \{\gamma \in GL(H_Q) | \gamma \text{ commutes with } Hg(h)\} \tag{10}$$

Clearly:

$$End_{Hg(h)}(H_Q) \subset End_{Hg(w)}(H_Q) \cap End_{Hg(g)}(H_Q) \tag{11}$$

Transposition with respect to the polarization form $<,>$ determines an *involution* on $End_{Hg(h)}(H_Q)$ and since on $J(w)$, $<,>$ corresponds to a positive $(1,1)$ form:

$$T_r({}^t\gamma \cdot \gamma) > 0, \text{ for nonzero } \gamma \in End_{Hg(h)}(H_Q) \tag{12}$$

Using now the fact the h is simple and $h^{3,0} = 1$, we have an isomorphism:

$$End_{Hg(h)}(H_Q) \to K \subset C \tag{13}$$

onto a number field K, sending γ to the eigenvalue of γ on $H^{3,0}$.

Let σ be the induced involution on K.

Lemma 2.1. *One of the following statements holds:*

1) *K is a totally real number field and the involution σ is the identity;*

2) *The invariant part K^σ is a totally real number field and K is a totally imaginary quadratic extension of K^σ. The involution σ is the complex conjugation.*

Proof. Follws from (12) and [6, p. 41] (see also [9, 1.5.1]).

Proposition 2.2. *Let $h \in D$ be a simple polarized rational Hodge structure of weight 3, with $h^{3,0} = 1$. The following statements are equivalent:*

(i) *h is of CM type i.e. $Hg(h)$ is a torus algebraic group;*

(ii) *$End_{Hg(h)}(H_Q)$ is isomorphic (via (13)) with a number field K of degree $[K : Q] = \dim_Q H_Q = 2(1 + h^{2,1})$.*

If this is the case, then K necessarily satisfies statement 2) in the preceding lemma and the intermediate Jacobians $J(w)$ and $J(g)$ are both of CM type; more precisely:

$J(w)$ is of CM type $(K, \{\varphi_1, \varphi_2, \cdots\})$, with φ_1 the inclusion of K in C and certain other embeddings $\varphi_i, 2 \leq i \leq 1 + h^{2,1}$, while

$J(g)$ is of CM type $(K, \{\overline{\varphi_1}, \varphi_2, \cdots\})$.

Proof. Assume (i). the centralizer of the algebraic torus $Hg(h)$ is defined over Q and contains a maximal torus of $GL(H_Q)$, defined over Q. It follows that $End_{Hg(h)}(H_Q)$ contains a commutative semi-simple Q-algebra of degree $\dim_Q H_Q$, hence (ii).

Assume (ii). By (11) and CM(ii), the Weil intermediate Jacobian $J(w)$ is of CM type $(K, \{\varphi_1, \varphi_2, \cdots\})$, with K totally imaginary. Clearly, $J(g)$ is then of CM type $(K, \{\overline{\varphi_1}, \varphi_2, \cdots\})$.

The rational points of $Hg(w)$ and $Hg(g)$ are thus in $End_{Hg(h)}(H_Q) \approx K$ (cf. [4,p.248] or [6]). As a consequence, $Hg(w)$ and $Hg(g)$ commute and the Q-closure of the group generated by $Hg(w)$ and $Hg(g)$ is commutative. Relation (8) shows that $Hg(h)$ lies in the latter group, hence (i).

Reverting now to the general case, we summarize our results in:

Theorem 2.3. *Let $h \in D$ be a rational polarized Hodge structure of weight 3, with $h^{3,0} = 1$:*

$$h \equiv H_C = H^{3,0} \oplus H^{2,1} \oplus H^{1,2} \oplus H^{0,3}.$$

The following conditions are equivalent:

(i) h is of CM type;

(ii) The Weil and Griffiths intermediate Jacobians: $J(w)$ and $J(g)$ are of CM type and $Hg(w)$ commutes with $Hg(g)$;

(iii) h is an isolated fixed point in the period domain D for the action of some rational transformation in the Hodge group $Hg(h)$;

(iv) h is the only common fixed point, in some neighbourhoold of h in D, for a set of elements in G_Q.

Proof.   The equivalence of (i), (iii) and (iv) holds in general and was proven in Proposition 1.1.

The equivalence of (i) and (ii) follows from Proposition 2.2. and the splitting (9): $h - T(h) \oplus T(h)^\perp$.

Corollary 2.4. *The CM points are dense in the period domain D.*

In fact, already the simple rational Hodge structures of CM type are dense in D. It suffices to find one such (e.g. produce the Hodge decomposition from convenient eigenspaces of the multiplication endomorphism defined by a primitive element of a field K as above, with K considered as a Q-vector space) and then "translate" by elements of G_Q.

Remark.   By definition, a Calabi-Yau threefold X will be of CM type when its "period point" $h(X) = H_C = H^{3,0} \oplus H^{2,1} \oplus H^{1,2} \oplus H^{0,3}$ is of CM type. (Conceive of H_Z as fixed: the integral cohomology of the underlying differential manifold.)

The versal deformation of X determines locally a submanifold of dimension $h^{2,1}$ in D, which is of dimension $h^{2,1} + \dfrac{(h^{2,1} + 1)(h^{2,1} + 2)}{2}$. It is not obvious a priori whether this submanifold carries CM points.

In the next section we shall see examples where CM points are dense on images of period maps.

3. Examples: Calabi-Yau threefolds of CM type

The Fermat quintic threefold: $X_5^3 = \{\sum_{i=0}^{4} Z_i^5 = 0\} \subset \mathbf{P}_4$.

This case is well known [7, 8]. Let $h - h(X_5^3)$ be the period point. The Hodge conjecture holds and the "transcendental" part $T(h)$ is of dimension 4 with algebra of rational Hodge endomorphisms isomorphic with the cyclotomic field generated over Q by fifth roots of unity. The latter field is also the CM type of all fifty simple factors in the "algebraic" part $T(h)^{\perp}$.

Let E_α, $\alpha = 1, 2, 3$ be elliptic curves.

Denote by ι_α the involution $z_\alpha \rightarrow (-z_\alpha)$ of E_α.

On $E_1 \times E_2 \times E_3$, consider the action of the group $G_4 \approx \mathbf{Z}_2 \times \mathbf{Z}_2$, obtained by letting involutions operate on two factors and the identity on the third. These transformations clearly preserve the holomorphic 3-form on the torus and the quotient

$$E_1 \times E_2 \times E_3/G_4 \tag{14}$$

has Calabi-Yau *resolutions* with Euler number $e = 96$ and $h^{2,1} = 3$.

Actually, the singularities in (14) are along a configuration of 48 rational curves that can be represented as a cubic network with 4^3 points of intersection (corresponding to the 2-torsion points on the 3-torus). A projective resolution may proceed by blowing up these curves, and there are, essentially, three choices for the sequence of blow-ups. The rational Hodge structure on the resulting Calabi-Yau threefold is nevertheless the same, namely: the G_4-invariant part of the weight 3 Hodge structure of the torus. This is precisely of $h_1 \otimes h_2 \otimes j_3$, where h_α stands for the weight 1 rational Hodge structure of E_α. Using Proposition 1.2, we have:

Proposition 3.1. *A Calabi-Yau resolution of $E_1 \times E_2 \times E_3/G_4$ is of CM type if and only if each elliptic curve E_α is of CM type.*

Considering first the Kummer surface $E_2 \times E_3/\iota_2 \times \iota_3$, with the involution induced by $\iota_2 \times id = id \times \iota_3$, certain resolutions of (14) correspond to the following construction: resolution

$E \times K3/\iota \times \sigma$, where E denotes an elliptic curve with involution ι, and $K3$ denotes and *algebraic* surface of that type with an involution σ reversing the sign of the holomorphic 2-form.

Take, for example, in the role of $K3$ above, the (minimal) resolution of a double covering of $\mathbf{P}_2$, ramified over 6 lines in general position. The fixed locus of the natural involution σ will consist of 6 rational curves and, on $H^2(K3)$, σ^* has the eigenvalue $(+1)$ with multiplicity 16, while (-1) comes with multiplicity 6.

Such surfaces clearly depend on 4 moduli (6 general points in the dual $\check{\mathbf{P}}_2$, of which 4 may be prescribed, due to the action of $PGL(3)$, leave 4 independent parameters) and the image of period map lies in the orthogonal of the $(+1)$-eigenspace. CM points are dense on this image, since already "singular" points (i.e. $H^{2,0} \oplus H^{0,2}$ is defined over $\mathbf{Q}$) are dense on it.

Now, $E \times K3/\iota \times \sigma$ has singularities along $4 \times 6 = 24$ disjoint rational curves, and a blow-up yields a Calabi-Yau threefold with Euler number $e = 72$ and $h^{2,1} = 5$. The rational Hodge structure is the invariant part of the Hodge structure on $H^3(E \times K3)$, hence CM points will be dense on the versal deformation of this threefold, corresponding to instances where both E and $K3$ are of CM type.

References

[1] C. Borcea, *K3 surfaces and complex multiplication*, Revue Roumaine Math. Pures et Appliq. 31 (1986) 499-505.

[2] P. Deligne, *La conjecture de Weil pour les surfaces K3*, Invent. Math. 15 (1972) 206-226.

[3] P. Deligne, *Variétés de Shimura: interprétation modulaire et techniques de construction de modèles canoniques*, Proc. Symp. Pure Math. 33 (1979) part 2, 247-290.

[4] D. Mumford, *Anote on Shimura's paper "Discontinuous groups and Abelian varieties"*, Math. Ann. 181 (1969) 345-351.

[5] J.P. Murre, *Abel-Jacobi equivalence versus incidence equivalence for algebraic cycles of codimension two*, Topology 24 (1985) 361-367.

[6] G. Shimura, Y. Taniyama, *Complex Multiplication of Abelian Varieties and its Applications to Number Theory*, Public. Math. Soc. Japan 6, Tokyo, 1961.

[7] T. Shioda, *What is known about the Hodge conjecture?* in: "Algebraic Varieties and Analytic Varieties", Adv. Studies Pure Math. 1 (1983) 55-68.

[8] T. Shioda, *Geometry of Fermat varieties*, in: "Number Theory Related to Fermat's Last Theorem", Progress in Math. 26 (1982) 45-56.

[9] Yu.G. Zarhin, *Hodge groups of K3 surfaces*, J. reine u. angew. Math. 341 (1983) 193-220.